AZUKI BEAN: BOTANY, PRODUCTION AND USES

AZUKI BEAN:
BOTANY, PRODUCTION AND USES

Thomas A. Lumpkin and Dean C. McClary

Department of Crop and Soil Sciences
Washington State University
USA

CAB INTERNATIONAL

CAB INTERNATIONAL
Wallingford
Oxon OX10 8DE
UK

Tel: Wallingford (01491) 832111
Telex: 847964 (COMAGG G)
E-mail: cabi@cabi.org
Fax: (01491) 833508

A catalogue record for this book is available from the British Library.

ISBN 0 85198 765 6

Printed and bound in the UK at the University Press, Cambridge

TABLE OF CONTENTS

FOREWORD

Azuki bean is a minor legume crop of Asian origin. Formerly it belonged to the genus *Phaseolus* and was a member of the Asian Phaseolus Group, but it has recently been transferred to the genus *Vigna* with relatives like mungbean, black gram, rice bean, etc. From ancient times azuki has been an important crop in "slash and burn" agriculture in hilly regions. It helps to maintain soil fertility in the cropping sequence and is a good vegetable protein source. Azuki also occupies a unique position in Asian "legume food culture" because of the beautiful red color of the testa. In Asia, red is the most important color associated with celebration. The red color of azuki is especially important for agricultural peoples of East Asia, who often have customs and beliefs that a spirit dwells within grain. In Japan, azuki is an essential ingredient of both *seki-han*, a festive red rice dish prepared by steaming glutinous rice together with red azuki beans, and *an*, a sweetened bean paste used for traditional cakes introduced from China between the 6th and 7th centuries AD. If we trace the origin of the azuki folklore common in Japan and the Korean peninsula, we arrive in southern China.

Despite its cultural importance to some countries, it can be inferred that azuki is a crop of little concern to grain legume scientists, even in Asia, especially when compared to soybean, another legume of Asian origin. For example, no statistics exist on the global production of azuki, although it is known to be very important in East Asia. Another indication is that a comprehensive monograph on azuki has not been written, even by Asian scientists.

I learned about the proposed publication of *Azuki Bean: Botany, Production and Uses* by Professor T. A. Lumpkin and Mr Dean McClary, Washington State University, USA, from a book catalog, and I confess, that at that moment, I felt that Asian grain legume agronomists had been "scooped" by "foreigners". Unexpectedly, I was asked to contribute a foreword to this book through my friend Dr K. Kyuma, Dean of the Faculty of Agriculture, Kyoto University, Japan, in March 1992.

As can be seen from the list of contents, the authors describe the botany, morphology, physiology, production, pests, diseases, breeding, genetics, uses, processing and marketing, based on a review of approximately 800 literature and journal sources. This is the first monograph on azuki bean in the world and will contribute greatly to enhancing academic interest in this legume by young students and agronomists, and to raising the stature of azuki as a food and crop, that is, to shift this small, red-seeded legume crop from the status of a minor crop to that of a major crop.

Kazumi Maeda, Professor
Crop Science, Faculty of Agriculture
Kochi University
Japan

ACKNOWLEDGEMENTS

We gratefully acknowledge some of those whose assistance made the completion of this book possible. We thank the IMPACT program at Washington State University for funding much of this effort. We especially thank Suzanne M. Farrow for her diligent editorial and publication assistance, and Virginia Elliot and Hu Jia for skillful translation of much of the cited information. We also appreciate the patience and assistance of our publisher, including Tim Hardwick and Anna Zwicky. Others who made important contributions include: AVDRC, Columbia Bean and Produce Co., Kyoto University, Marubeni of America Corp., National Science Foundation, U.S. Plant Germplasm System, H. Aguilar, G. Benton, L. Elberson, A. Engquist, D. Evans, G. Gillilard, A. Hang, R. Hannan, D. Helm, W. M. Hsieh, W. L. Jin, W. Kaiser, M. Kobayashi, J. Konovsky, K. Kyuma, T. Matoh, K. Nelson, S. Niwa, D. O'Rourke, Y. Pomeranz, M. Silbernagel, M. Stein, R. Summerfield, B. Swanson, L. Tanigoshi and G. Waite.

ABOUT THE AUTHORS

Dr Thomas A. Lumpkin received his undergraduate education at Washington State University and his graduate education at the University of Hawaii. His career in the study of Asian agriculture spans 22 years, including two years each at the Zhejiang Academy of Agricultural Sciences, China and at Kyoto University, Japan. His interest in East Asian crop development has fostered research on Azuki, Edamame soybean, Wasabi radish and Chinese Milkvetch (*Astragalus sinicus*). Dr Lumpkin, currently a professor of Agronomy and Asian Studies, teaches world crops and agricultural systems and leads a research project on East Asian Crop Development in the Department of Crop and Soil Sciences, Washington State University, Pullman, WA, 99164-6420; E-mail: lumpkin@wsuvm1.csc.wsu.edu.

Dean McClary received his Bachelor of Science degree in Agriculture from Washington State University in 1987 and completed his Master of Science in Agronomy with Dr Lumpkin in 1990. His graduate research project consisted of an assessment of azuki production potential in Eastern Washington and included several extended periods of study in Japan. Mr McClary worked in Dr. Lumpkin's East Asian Crop Development Program as an agricultural technician until 1992 and currently teaches agriscience courses at Ferndale High School, Washington.

to the mothers of the authors

Alma R. Lumpkin

and

Mary McClary

1

INTRODUCTION

THE PLANT'S NAME

"Azuki" is the way Japanese read and Romanize the Chinese characters meaning "small bean" (Table 1.1). This name, azuki, will be used in this book instead of "adzuki" or "adsuki" which are normally used in English-language publications. Thus, this book adopts the original Japanese name. Also "azuki bean" will not be used in this text since the literal meaning, "small bean bean", is obviously a redundancy in terms. Azuki is often referred to as "Red Diamond" in the Japanese commodity business because of its high value. There are also many other colloquial names for azuki, the most common being *shouzu*. Other colloquial names are often associated with one or a few prefectures in Japan, e.g. *akamame* in Oita; *ankomame* in Hokkaido, Shiga and Kochi; *anmame* in Yamanashi; *antoki* in Saitama and Aichi; *irakuri* in Mie; *shoumame* in Tokushima; *komame* in Shizuoka and Hyogo; *kannome* in Yamaguchi; *narazu* in Ehime; *omame* in Shizuoka; *konaremame* in Aichi; and, *gururimame* in Kumamoto (Stat. Research ..., 1951).

The Chinese have numerous written names for azuki, most of which are obsolete or not in common usage. The two and three character names listed in Table 1.1 are from the Chinese book, *Edible Bean Crops* (see Duan, 1989). The Romanized (Hanyu Pinyin) and literal English translation of these names are also listed. Many of the Chinese names for azuki include one of several Chinese characters signifying some form of the color red. These Chinese characters can be identified in the list by the Hanyu Pinyin Romanizations of *chi*, *hong* and *zhu*. Both Koreans and Japanese usually write the name "azuki" using Chinese characters, especially the first Chinese name in the list (小豆) and occasionally the second name (赤豆), which normally refers to a dark red azuki used for medicinal purposes. Koreans and Japanese also write the name for azuki with their national alphabets (Table 1.1). In the two Japanese alphabets, the name is pronounced as "azuki", in the same way that they pronounce the first Chinese name. Koreans call azuki "pat".

English names for azuki include azuki, adzuki, adsuki, atsuki, atzuki, red bean, small bean, small red bean, crimson bean and small crimson bean (Ag. Lexicon ..., 1979; Duke, 1981; Ci Hai ..., 1983; Duan, 1989). The bean is called *rains* or *guruns* in Hindi, *adzukibohne* in German, *fagiolo adzuki* in Italian, *frijol adzuki* in Spanish, *frijol diablito* in Cuban, *haricot adzuki* or *haricot a feuilles angulaires* in French, *poroto arroz* in Argentinian (Kay, 1979), *judia adzuki* in Mexican (Arias, 1975), *feijao adzuki* in Angolan and *nirikia* in Zairean (Plant Production ..., 1959).

Table 1.1. Chinese character names for azuki, along with their Romanization by the Hanyu Pinyin system used in mainland China and their meaning in English. At the bottom of the list, the name "azuki" is written in the two Japanese alphabets and the Korean name for azuki, "pat", is written in the Korean alphabet. The first Chinese name for azuki (小豆) is the most common or formal way of writing the name in Chinese, Japanese and Korean.

CHINESE HANZI	CHINESE HANYU PINYIN	ENGLISH TRANSLATION
小豆	*Xiaodou*	Small bean
赤豆	*Chidou*	Dark red bean
赤小豆	*Chixiaodou*	Dark red small bean
红豆	*Hongdou*	Red bean
红小豆	*Hongxiaodou*	Red small bean
金豆	*Jindou*	Golden bean
金红豆	*Jinhongdou*	Golden red bean
米赤豆	*Michidou*	Rice dark red bean
杜赤豆	*Duchidou*	*Du* dark red bean
朱豆	*Zhudou*	Vermilion bean
竹豆	*Zhúdou*	Bamboo bean
虱牳豆	*Shimudou*	Louse bean
アズキ	*Azuki*	(Japanese Katakana alphabet)
あずき	*Azuki*	(Japanese Hiragana alphabet)
팥	*Pat*	(Korean alphabet)

THE CROP AND WORLD LITERATURE

Hymowitz (1987) has called azuki one of the 12 most important grain legumes in the world. His statement is based on azuki's co-evolution with rice over wide areas of Far East and Southeast Asia, trade in international markets, large germplasm collections and national research attention. Another characteristic of major world crops is an extensive body of scientific literature, most of which, published in the producing countries' languages. Azuki is primarily produced and studied in the countries of northeast Asia and most research has been published in Japanese, Chinese and Korean. The lack of significant research interest outside of northeast Asia and the difficulty of translation from these languages has constrained the development and compass of azuki. Part of this constraint is the lack of a comprehensive review of azuki literature in the English language. The task of compiling, translating and reviewing scientific literature from Asia and elsewhere about azuki has been complicated and time consuming, but it has now been accomplished in this book.

There are four previous reviews of azuki literature in the English language. Sacks (1977) incorrectly stated that Piper and Morse (1914) was the earliest Western language literature review of azuki. The latter work was a brief introduction to azuki with a few observations about U.S. adaptation and without references. Sacks published the first true review, but it was short and had an ethnobotanical emphasis. In more recent efforts, Fery (1980) published an introductory review of azuki genetics while Kay (1979) and Duke (1981) both prepared short, general reviews based on small bodies of azuki literature.

Comprehensive literature reviews would be expected in countries with a long tradition of production and use. However, the authors were unable to identify a single source in the Japanese, Korean or Chinese languages. Japanese researchers frequently cite Kawahara's (1962) chapter on azuki in his college text, as the only literature review in their country. Hoshikawa's (1985) chapter in *Edible Crops*, provides updated information on Kawahara's work. Duan's (1989), 'Small Bean' chapter in *Edible Bean Crops*, is the only known summary of azuki ethnobotany, morphology, physiology, production, pests and uses currently available in the Chinese language.

2

BOTANY OF AZUKI

Botanical descriptions of azuki have been published for hundreds of years in China (Fig. 2.1) and more recently by the Japan Bean Fund Association (1977, many photos), Ohwi (1984, *Flora of Japan*), Yamaguchi (1992, wild azuki with illustrations) and others.

TAXONOMY

Azuki belongs to the family Fabaceae, tribe Phaseoleae, subtribe Phaseoliinae and has had 15 different genus-species names since it was first described by Englebert Kaempfer (1712, also see Sacks, 1977). The most recent consensus is genus *Vigna* Savi, subgenus *Ceratopteris* (Piper) Verdc., species *Vigna angularis* (Willd.) Ohwi and Ohashi (Ohwi and Ohashi, 1969; Lawn and Ahn, 1985; Smartt and Hymowitz, 1985). *Vigna angularis* var. *angularis* and the wild conspecific form *V. angularis* var. *nipponensis* are two generally recognized azuki subspecies (Smartt and Hymowitz, 1985; Baudoin and Maréchal, 1988; Tomooka, 1991; Yamaguchi, 1992) although Ohwi and Ohashi (1969) listed a total of 15.

The inclusion of azuki in *Vigna* is a recent taxonomic decision. Azuki was originally classified as a member of the genus *Phaseolus* based on morphological characteristics of the stigma and style, etc. (Piper and Morse, 1914; Hoshikawa, 1985). The creation of a new genus called *Azukia* (see Table 2.1), on the basis of stipule and flower characteristics, has been repeatedly suggested but has not been generally accepted for the Asiatic *Phaseolus* and now *Vigna* members (Takahashi, 1909; Ohwi, 1953, 1984; Smartt, 1980, 1981; Maréchal, 1982; Chandel *et al.*, 1988; Tateishi and Ohashi, 1990; Tomooka *et al.*, 1991). Other taxonomists have placed *Azukia* species in the genus *Dolichos*. Later, azuki and other Asiatic members of *Phaseolus* were reclassified into the genus *Vigna* by Verdcourt (1970) and confirmed by Maréchal *et al.* (1978) based on early vegetative characteristics and the morphology of the flower and stipule. Recent results from morphological, physiological and genetics research has supported this taxonomic change. For example, pollen characteristics of azuki more closely resemble those of *Vigna* than *Phaseolus* (Ohashi and Takahashi, 1981). Azuki seeds differ from *Phaseolus* by the absence of pipecolic acid and the acidic peptide γ-glutamyl-S-methylcysteine (Casimir and Le Marchand, 1966; Bell, 1971). An isoelectric focusing and electrophoresis study by Jaaska and Jaaska (1988) using aspartate superoxide dismutase and aminotransferase isoenzymes also supports the inclusion of azuki in *Vigna*. However, fluorescent banding pattern research by Zheng *et al.* (1993) groups azuki with two species of *Phaseolus*.

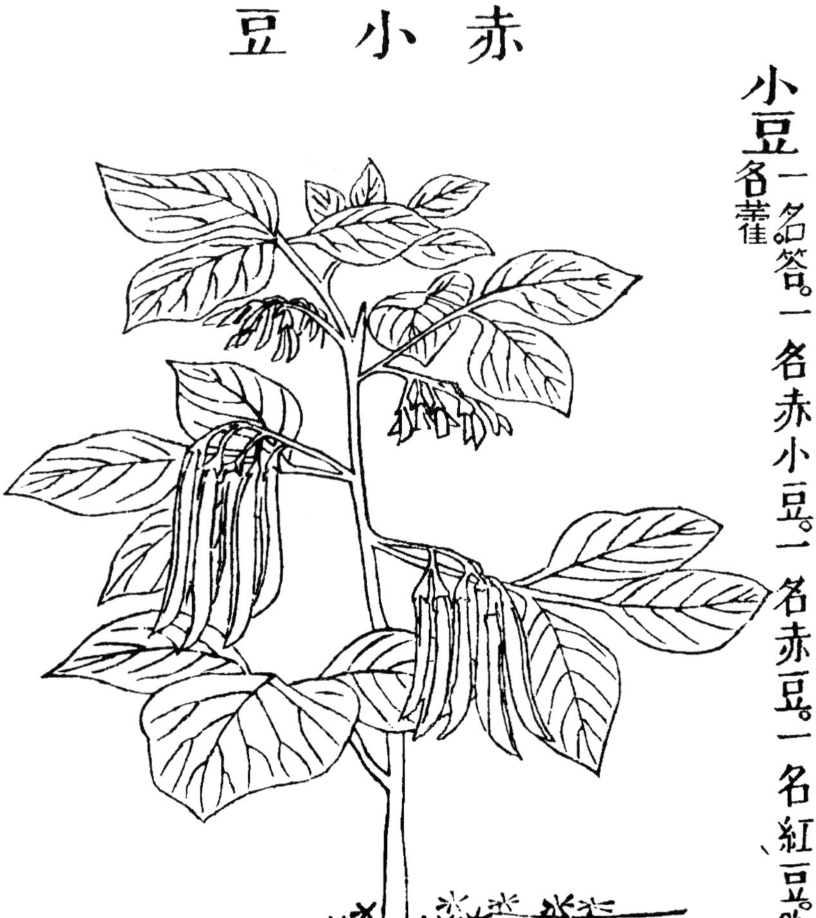

Fig. 2.1. A Qing Dynasty period illustration of azuki (sometimes confused with rice bean), with synonyms vertical on the right, from the book *Shou Shi Tong Kao* (*Compendium of Works and Days*) (授時通考) published in 1742 (adapted from Bray, 1984). Clarity has been enhanced by computer.

The Subgenus Ceratotropis

Azuki is a member of the highly homogeneous Asian subgenus *Ceratotropis* which comprises about 17 species, including seven cultivated species and their wild progenitors (Verdcourt, 1970; Maréchal *et al.*, 1978, 1981; Arora, 1985; Baudoin and Maréchal, 1988; Egawa, 1988; van Oers, 1989; Tateishi and Ohashi, 1990). Members of *Ceratotropis* are differentiated from each other on the basis of unique floral morphology, stipule characteristics and region(s) of domestication (Maréchal *et al.*, 1981; Smartt, 1981, 1984; Tateishi and Ohashi, 1990). The key morphological characteristics of the subgenus are peltate stipules, yellow asymmetric flowers, a style that extends beyond the stigma, and triporate pollen with deep-walled large retriculations and a granulate infratectum structure (Verdcourt, 1970; Baudoin and Maréchal, 1988; Tateishi and Ohashi, 1990). In the flowers, the keel and style are curved to the left side about 360° with their tips facing a sharp pointed unilateral pocket on the left side of the keel. Complete unanimity is still lacking so taxonomy within the subgenus remains quite confused (Table 2.1).

Verdcourt (1970) classified five *Ceratotropis* cultigen species into a mungbean group which included azuki, *Vigna aconitifolia* (moth bean, mat bean), *Vigna mungo* (blackgram, black mapte, *mash*, *mungo*, *urd*, *urid*, woolly pyrol), *Vigna radiata* (Chickasaw pea, *bundo,* goldengram, greengram, *moong*, *mung*, mungbean, *mungo*, Oregon pea, *yae-nari*) and *Vigna umbellata* (rice bean, climbing mountain bean, Jerusalem pea, Mambi bean, red bean, oriental bean, Pegin bean, *tsuru-azuki, kanime*). For convenience, Tomooka *et al.* (1991) divide *Ceratotropis* into two groups based on seedling characteristics (see Maekawa, 1955). Their mungbean group (Tomooka *et al.*, 1991) is centered on the Indian subcontinent and is composed of *V. radiata*, *V. mungo* and *V. aconitifolia* which have seedlings with epigeal germination and a short petiole on the primary leaf. Their azuki group is centered in East Asia and is composed of *V. angularis* and *V. umbellata* which have seedlings with hypogeal germination and a long petiole on the primary leaf. Tateishi (1989) and Tateishi and Ohashi (1990) have described an azuki bean group composed of *V. aconitifolia*, *V. angularis* var. *angularis*, *V. mungo* var. *mungo*, *V. radiata* var. *radiata*, *V. umbellata* and their wild forms primarily based on the key characteristics of the subgenus, while Yasui *et al.* (1985) used differences in the low molecular weight carbohydrate composition of seeds to divide *Ceratotropis* in to a mungo-radiata group and an angularis group homogenously consisting of *V. angularis*, *Vigna nakashimae* (*hime-tsuru-azuki*), *Vigna reflexo-pilosa* (*o-yabu-tsuru-azuki*), *V. umbellata* and a slightly more remote *V. aconitifolia*.

Table 2.1. Synonyms for some members of the subgenus *Ceratotropis*. Known cultivated (C) and wild (W) forms are indicated in the second column. No attempt has been made to correct the confusion of different classifications in the literature (See Verdcourt, 1970; Maréchal *et al.*, 1978; Baudoin and Maréchal, 1988; Tateishi, 1984, 1989 and Tomooka *et al.*, 1991).

Current Name	CW	2n=	Synonyms
Vigna aconitifolia (Jacq.) Maréchal	CW	22	*Phaseolus aconitofolius* Jacq.; *P. palmatus* Forsk.; *Dolichos dissectus* Lam.
Vigna adenantha (G.F. Meyer) Maréchal, Mascherpa & Stainier			*Phaseolus adenantha* (G.F.) Meyer
Vigna angularis (Willd.) Ohwi and Ohashi	CW	22	*Dolichos angularis* Willd.; *Phaseolus angularis* (Willd.) Wight; *Azukia angularis* (Willd.) Ohwi
Vigna angularis var. *angularis*	C	22	
Vigna angularis var. *nipponensis* (Ohwi) Ohwi and Ohashi	W	22	*Phaseolus trilobus* Ait. *sensu* auctt. Jap., *P. nipponensis* Ohwi; *P. angularis* var. *nipponensis* (Ohwi) Ohwi; *P. angularis* f. *nipponensis* (Ohwi) Kitamura; *Azukia angularis* var. *nipponensis* (Ohwi) Ohwi.
Vigna bourneae Gamble	W		
Vigna dalzelliana (Kuntzze) Verdc.	W		*Phaseolus pauciflorus* Dalz.; *P. dalzelliana* Kuntzze; *P. dalzelli* Cooke
Vigna glabrescens M. M. & S.	CW	44	*Phaseolus glaber* Roxb.; *P. glabrescens* Steudel; *Vigna mungo* var. *glaber* (Roxb.) Baker; *V. radiata* var. *glabra* (Roxb.) Verdc.; *V. reflexo-pilosa* Hayata
Vigna grandiflora			*Phaseolus sublobata* var. *grandiflorus*; See Tomooka *et al.*, 1991.
Vigna grandis (Dalz. and Gibs.) Verdc.			*Phaseolus grandis* Dalz. and Gibs.
Vigna hirtella Ridley			
Vigna hosei (Craib) Backe			*Dolichos hosei* Craib
Vigna khandalensis (Santapau) Raghava and Wadhwa	W		*Phaseolus grandis* Dalz. and Gibs.; *P. khandalensis* Santapau; *Vigna grandis* (Dalz. and Gibs.) Verdc.
Vigna malayana M. R. Henderson			*V. parviflora* Ridley
Vigna minima (Roxb.) Ohwi and Ohashi	W		*Phaseolus minimus* Roxb.; *Azukia minima* (Roxb.) Ohwi
Vigna minima var. *minima* Tateishi			*Phaseolus minimus* Roxb.; *P. calcaratus* var. *gracilis* Prain; *P. calcaratus* Roxb. var. *gracilis* Prain; *P. calcaratus* Roxb. var. *glabra* (Roxb.) Prain; *P. trilobus* Ait. *sensu* Matsum.; *P. heterophyllus* Hayata; *P. gracilicaulis* Ohwi; *Azukia minima* (Roxb.) Ohwi; *A. heterophylla* (Hosokawa) Masamune; *V. gracilicaulis* (Ohwi) Ohwi & Ohashi; *V. umbellata* (Thunb.) Ohwi & Ohashi var. *gracilis* (Prain) Maréchal, Mascherpa & Stainier
Vigna minima subsp. *gracilis*			See Tomooka *et al.*, 1991.

Table 2.1. Continued.

Vigna minima var. *minor* (Matsum.) Tateishi			*Azukia riukiuensis* (Ohwi) Ohwi; *Vigna riukiuensis* (Ohwi) Ohwi and Ohashi; *V. lutea* A. Gray var. *minor* Matsum.; *V. riukiuensis* (Ohwi) Ohwi & Ohashi; *Phaseolus minimus* Roxb.; *P. minima* (Roxb.) Ohwi and Ohashi *sensu* Walker
Vigna minima subsp. *nakashimae*			See Tomooka *et al.*, 1991.
Vigna minimus Ohwi and Ohashi			
Vigna mungo (L.) Hepper	CW	22	*Phaseolus mungo* L.; *P. viridissimus* Ten.; *Azukia mungo* (L.) Masamune
Vigna mungo var. *mungo*	C	22	*Phaseolus mungo* L.
Vigna mungo var. *silvestris* Lukoki, Maréchal and Otoul	W	22	
Vigna nakashimae (Ohwi) Ohwi and Ohashi	W	22	*Phaseolus nakashimae* Ohwi; *Azukia nakashimae* (Ohwi) Ohwi; [*P. minimus sensu auct*, Japon., non Roxb.]; *Vigna minima* subsp. *nakashimae* (Ohwi) Tateishi
Vigna papuana Bak. f.			
Vigna radiata (L.) Wilczek	CW	22	*Phaseolus radiatus* L.; *P. radiatus* var. *typica* Matsum.; *P. aureus* Roxb.; *Azukia radiata* (L.) Ohwi; *Rudua aurea* (Roxb.) Maekawa; *Vigna aureus* (Roxb.) Hepper; [*Vigna mungo* sensu Hepper, non (L.)]; [*Phaseolus mungo sensu* F.B.I. *non* L.]
Vigna radiata var. *radiata*	C	22	*Phaseolus radiatus* L.; *P. aureus* Roxb.; *Azukia radiata* (L.) Owhi; *Rudua aurea* (Roxb.) F. Maekawa
Vigna radiata var. *sublobata* (Roxb.) Verdc.	W	22	*Phaseolus sublobata* Roxb.; *P. trinervius* Wight and Arn.; *Vigna opisotricha* A. Rich.; *V. brachycarpa* Kurz; *V. perrierianna* R. Viguier; [*Phaseolus radiatus* auctt. mult. *non* L. *sensu stricto.*]
Vigna radiata var. *glabra* (Roxb.) Verdc.		44	*Phaseolus glaber* Roxb.; *P. glabrescens* Steudel; *Vigna mungo* var. *glaber* (Roxb.) Baker; *P. calcaratus* var. *glaber* (Roxb.) Prain
Vigna radiata var. *setulosa* (Dalz.) Ohwi and Ohashi			*Phaseolus setulosa* Dalz.
Vigna reflexo-pilosa Hayata	CW	44	*Phaseolus reflexo-pilosus* (Hayata) Ohwi; *P. mungo* L. *sensu* Forbes & Hemsley; *P. neocaledonicus* Baker f.; *Azukia reflexo-pilosa* (Hayata) Ohwi; *Vigna glabrescens* M. M. & S; *V. Catjang* Endl. var. *sinensis* King *sensu* Matsum.
Vigna riukiuensis (Ohwi) Ohwi and Ohashi	W	22	*Phaseolus riukiuensis* Ohwi; *Azukia riukiuensis* (Ohwi) Ohwi; *Vigna minima* var. *minor* Tateishi
Vigna stipulacea			See Tateishi and Ohashi, 1990.
Vigna trilobata (L.) Verdc.			*Dolichos trilobata* L.; *Phaseolus trilobatus* (L.) Schreb.; [*P. trilobus sensu* Aiton et auctt. mult., *non Dolichos trilobata* L.]
Vigna umbellata (Thunb.) Ohwi and Ohashi	CW	22	*Dolichos umbellatus* Thunb.; *Phaseolus pubescens* Blume; *P. calcaratus* Roxb.; *P. chrysanthus* Savi; *P. torosus* Roxb.; *P. ricciardianus* Tenora; *V. calcarata* (Roxb.) Kurz; *Azukia umbellata* (Thunb.) Ohwi
Vigna umbellata var. *umbellata*	C		*Phaseolus calcaratus* var. *major* Prain; *Phaseolus calcaratus* var. *rumbaiya* Prain
Vigna umbellata var. *gracilis* (Prain) M.M.&S.	W		*Phaseolus calcaratus* var. *gracilis* Prain

Research has demonstrated many similarities among the *Ceratotropis* cultigens (Poehlman, 1991) (Table 2.2). Danimihardja and Lester (1974) found that azuki, *V. umbellata* and *V. radiata* have similar protein bands. Turkova and Klozova (1985) used immunoelectrophoresis on the seed proteins of 10 species of *Vigna*. They concluded that azuki, *V. mungo*, *V. radiata*, *Vigna trilobata* (pillipesara bean, jungli bean) and *V. umbellata* probably constitute a taxonomic system group based on the similarities of pattern. Starch gel electrophoresis procedures were used to compare peroxidase, acid phosphates and esterase isozymes from the primary leaves of azuki, *V. mungo* and *V. umbellata* and revealed that these three species had almost identical acid phosphates banding patterns (Bassiri and Adams, 1978). *V. mungo* and *V. radiata* were reported to have an almost identical composition of low molecular weight carbohydrates (LMWC) in the seed, while azuki and *V. umbellata* were part of a different LMWC pattern group (Yasui *et al.*, 1985). DNA sequencing has shown that azuki and *V. radiata* are closely related but *V. radiata* has tandemly repeated non-rDNA 174-bp elements as prominent independent satellite DNA within the genome that are not found in the azuki genome (Unfried *et al.*, 1991). Lectins have been investigated as possible taxonomic tools, however their usefulness within Papilionoideae has not provided results that support the results of other tools. For example, extracts from the seeds of various Papilionoideae including *V. umbellata*, yielded several different types of haemagglutinations, while azuki and some other seemingly unrelated legumes lacked haemagglutination (Toms, 1981). Additional information about the relationships of species in this subgenus can be found in the section on interspecific hybridization in Chapter 7.

MORPHOLOGY AND PHENOLOGY

Seed

Azuki seeds are subcylindric with subtruncated ends (Piper and Morse, 1914; Sacks, 1977) and vary in length (5.0 to 9.1 mm), width (4.0 to 6.3 mm), thickness (4.11 to 6.04 mm) and weight (30 to 250 mg per seed) (Leese, 1958; Sacks, 1977; Kay, 1979; Chiba, 1980; Duke, 1981; Nakaseko and Gotoh, 1982; Nakaseko, 1983; Hoshikawa, 1985; Chinese Ag. Resources ..., 1987; Yoshida, 1987). Thickness measurements, taken from both sides of the hilum, are highly correlated to seed weight, and seed weight is correlated to specific gravity, seed length and width (Tanda, 1958; Yoshida, 1987). Much of the size variability among cultivars can be attributed to the development of two distinct market classes in Japan; the regular type (>4.2 mm length) and a larger *Dainagon* type (>4.8 mm length) (Hoshikawa, 1985). A model has been devised by Terai and Horie (1991) to separate white azuki seeds into grain-size type, medium type and grain-number type.

Azuki seed is characterized by a smooth coat, a strongly defined cotyledonary ridge and an elevated, countersunk micropyle (see cover photo) (Piper and Morse, 1914; Leese, 1958; Duke, 1981; Duan, 1989). The micropyle is the primary region of water imbibition during seed germination (Hoshikawa, 1985). The hilum is white and approximately 2.4 to 3.3 mm long and 0.6 to 0.8 mm wide (Duke, 1981). A thin caruncle encompasses the hilum cushion, and the hilum floor is concave from the surrounding seed tissue. Piper and Morse (1914) contend that the hilum is narrow, linear and not concave (Leese, 1958; Ag. Lexicon ..., 1979; Ci Hai ..., 1983; Ohwi, 1984). The seed testa as a percentage of total seed weight ranges from 9.2 to 10.3% for some cultivars (Hsu, 1989b).

Table 2.2. Morphological key for the identification of some *Vigna* species. Adapted from Jain and Mehra (1980) and extended by use of Tomooka *et al.* (1991).

Leaflets deeply lobed with 3 to 5 narrow lobes; pods glabrous or nearly so:	
Stipules small, linear to lanceolate	*V. aconitifolia*
Stipules large, oval	*V. trilobata*
Leaflets entire or very occasionally with 2 to 3 shallow broad lobes:	
Plants and pods hairy; seeds usually dull; germination epigeal; a short petiole on the primary leaf of the seedling:	
Pods point upwards, covered with long hairs; seeds oblong, usually blackish; hilum concave	*V. mungo*
Pods point sideways or down, covered with short hairs; seeds nearly or quite globular, brown to black; hilum not concave	*V. radiata*
Plants and pods glabrous or nearly so; seed smooth and shiny; germination hypogeal; a long petiole on the primary leaf of the seedling:	
Pods constricted between seeds, hilum not concave	*V. angularis*
Pods not constricted between seeds; hilum concave	*V. umbellata*

Seed colors (see cover) range from solid maroon to solid black, blue-black, green, ash gray, green-gray, brown, reddish-brown, straw yellow, white and various mottled combinations (Piper and Morse, 1914; Liu and Wang, 1959; Zhou *et al.*, 1964; Chang and Han, 1965; Thomas *et al.*, 1974; Sacks, 1977; Murata, 1980; Duke, 1981; Zheng and Zhang, 1983ab; Zhou, 1983; Chinese Ag. Resources ..., 1987; Duan, 1989; van Oers, 1989). The seed testa colors of over 1000 Chinese landraces were about 40% red, 36%

white, 12% striped in various colors and 8% green (Hu, 1984). Normally red is the preferred color for consumption, but in China's Shaanxi Province more white landraces were collected (Zhou, 1983). Yoshida *et al.* (1989b) quantified maroon azuki seed testa differences among seven cultivars. A highly significant correlation was found between seed testa chroma and the degree of seed testa yellowing with time of harvest; both features intensified with later harvest dates. Differences in seed testa color in some cultivars can be due to differences in physiological maturity (Hokkaido Agricultural Affairs Bureau, 1987).

Anatomical studies of the azuki seed have revealed that the seed testa epidermis is underlaid by a thick palisade cell layer over compactly arranged parenchyma cells; the subepidermis lacks the hourglass-shaped cells found in some *Phaseolus* spp.; starch granules are spherically shaped and loosely organized in the cotyledons; and, the hilum region has a double layer of palisade cells and a sub-hilar region plug (Takeda and Shibaoka, 1978; Chilukuri and Swanson, 1991). The starch granules were noted to be larger (20 to 25 μm wide; 20 to 40 μm long) than those observed in some other legumes (Sefa-dedeh and Stanley, 1979).

Vegetative features

Azuki seedlings begin life with hypogeal emergence (Fig. 2.2) and their first leaves are simple, opposite, cordate and long-petioled (Maekawa, 1955; Yarnell, 1965; Ag. Lexicon ..., 1979; Kay, 1979; Ohwi, 1984; Hardman, 1985; Chandel *et al.*, 1988; Duan, 1989; Dana and Karmakar, 1990). The vegetative plant is a bushy, usually erect glabrous to slightly pubescent annual that normally grows from 20 to 90 cm in length, depending upon maturity type (Piper and Morse, 1914; Matlock and Oswalt, 1963; Sacks, 1977; Kay, 1979; Chiba, 1980; Duke, 1981; Northwest ..., 1981; Ci Hai ..., 1983; Hu, 1984; Hoshikawa, 1985; Duan, 1989; van Oers, 1989). Some azuki cultivars exhibit viney growth such as *Kigane dainagon*, while others climb, with support, 1 to 3 m (Strand, 1943; Noda, 1952; Duke, 1981; Nakaseko, 1983; Hu, 1984; Ohwi, 1984; Hoshikawa, 1985; Jin and Jiang, 1993). Chinese landrace collections have growth habit ranges from 42 to 84% semi-prostrate to prostrate (Zheng and Zhang, 1983b; Zhou, 1983; Jiang and Chen, 1984).

After the first simple pair, subsequent leaves are pinnately trifoliate. Leaflets measure 5 to 10 cm long and 5 to 8 cm wide (Matlock and Oswalt, 1963; Duke, 1981; Ohwi, 1984, Chandel *et al.*, 1988) and range in shape from lanceolate to broadly ovate (Maekawa, 1955; Hoshikawa, 1985; Duan, 1989). Leaflet margins range from entire to three-lobed (Figs. 2.3 and 2.4). Lateral leaflets may have one lateral lobe on the stem side while the terminal leaflet may have a lateral lobe on each side. The middle leaflet is attached to the petiole by a long petiolule. The tips of leaflets range from acute to acuminate and when lateral lobes are present, they have tips ranging from blunt to acuminate. Most plants exhibit a progressively wider angle between the petiole and the main stem from the top of the plant to the bottom (Nakaseko *et al.*, 1979b; Nakaseko, 1983).

Azuki stipules are small, peltate, lanceolate-subulate, acuminate, often bifid, two per node, with basal appendages (Piper and Morse, 1914; Maekawa, 1955; Ohashi, 1980; Ohwi, 1984; Hoshikawa, 1985; Baudoin and Maréchal, 1988; Tateishi and Ohashi, 1990). Ohwi and Ohashi (1969) reported that the stipules attach to the stems in the middle, but Ohwi (1984) later suggested they were basifixed. The best explanation for this apparent difference was provided by Tateishi and Ohashi (1990) who reported that stipule attachment depends upon physiological age; azuki stipules are initially basifixed with truncate bases, later basifixed with auriculate bases and finally peltate. The stipules of some azuki cultivars sometimes contain sufficient anthocyanin to exhibit a purplish color.

Fig. 2.2. Ontogeny of an azuki plant. Adapted from Murata, 1983b.

The stems of some cultivars can function like stolons when they come in contact with the damp soil, producing roots and secondary plant development (Ci Hai ..., 1983; Duan, 1989). Stem color is normally green but some cultivars exhibit purplish coloration (Hoshikawa, 1985; Duan, 1989). The main stem has about 20 nodes per plant, depending upon the cultivar, with four to six branches developing between the fourth to ninth main stem nodes (Hu, 1981, 1984; Hoshikawa, 1985; Duan, 1989). Chiba (1980) recorded an average of 13.6 main stem nodes and 2.7 branches per stem, while Tasaki (1963) recorded averages of 16 to 36 cm per internodes, 2 to 15 branches per stem, 6 to 14 mm width of stem base and 21 to 105 cm length of branches at maturity. Terai and Horie (1991) found that their category of grain number type plants had relatively large numbers of branches. Secondary branching occurs (Hoshikawa, 1985), but not under normal planting densities (Nakaseko *et al.*, 1979a; Nakaseko, 1983). Jin and Jiang (1989) found two patterns of branching within azuki cultivars, i.e. multiple and single branching. The position of the first branching node is an important consideration for mechanical harvesting. Hsu *et al.* (1989b) found that the average position of the first node was 12.9 cm above ground level in seven advanced azuki breeding lines and *Kaohsiung* No. 3.

Fig. 2.3. Trifoliate leaves and fertile nodes of azuki. Adapted from Botanical Institute ..., 1972 and Duan, 1989.

Azuki has a taproot type of root system that extends spherically 40 to 50 cm from the point of seed germination. Secondary root branch development occurs later in the growing season about 10 to 20 cm below the soil surface where individual branches can reach 40 cm (Kakuta and Satoh, 1953a; Tanaka, 1968ab; Hoshikawa, 1985; Duan, 1989). Secondary roots develop from both the taproot and the hypocotyl (Tanaka, 1968ab). Root weight at maturity has been found to have a highly significant correlation with seed yield (g/m^2) (Sawaguchi and Nomura, 1978).

Root nodules, resulting from cowpea group *Rhizobium* infections, are spherical, 4 to 10 mm in diameter and begin developing one to three weeks after germination (Hoshikawa, 1985; Tsai, 1988). Maximum root nodule weight per plant is reached about two weeks after anthesis (Duan, 1989). Nitrogen fixation by *Rhizobium* decreases as soil fertility and/or the amount of applied nitrogen fertilizer increases (Tsai, 1988). The amount of nitrogen fixed by azuki's rhizobial symbiont is also affected by soil moisture content and a soil pH less than 5 (Nishimune, 1984; Wu and Hsu, 1986). Nitrogen fixation levels of 41 to 100 kg N/ha are reported for azuki in one review (Liang and Zhang, 1991). In a five-year study on three soil types, a range of 1 to 60 kg N/ha was fixed (Nishimune, 1984).

Reproductive features

Azuki can be classified by its relative degree of determinacy or indeterminacy (Duan, 1989). Azuki was classified as indeterminate, but with 85% of its pods maturing at approximately the same time, by Matlock and Oswalt (1963), Robinson (1964), Ahn *et al.* (1981) and Kim *et al.* (1981).

Fig. 2.4. Leaf types found in azuki. Leaflets range from lanceolate to broadly ovate and have margins ranging from entire to 3-lobed. Lateral leaflets on the lower leaf have one lateral lobe on the stem side while the terminal leaflet has a lateral lobe on each side. The tips of leaflets range from acute to acuminate.

Others (Nomura, 1969ab; Oikawa and Ogawa, 1969) have reported a low mature-pod ratio (mature pods/all pods harvested) of only 25 to 51%. Lawn (1979ab, 1983) found azuki to be less indeterminate than *V. radiata*, *V. mungo* and *V. umbellata*. Maturity time and environmental conditions during the growing season are reported to influence the degree of indeterminacy. Early-maturing types of azuki are normally determinate (Kay, 1979), while late-maturing cultivars are semi-determinate (Sawaguchi and Nomura, 1980). Seven Japanese azuki cultivars introduced to India were found to be determinate (Thomas *et al.*, 1974; Gill and Sharma, 1980). Twenty-three azuki cultivars were observed to be determinant during the Philippines wet season, yet were indeterminate when grown during the dry season (Hernandez *et al.*, 1978).

The inflorescences and flowers of species in the *Ceratotropis* subgenus are homogeneous (Tateishi and Ohashi, 1990). Azuki inflorescences (Figs. 2.5 and 2.6) are axillary racemes of 2 to 20 flowers. The racemes are borne on long pedicels at lower nodes of the stem or branches and on short to subsessile pedicels on higher nodes (Piper and Morse, 1914; Kay, 1979; Duke, 1981; Ohwi, 1984; Hoshikawa, 1985; Baudoin and Maréchal, 1988; van Oers, 1989; Duan, 1989). The azuki raceme possesses an unusual extrafloral nectary within a swollen part of every node on the peduncle (Takahashi, 1917). The flowers (Figs. 2.5 and 2.6) are 15 to 18 mm in length and are only yellow with color intensity variable by cultivar. The standard petal is asymmetric with a sinistral twist, prominent and has a central inner surface protuberance. The keel petals are fused along margins and asymmetrical with a sinistral curve. The left keel petal has a convex region, hornlike, called the keel pocket. The style is flattened, abruptly geniculate in the middle to upper portion, fimbriate on one side and the end has a subulate point (beak) that extends beyond the stigma. The stigma is discoid and lateral with a geniculated tip. Stamens are diadelphous (9 + 1). Bracteoles are 6 to 8 mm across, about twice as long as the calyx tube, narrowly ovate with few nerves and have an acute tip (Liu and Wang, 1959; Ohwi and Ohashi 1969; Ohashi, 1980; Ci Hai ..., 1983; Ohwi, 1984; Hoshikawa, 1985; Duan, 1989; Tateishi and Ohashi, 1990).

Fig. 2.5. Stages in the maturation of an azuki inflorescence. Note the twisted banners of the flowers, the indeterminate nature of the inflorescences, nodes where flowers have abscised, the retention of the corolla on the apex of young pods and the calyx at the base of each pod.

Azuki pollen is predominately suboblate in shape and has an average size of 37 by 41 μm (Alberto and Taylor, 1966). Chang (1988a) reported that azuki pollen is spherical and has an average diameter of 43.6 μm while Horvat and Stainier (1980) stated that azuki pollen could be as large as 55 μm. Azuki pollen is also characterized by large reticulations and triporate, subgranulate infrareticulum structures (Alberto and Taylor, 1966; Ohashi, 1980; Baudoin and Maréchal, 1988; Tateishi and Ohashi, 1990). Azuki pollen wall and pore area characteristics are described by Horvat and Stainier (1980). The presence of twisted reticulations has been used to support the inclusion of azuki in the genus *Vigna*, rather than *Phaseolus*, and the subgenus *Ceratotropis* (Ohwi and Ohashi, 1969; Horvat and Stainier, 1980). Alberto and Taylor (1966), for example, reported that the morphology of azuki pollen is similar to that of pollen from *V. aconitifolius*. The average stainability of azuki pollen produced by four different cultivars was 82.5% (Ahn, 1988). The optimum temperature for pollen germination is approximately 25°C (Tasaki, 1957b). *In vitro* germination of azuki pollen is best achieved with a 1% agar media having 30% sucrose and a pH of 5.6 (Tasaki, 1957b).

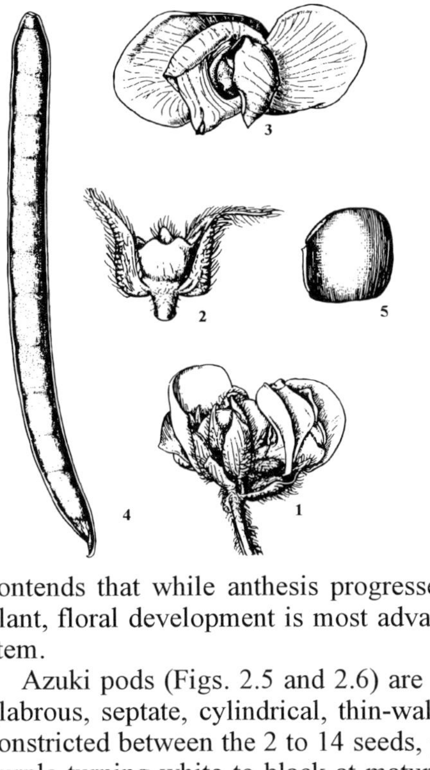

Fig. 2.6. Inflorescence (1), calyx and bracteoles (2), flower (3), pod (4) and seed (5) of azuki. Adapted from Yamaguchi, 1992.

Azuki inflorescences start developing 18 to 23 days before anthesis and flower primordium 21 days before anthesis. Anthesis normally begins in the morning and can continue for up to 40 days (Takahashi, 1917; Hoshikawa, 1985; Duan, 1989). Floral development progresses acropetally with anthesis beginning on lower main stem nodes and branches first (Nomura and Asanuma, 1970; Satoh, 1979). Harada (1953) contends that while anthesis progresses from the bottom to the top of the plant, floral development is most advanced at the middle nodes of the main stem.

Azuki pods (Figs. 2.5 and 2.6) are 5 to 13 cm long, 0.5 cm in diameter, glabrous, septate, cylindrical, thin-walled with a short blunt terminal beak, constricted between the 2 to 14 seeds, pendulous at maturity, green or red to purple turning white to black at maturity and in clusters of two to six pods per pedicel and 4 to 195 pods per plant (Georgeson *et al.*, 1891; Piper and Morse, 1914; Liu and Wang, 1959; Matlock and Oswalt, 1963; Zhou *et al.*, 1964; Sacks, 1977; Ag. Lexicon ..., 1979; Kay, 1979; Duke, 1981; Hu, 1981, 1984; Ci Hai ..., 1983; Nakaseko, 1983; Zheng and Zhang, 1983b; Hoshikawa, 1985; Duan, 1989; van Oers, 1989; Dana and Karmakar, 1990). Seeds per pod averaged 6.72 for the cultivar *Chagarawase*, 6.64 for *Sakae shouzu* and 4.77 for *Akane dainagon* according to Murata and Narugawa (1978). An average of 6.61 seeds per pod for 190 cultivars was reported by Chiba (1980).

Researchers at the Hokkaido Prefectural Tokachi Agricultural Experiment Station (1978) observed the pattern of flower and pod development for four cultivars of azuki. Flowering duration was about 30 days, the average number of flowers per plant ranged from 48 to 65 and the number of mature pods ranged from 26 to 36. The percentage of mature pods from flowers ranged from 47 to 55. Similar research was conducted by Kim *et al.* (1981) in Korea. They found a range of 60 to 133 flowers and 13 to 45 pods per plant based on planting date, and reported that approximately 57% of all flowers produced by azuki were fertilized, with about 46% of those developing into mature pods. They also found that pod abscission, rather than flower abortion most affected the number of mature pods (also see Lee *et al.*, 1991). Floral and pod abortion rates as high as 69 to 76% have been observed in azuki (Lu and Ma, 1985).

Azuki blooms two to three times when planted early in the growing season or once if planting is delayed (Harada, 1953; Kim *et al.*, 1981). Flowers that develop early in each bloom produce a higher average seed weight than flowers that appeared later in the bloom (Nomura, 1969ab). Matlock and Oswalt (1963) reported that seeds contributed about 50% of mature pod weight. Dry seed weight per plant ranges form 3 to 78 g (Nakaseko, 1983; Hu, 1984). A harvest index (HI; seed weight/biological weight) average of about 28% can be calculated from Chiba's (1980) data on 190 cultivars although HI values of over 50% are not uncommon.

3

PHYSIOLOGICAL CHARACTERISTICS

GROWTH AND DEVELOPMENT

From planting until seedling establishment, azuki is recognized as a slow growing plant, especially in temperate regions (Piper and Morse, 1914; Yamanouchi, 1964b; Alford and Tibbitts, 1971; White, 1972; Japan Bean Fund ..., 1987; Sawaguchi, 1987). Emergence rates are 2 to 3 days slower for azuki than that for mungbean, green gram and rice bean (Lawn, 1979ab). Seed germination (Shida and Danjo, 1971; Lee and Karunanithy, 1990) and seedling emergence varies between 7 and 20 days depending upon soil temperature and the soil moisture (Zhou *et al.*, 1964; White, 1972; Kay, 1979; Duke, 1981; Dai and Zhang, 1982; Zhou, 1983; Duan, 1989; Durgan *et al.*, 1989). The seeds require a minimum germination temperature of 6 to 10°C, an optimum of 30 to 34°C; and a maximum of 42 to 44°C (Hoshikawa, 1985; Ushirogi *et al.*, 1986; Wu and Hsu, 1986; Duan, 1989). A soil temperature change of 1°C at 10 cm will cause a corresponding 2-day germination time difference (Ushirogi *et al.*, 1986). The new Chinese azuki cultivar *Longxiao dou* No. 1, has an optimal germination temperature requirement of only 25°C (Zhang *et al.*, 1981).

Cultivars grown in subtropical Taiwan are reported to perform best if soil moisture and air temperatures are high from planting to flowering, but cool and dry from pod development through physiological maturity (Wu and Hsu, 1986). In contrast, the seedlings of cultivars such as *Erimo* grown in temperate Washington State, do best in drier soil, which is warmer under cool spring growing conditions.

Azuki seedling dry weight and leaf area index (LAI) 10 days after emergence is highly correlated to initial seed dry weight (Nakaseko and Gotoh, 1982), while early growth rate in azuki is positively correlated to leaf area (Nakaseko, 1983). Azuki growth rates 30 days after emergence measured as grams per plant per day were found to be less than those for kidney bean and soybean (Nakaseko and Gotoh, 1982, 1983; Nakaseko, 1983). Azuki was recorded to have a slower rate of overall growth as compared to black gram and mungbean (Lawn, 1983). Yamanouchi (1964b) contends that azuki vegetative development is not only inferior to soybean and kidney bean but is characterized by shorter periods of vigorous reproductive development. Some Japanese azuki cultivars exhibited a net assimilation rate (NAR) peak, either at the beginning of primary branch elongation and/or at pod-filling, when LAI and E_{PAR} (crop growth rate/photosynthetically active radiation) are at maximum (Nakaseko *et al.*, 1979a; Nakaseko, 1983; Nakaseko and Gotoh, 1983). LAI reached a maximum of 5 at 95% light interception for the cultivar *Takara* (Nakaseko and Gotoh, 1983). Singh and Singh (1983) developed methods for determining leaf area from linear measurements.

Sakuratani (1974) studied the temperature environment in the air layer over a young azuki crop with an LAI of 1. Some of his discoveries included: the value of the zero-plane displacement decreased with increasing wind velocity; with a wind velocity between about 0.5 and 2.7 m/s, exchange velocity was between 7 and 14 cm/s; and, the LAI of 1 had values for the Bowen ratio of 0.35 to 0.45.

A growth manipulation study determined that the suppression or the physical removal of the main vegetative stem caused growth compensation through branching and increased seed yields (Yoshida *et al.*, 1989a). In contrast, removal of branches or restriction of anthesis to a two-week period reduced seed yield. Breaking but not removing the lower branches of azuki at flowering caused a dramatic increase in the number of pods per plant and seed yield as compared to the control (Wang, 1990).

Lawn (1979ab, 1983) determined that there was only a marginal increase in LAI from the beginning until the end of flowering. Petiole length correlated positively with the LAI at each node; the longer the petioles, the larger the LAI. Highest total leaf dry matter occurs between the seventh and ninth nodes for most cultivars (Nakaseko, 1983; Yamanouchi, 1964b). Nakaseko and Gotoh (1981) found that the LAI at 95% light inception was 5.0 for *Takara shouzu* and 3.3 for *Akatsuaki dainagon*. Sakuratani (1974) demonstrated that an azuki field canopy has more intensive thermal convection, higher heat flux, malle exchange velocity and a different air temperature than a grass field, and these can be attributed to differences in LAI and thermal properties. Among several *Vigna* species, azuki has been found to have a lower crop growth rate (CGR) than black gram, mungbean and rice bean across 17 planting dates, but a higher harvest index (HI) (Lawn, 1979ab). CGR in azuki is highly correlated with the duration of total leaf area (Tsai, 1982).

Sawaguchi and Nomura (1980) looked for differences in CGR between cultivars. They could not find significant differences during early growth stages but with the commencement of anthesis, total plant dry weight increased dramatically and cultivar differences became apparent. The differences in CGR for early maturing and later maturing cultivars were found to be mainly due to increases in main stem height, the degree of primary branching, pods per plant and seeds per pod (Hu, 1984).

Azuki is traditionally classified as a quantitative short-day plant with varietal variation in daylength sensitivity (Miyagi and Yasukawa, 1934; Allard and Zaumeyer, 1944; Kakuta and Satoh, 1953b; Tasaki, 1957a, 1965; Nomura *et al.*, 1965; Purseglove, 1974; MacKenzie *et al.*, 1975; Lawn, 1979ab; Kay, 1979; Summerfield and Wien, 1980; Japan Bean Fund ..., 1987; Duan, 1989; van der Maesen and Somaatmadja, 1989). One Indian, one Japanese and 10 Korean accessions were found to be day-neutral when grown in Hawaii (Hartmann, 1969; Duke, 1981). The Asian Vegetable Research and Development Center (AVRDC) tested 103 accessions for photoperiod sensitivity and found that 35% were day-neutral and 61% did not begin anthesis until a critical daylength of 12 hours was reached; accessions collected from higher latitudes were day-neutral (MacKenzie *et al.*, 1975; Ahn *et al.*, 1981). Strong correlations are found between daylength sensitivity and maturity type. Early maturing azuki cultivars are less sensitive to daylength while late-maturing types are very short-day sensitive (Tasaki, 1965). Three azuki cultivars tested remained in a vegetative state when maintained under 24 hours of illumination, beginning one month after emergence (Kakuta and Satoh, 1953b). Azuki subjected in the vegetative stage to light periods shorter than the critical daylength required for anthesis initiation results in earlier flowering, fewer main stem nodes, shorter plants and reduced yields (Duan, 1989). Temperature strongly interacts with photoperiod, e.g. short days and high temperatures will make any sensitive cultivar flower by 30 days after planting (Tasaki, 1960).

Azuki cultivars can be classified as early, intermediate or late based on the number of days required for maturation (Nomura and Satoh, 1977). Early maturing cultivars were found to be 7.1 cm shorter than intermediate maturing cultivars, which were 5.8 cm shorter than late maturing cultivars. Early maturing cultivars produced an average seed yield of 1610 kg/ha and 47.8% of all pods were less than 10 cm from the ground. Intermediate maturing cultivars yielded 2220 kg/ha and had 36.6% of all pods less than 10 cm from the ground; late maturing cultivars yielded 2470 kg/ha and had 21.7% of all pods less than 10 cm from the ground.

Air temperatures as well as photoperiod effect the rate of azuki growth and development (Tasaki, 1957a). Lawn (1979ab) hypothesized that in regions with large diurnal temperature variations during the growing season, azuki is influenced more by ambient temperature than photoperiod. Lawn reported that 75% of final azuki dry matter weight variation spanning 17 planting dates was attributable to differences in temperature. The cultivars tested by Park and Hong (1970) and Lawn (1979ab) might have been classified by Tasaki (1965) as intermediate maturity types which are sensitive to both temperature and daylength, especially at floral initiation. Azuki cultivars classified as early, intermediate and late maturing all required increasing temperatures for floral initiation as daylength increased from 8 to 15.5 hours (Tasaki, 1965). A negative correlation was found between daylength sensitivity and temperature in azuki cultivars; early maturing, daylength insensitive cultivars were temperature sensitive, while late maturing, daylength sensitive cultivars were temperature insensitive (Tasaki and Honma, 1965). An optimum temperature for azuki flowering is 26°C (Harada, 1953; Wu and Hsu, 1986). Duan (1989) stated that the best temperature for floral induction, initiation and development is 24°C; temperatures below 16°C during these physiological stages can result in reduced flower and pod numbers.

The development sequences of azuki pods and seeds have been found to be sigmoidal in nature. Murata and Narugawa (1978) found that seed weight increases rapidly about 24 days after anthesis for all cultivars tested and the maturity characteristics of each cultivar determined when the highest seed weight was achieved.

Nyctinastic movement of azuki leaves is controlled by changes in turgor pressure as affected by light intensity and wavelength (Tasaki, 1961). There are periodic rotational movements of the two unifoliate seed leaves of azuki when exposed to continuous light (Alford, 1970; Alford and Tibbitts, 1971; Nakaseko, 1984a). Rotational movements began at the midvein rotational axis three days after the leaves unfolded; amplitude and time duration remained constant under most conditions for more than six days after initiation (Alford and Tibbitts, 1971). The two leaves of plant were found to be closely synchronized in their individual rotational movements, but there was no synchronization of movement between plants. The spontaneous commencement and progression of circadian rhythms of azuki unifoliate leaves have been studied (Alford and Tibbitts, 1970). These researchers determined a mean rhythm cycle of 27.3 hours which did not change with leaf aging, and observed that unifoliate leaves on the same plant and on different plants were not always synchronized.

A two-dimensional photon counting system was able to detect emission patterns of ultraweak light originating from the wound sites of physically injured azuki seedlings, and was used to measure the response of the plants to H_2O_2 (Suzuki *et al.*, 1991). In microgravity created by a three-dimensional clinostat, azuki root grew in random directions, while shoots continued some directional growth (Hoson *et al.*, 1992). When lateral roots are about to emerge, the normal electric potential of the root has been found to drop at the point of emergence (Hamada *et al.*, 1992).

BIOCHEMISTRY OF GROWTH

Various chemical isolation and characterization studies have been conducted using azuki. The amino acid sequence of the azuki protease inhibitors AB-I, AB-IIa and AB-IIc have been isolated and determined (Ishikawa *et al.*, 1985). Crystallization of AB-IIa, AB-I and AB-I bound with trypsin has been conducted, while the nature of the AB-I binding site for trypsin was determined later (Tanaka *et al.*, 1983; Tsunogae *et al.*, 1986). Azuki trypsin-chymotrypsin inhibitors ABI I, II and II' were found to be 74 to 77% homologous with those of *Phaseolus vulgaris* var. *nanus* cv. Borlotto (Funk *et al.*, 1993). Protease inhibitors I-A, I-A' and I-B, all which have two reaction sites consisting of arginine-serine and lysine-serine peptide bonds, have been identified and characterized (Yokota *et al.*, 1983). Proteases, glucosidases and glucuronidases could not be detected in azuki seedlings using histochemistry, although other acid and neutral hydrolases were found (Gossrau, 1991).

The flavonol glycoside, 3-robinoside 7-rhamnoside, has been isolated and identified in azuki (Harborne, 1971), and its urease characterized (Bailey and Boulter, 1971). Suspensions of azuki cells have been treated with actinomycin D and found to produce three different isoflavone glycosides (Kobayashi and Ohta, 1983). Kojima *et al.* (1991bf) found the novel diglycosyldiacylglycerol (DGDG), galactosyl(β1 leads to 6)-galactosyl(β1 leads to 3')-diacylglycerol (B isomer) in azuki beans and traced its concentrations from the seed through different levels of plant development. Concentrations of the B isomer were highest in seeds and least in plant parts. Nozawa *et al.* (1992, 1993) have found that the reconstitution of an azuki subtilisin inhibitor (see Yoshikawa *et al.*, 1985) from two fragments is temperature dependent.

Peroxidase activity in the roots and shoots of azuki seedling was found to increase from one to three days after germination, reach a maximum level at four days and then decline (Jun *et al.*, 1983). An electrophoresis study demonstrated that stems of seedlings exhibited less peroxidase and more phosphates banding than primary leaves while esterase activity was found in primary leaves and stems but not roots (Bassiri and Adams, 1978). Bound and soluble peroxidases have been found in the cell walls of azuki epicotyls and ascorbic acid has been found to regulate their oxidation of phenolics (Takahama, 1979). Hydrogen peroxide content and catalase activity were correlated in azuki seedling tissue; both decreasing, for example, in cotyledons, roots and the epicotyl five days after germination (Jun *et al.*, 1983).

The purification product of a *Lentinus edodes* mycelium culture was found to promote adventitious root development from epicotyl cuttings (Mitsuhashi-Kato *et al.*, 1985). Others have found an epidermal growth factor that promotes adventitious root formation in epicotyl cuttings (Kato *et al.*, 1993). Iwanami (1982) found four different acids that would negate the normal negative geotropism and positive phototropism of azuki seedling cuttings. The surface electrical potentials that develop along the epicotyl as a result of geotropism increase along the upper side and decrease along the lower side when the epicotyl is horizontal (Imagawa *et al.*, 1991). Actinomycin D and 2,4-D have been studied for their effects on root development on stem cuttings (Mitsuhashi and Shibaoka, 1981).

Nishitani and Masuda (1979) determined the upper portion of intact azuki epicotyls contained more pectic substances but less α-cellulose than lower portions. These researchers also demonstrated a similar demarcation of lower versus upper epicotyl tissue based on the percentages of two neutral sugars, xylose and galactose.

Mannitol, which reduces turgor pressure (Tatt, 1980), was found to inhibit polysaccharide increases in azuki cell walls induced by auxin and sucrose (Nishitani *et al.*, 1979).

Methyl viologen-linked nitrite reductase (EC 1.7.7.1), which catalyzes the six electron reduction of nitrite to ammonia, has been isolated from azuki roots. The enzyme has an estimated total molecular weight of 62,000 for two sub-units and absorbance ratios of 0.39 (A_{573}/A_{397}) and 0.30 (A_{397}/A_{280}) (Nagaoka *et al.*, 1984; Ishiyama and Tamura, 1985; Ishiyama *et al.*, 1985). Three different phytoalexin elicitors and vanadate were determined to cause an elevation in phenylalanine ammonia lyase (PAL) activity and isoflavone glycosides concentrations in suspension-cultured azuki cells (Hattori and Ohta, 1985).

The highest carbohydrate concentration in five different azuki plant parts occurred at different times in the growth cycle with the maximum accumulation sequence over time being leafblade-petiole-stem tissue-pod-seed. (Sawaguchi and Nomura, 1981). Carbohydrates are accumulated in the leaves until flowering, in the petioles during rapid vegetative growth and in the pods and seeds during pod-filling (Sawaguchi, 1986). Natural ^{13}C values in azuki were least depleted in root nodules and most depleted in leaves, with levels in mature seeds being intermediate to the two (Yoneyama and Ohtani, 1983).

GROWTH REGULATOR EFFECTS

Oikawa and Ogawa (1969) hypothesized that applications of the auxin transport inhibitor (TIBA) (2,3,5-triiodobenzoic acid compound with N-methylmethanamine [1:1]) could increase the harvest index of azuki by reducing vegetative growth and improving flower and pod development. Azuki plants treated with 25, 50 and 100 ppm TIBA were shorter but displayed increased branching. Azuki treated with 25 ppm TIBA had higher yields, and 50 and 100 ppm treated plants had more mature pods at harvest than the controls. These results may have been influenced by the unusual weather during growing season and plant height per plant density interactions. Yoshida *et al.* (1986a) conducted a similar TIBA experiment and found that plants treated with 50 and 25 g/ha had a lower LAI during the reproductive period, produced fewer pods and seeds per pod and had reduced yields as compared to untreated azuki.

Azuki epicotyl growth is influenced by brassinosteroid (Cerana *et al.*, 1983) and was enhanced by the plant hormone auxin (IAA) which induced polysaccharide synthesis in cell walls (Masuda, 1985; Masuda *et al.*, 1989). When in the presence of a sucrose or glucose substrate and adequate turgor pressure, the polysaccharides xylose, galactose and cellulose increased significantly while noncellulosic sugar content changed only slightly (Nishitani and Masuda, 1979, 1981, 1982a; Nishitani *et al.*, 1979). Galactose does not inhibit IAA-induced growth in azuki but did inhibit an IAA-induced decrease of the pH in the culture solution (Kokubo *et al.*, 1993) and does not inhibit UDP-glucose in azuki hypocotyls or auxin-induced elongation of azuki epicotyls (Yamamoto *et al.*, 1988).

Xyloglucan, which has been proposed as an antiauxin and has been found to inhibit 2,4-D induced growth of pea, did not inhibit IAA-induced growth of azuki epicotyls, even when the epicotyls were abraded; fucosyl-lactose also did not inhibit IAA-induced growth (Hoson and Masuda, 1991ab). However actinomycin D inhibits epicotyl elongation induced by IAA or IAA with GA_3 (Kaneta *et al.*, 1993). Xyloglucan antibodies specific for hepta- and octasaccharides also inhibited auxin-induced elongation of epicotyls and cell wall loosening in azuki but not oat (Hoson *et al.*, 1991b).

Nishitani (1992) has developed a novel way to monitor endo-xyloglucan transferase activity by observing changes in the molecular weight distribution of substrates. Nishitani and Tominaga (1992) isolated an endo-type glycosyltransferase from azuki epicotyls which is a glycoprotein with a mass of about 33 kDa. Induced structural changes in cell wall xyloglucans of azuki have been studied by Nishitani and Masuda (1982a, 1983).

The epidermis of azuki is highly sensitive to auxin and responsive to fusicoccin (Branca *et al.*, 1988). These reseachers suggested that the epidermis was the target for auxin activity and epidermal integrity was necessary for auxin-induced elongation. An extensive turnover occurs in two pectin fraction components of azuki cell walls, galactans and polyuronides, that coincides with auxin-induced cell wall extension. This suggests that galactans and polyuronides have major roles in the extension of azuki cell walls (Nishitani and Masuda, 1980).

The promotion of azuki epicotyl growth by auxin was demonstrated to be enhanced by gibberellins (GA) which promoted cell elongation and prepared the tissues for the auxin action (Shibaoka, 1972; Nishitani and Masuda, 1982b). Colchicine reversed the GA_3 effect almost completely but not the effect of auxin. The basal portions of azuki epicotyl sections were nonresponsive to IAA or GA treatments alone, but a GA and sucrose treatment promoted xylan and cellulose synthesis (Shibaoka *et al.*, 1977; Nishitani and Masuda, 1982b). Gibberellin applied alone was unable to induce the synthesis of cell walls or tissue growth (Shibaoka, 1972; Nishitani and Masuda, 1982b; also see Takeda and Shibaoka, 1978). The positive growth effect of GA_3 and the negative growth effect of kinetin on auxin-induced azuki epicotyl elongation were reversed by the presence of coumarin (Hogetsu *et al.*, 1974). Microtubules in the epidermal cell wall and the effect of hormones on their development have been studied by Shibaoka and Hogetsu (1977), Takeda and Shibaoka (1981ab) and Mizuno and Shibaoka (1982).

IAA-induced elongation of azuki epicotyls has been found to be suppressed by a fucose-binding lectin from *Tetragonolobus purpureus* and polyclonal antibodies raised against xylogulcan heptasaccharide when the cuticle has been abraded (Hoson and Masuda, 1991a; Hoson *et al.*, 1993)

The growth retardant, daminozide (butanedioic acid mono [2,2-dimethyl hydrazide]), was applied at 1125 kg/ha in a solution of 2500 to 5000 ppm to Chinese azuki cultivar *Ji* No. 1 just before anthesis. Plant height and the days to maturity were reduced, the number of branches per plant was unaffected and 100-seed weight and yields per plant were increased (Dou *et al.*, 1989; Wang, 1990). Ethylene (3 to 10 ppm) and 2,4-D (10 to 50 μM) applied after seed imbibition, shortened the roots and hypocotyls while the ethylene-releasing material ethephon (2-chloroethyl phosphonic acid) was ineffective. Azuki hypocotyls were more sensitive to these compounds than the roots (Chang, 1986). Imazeki (1993) has obtained a patent for the cDNA of ethylene-induced acidic chitinase, which was described by Ishige *et al.* (1993b). Ishige *et al.* (1991, 1993a) also found that ethylene caused the accumulation of seven different proteins in leaves of azuki. One of these was characterized after cloning and found to be a cationic isozyme of peroxidase.

Chlorophyll content of azuki demonstrates three peaks in concentration during its growth period (Jin and Gu, 1990). Nuclear-encoded subunits of PSI in chlorophyll, except psaG and psaK, were generally found to have two isoforms (Obokata *et al.*, 1993). RuBP carboxylase activity, photosynthesis and soluble protein content were all increased by treating the seeds of azuki with ß-pyridin propyl alcohol (Zhu, 1991). Wang (1990) recommends the spray application of an unidentified Chinese growth substance called *Bijou* which nearly doubled the yield of azuki in research plots. This substance caused the plants to be shorter, increased pods and seeds per plant and slightly increased seed weight. Pods are a strong sink for photosynthate (Kogure *et al.*, 1980).

MINERAL NUTRITION

Macronutrient concentration changes over the lifecycle of azuki and responses to different minerals have been studied. Kushizaki *et al.* (1961) made weekly observations and graphed the accumulation of dry matter and composition of N, P, K, Ca and Mg in azuki over its growing season (Figs. 3.1 and 3.2).

Nitrogen in azuki has been categorized into three source of origins by Nishimune (1984), i.e. nitrogen fertilizer, which is the dominant nitrogen source for early growth, and nitrogen derived from the soil and from symbiotic fixation which are the primary sources for the reproductive and physiological maturity phases of growth and development. Azuki uptake

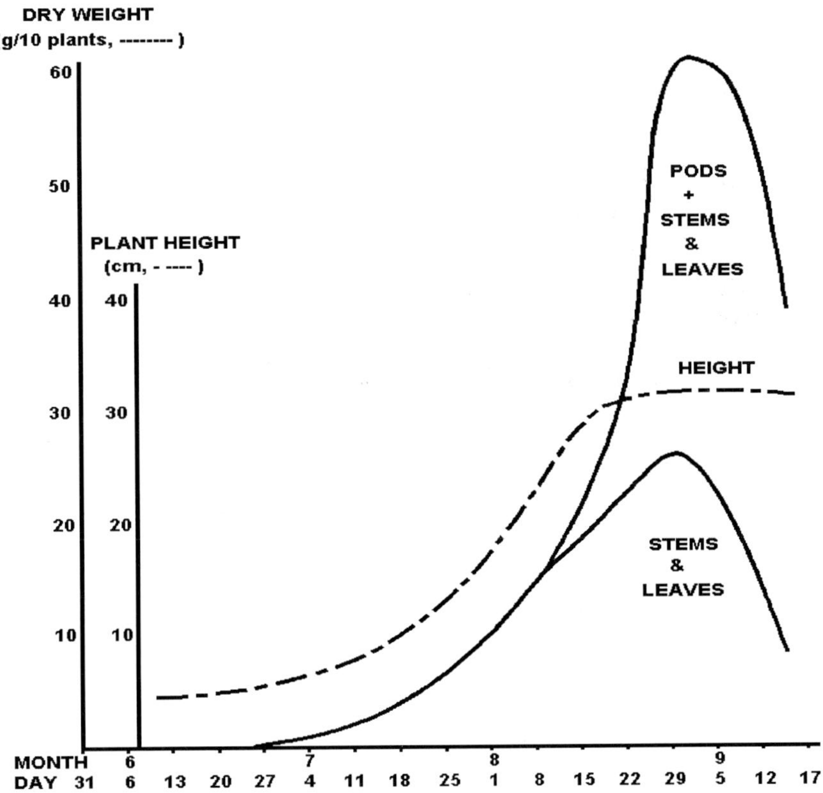

Fig. 3.1. Accumulation of weight and height during the maturation of a dainagon azuki cultivar on Hokkaido, Japan (adapted from Kushizaki *et al.*, 1961). The plants germinated on 2 June.

of nitrogen is slow until anthesis, after which it increases dramatically (Sawaguchi and Nomura, 1980; Sawaguchi, 1987). Maximum nitrogen content of leaves is reached a week before flowering, then drops by about 50% by flowering (Yamanouchi, 1964b). Nitrogen absorption levels have a high correlation with azuki root weight, and root absorption and azuki dry matter increase with nitrogen levels up to 200 kg/ha (Sawaguchi and Nomura, 1978). Nitrogen has also been shown to be positively correlated to mean seed size (Yoshida and Sato, 1990). Increasing nitrogen applications from 23 to 46 kg/ha resulted in an increase in the number of mature pods but caused a decline in the number of seeds per pod and overall yield (Furusawa *et al.*, 1974). Others (Hokkaido Prefecture ..., 1972b) found that increased fertilizer generally increased plant height, number of branches and pods, gross plant weight, yield and bean weight. On-farm testing has shown

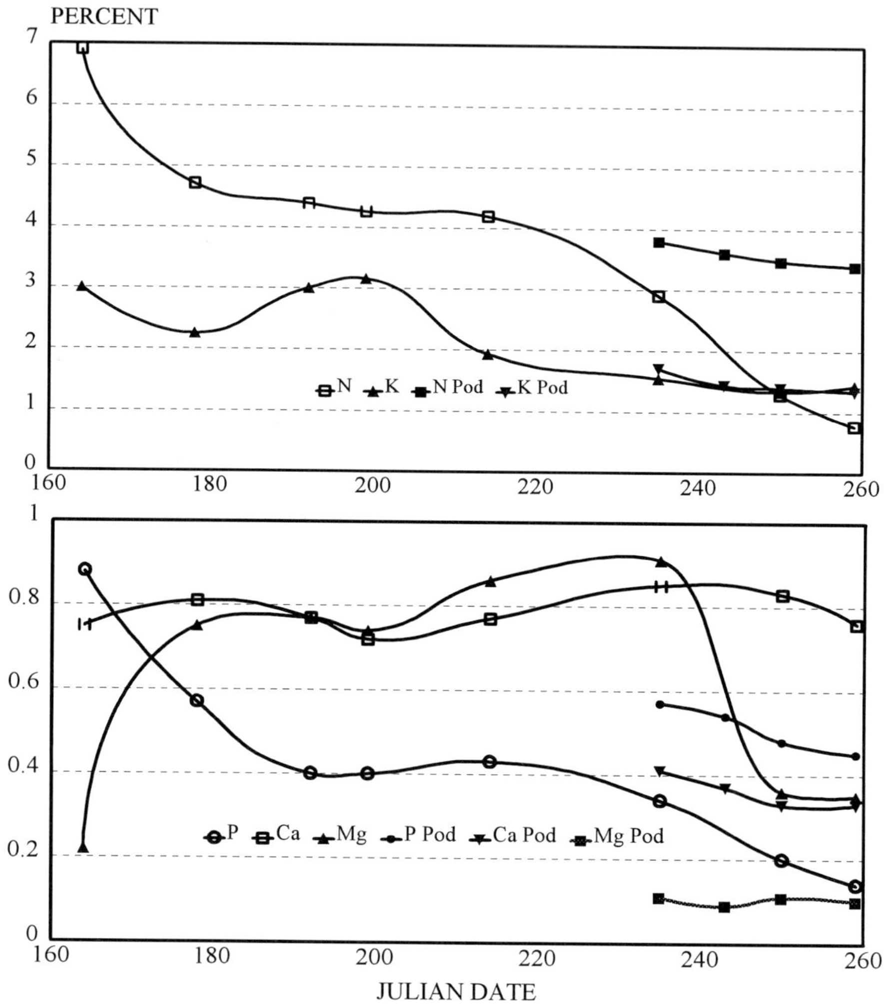

Fig. 3.2. Changes in the percent nutrient composition on a dry weight basis of dainagon azuki plants during their life cycle on Hokkaido, Japan (adapted from Kushizaki *et al.*, 1961). The long lines represent the elemental composition of stems and leaves, while the short lines on the right represent pods. From the left, data points represent the following sampling dates: 6/13, 6/27, 7/11, 7/18, 8/2, 8/23, 8/31, 9/7 and 9/16.

that nitrogen applications of over 30 kg/ha fail to improve yields and decrease the effectiveness of *Rhizobium* (Japan Bean Fund ..., 1987).

Osaki *et al.* (1992) developed a carbon-nitrogen interaction model for the growth of several crops including azuki. The model used for azuki was DM (plant dry weight) + DM0 (starting dry weight) + CNI' (carbon-nitrogen index). CNI' changed with the amount of nitrogen absorbed by harvest (Nh). Nh had a hyperbolic curve fitting the equation CNI' + 1/(aNh + b). The coefficient 'a' was nearly zero for azuki, which kept CNI' almost constant regardless of Nh.

Nitrogen fixation reaches a maximum at the beginning of pod-filling and then decreases rapidly (Sawaguchi, 1986). Allantoin-N is the principal form of N, originating from nitrogen fixation, in xylem transport (up to 80%) while amino acids, amides and NO_3-N (up to 90%), especially asparagine, are found in the xylem when nitrogen fertilizer is applied. Ureides are also nitrogen compounds produced in nodules and transported in azuki. Vance *et al.* (1985) traced carbon from $^{14}CO_2$ label (see Tazaki, 1979) in azuki and found 12% in the ureides xylem sap. The rest of the carbon was found in labeled TCA cycle intermediates that balance inorganic charged cations in the xylem. These researchers suggested that CO_2 fixation in azuki nodules provides little of the required carbon for the assimilation and transportation of fixed nitrogen. The bacteroid fraction of azuki nodules was found to have higher concentrations of ^{15}N than concentrations found in leaves or the atmosphere (Yoneyama *et al.*, 1984, 1987).

Phosphorus concentrations were highest in azuki leaves about 4 to 10 days before flowering and then gradually decreased (Yamanouchi, 1964b). The application of superphosphate at 100 to 200 kg/ha with 2 metric tons (mt) manure/ha resulted in the most azuki dry matter accumulation and yields as compared to eight other soil treatments (Sawaguchi and Nomura, 1978). During the early growth stage when growth is most dependent upon phosphorus availability, CGR is closely correlated with seed yield (Sawaguchi, 1986).

Phosphorus applications of up to 880 kg/ha increased azuki yields on volcanic soils, but began to peak at 220 kg, and most applied about 66 kg (Japan Bean Fund ..., 1987). Near Tianjin, China, phosphorus was applied in an experiment at rates between 15 and 165 kg/ha. Seed yield and economic yields were highest at 100 kg P/ha (2600 kg seed/ha), but seed yield per unit of P was highest at 15 kg P/ha (Mou and He, 1985). Tanaka (1982) studied the response of azuki to phosphorus in culture solution and found that azuki growth peaked at 0.05 ppm P. This response was similar to rice and maize. These three crops were more tolerant to low P than wheat, barley, sugar beet, tomato and Chinese cabbage. In contrast, Tadano and Sakai (1991) studied acid phosphatase activity in the roots of azuki and other legumes under conditions of phosphorus sufficiency and deficiency. All legumes showed an increase in phosphatase activity under deficiency, but azuki showed the least increase of activity, only 1.5 times, while lupin increased 19.9 times.

Azuki requires high levels of calcium, especially after anthesis (Japan Bean Fund ..., 1987). Calcium translocation was determined to be highest in azuki stem tissue and accumulated in leaf tissue with little translocated to other plant parts such as developing seeds (Yamanouchi, 1964a). Azuki is susceptible to nickel toxicity, which is usually concentrated in soils of volcanic origin (Duke, 1981), and lead (Pb) toxicity (Wagatsuma *et al.*, 1985) when produced in drained rice paddy fields. Azuki seed yields decreased more than 33% when plants were grown with 300 mg Pb/kg soil as compared to production in soil with no lead (Kim *et al.*, 1988). Application of more than 4 kg/ha of boric acid for sugar beet production on volcanic soils in the Tokachi district of Hokkaido was determined to cause toxicty symptoms in a subsequent crop of azuki (Mori *et al.*, 1966; Kay, 1979).

Azuki was injured when a minimum of 1.2 ppm of arsenic in a solution with 0.013 molar (M) sulfuric acid and 0.05 M hydrochloric acid was applied to orchard soils (Hanada *et al.*, 1975). These researchers found that arsenic was bound when iron-derived sulfuric acid and calcium carbonate were applied together, while separate applications of either of the same materials or zinc-derived sulfuric acid, or repeated applications of phosphoric acid did not tie up arsenic in enriched soils (Hanada *et al.*, 1975). A pH at or higher than 7.8 caused zinc and iron deficiency chlorosis in azuki (Sperbeck,1981; Hardman, 1985).

WATER USE REQUIREMENTS

The water demands of azuki vary with the stages of growth and development. Tasaki (1959) observed the effect of different levels of sub-irrigation and soil moisture on plant height, dry weight, leaf drop, branch length and numbers of leaves, nodes and branches. He reported that early azuki growth and development up to 40 days after planting proceeded normally at soil moisture contents as low as 40% of field capacity while a soil moisture range of 80 to 90% field capacity was required for optimal azuki seed production. Adequate soil moisture is especially critical for azuki at anthesis (Ag. Lexicon ..., 1979; Ci Hai ..., 1983). Total azuki seed dry matter and nitrogen per unit area was not affected by a cyclical soil moisture treatment of 2.8 pF from the onset of the first trifoliate to physiological maturity (Trung *et al.*, 1985). Azuki does not perform as well as soybeans when produced on highly water-saturated and poorly drained soils of Taiwan; any soil with a high percentage of moisture at planting can encourage azuki seed rot and poor stand establishment (Wu and Hsu, 1986). The presence of a water-saturated zone 10 cm below the soil surface during the complete growth cycle resulted in low total dry matter and pods per plant. Tasaki (1959) hypothesized that early maturing azuki cultivars would achieve maximum seed yields if the water-saturated zone was at least 30 cm below the soil surface and 50 to 70 cm for late maturing cultivars.

Azuki has a comparatively greater tolerance for heavy rainfall than other grain legumes, although azuki yields are suboptimal on heavy, insufficiently drained soils (Kay, 1979; Wu and Hsu, 1986; Breene and Hardman, 1987). Under East Asia's monsoon growing conditions, flooding from typhoons and other disturbances are frequent. Sunada *et al.* (1975) states that azuki is sensitive to excessive moisture. Goto *et al.* (1983, 1985) tested azuki's ability to withstand 3 cm of standing water at various growth stages over a range of duration from 3 to 36 hours; some pots were also subjected to 48 hours of waterlogging. Damage generally increased with flooding duration. The flowering stage was generally the most affected (wilting, abortions, etc.) although flooding at the fourth trifoliate growth stage eventually resulted in the largest yield losses (20 to 74%). Flooding at all but the most mature stage caused a delay in maturity. Dew accumulation during the winter on fall-planted azuki was recorded by Hsu (1983) in order to determine its importance as a component of the total water requirements of azuki produced in southwestern Taiwan.

ENVIRONMENTAL RESPONSES

Substantial research has demonstrated the effects of air temperature and various planting times and densities on azuki vegetative and reproductive performance. Differences in cultivation practices and the application of growth regulators have also been studied in detail.

Temperature experiments

Azuki performs best under warm and dry conditions; it tolerates high temperatures but is sensitive to frost or cold weather conditions (Motomiya and Ito, 1972; Purseglove, 1974; Sunada *et al.*, 1975; Ag. Lexicon ..., 1979; Kay, 1979; Lawn, 1979ab; Duke, 1981; Ci Hai ..., 1983; Japan Bean Fund ..., 1987). After planting, azuki germination rate has a significant correlation with soil temperature; as temperatures drop from 14 to 11°C, the germination time increases from 12 to 22 days (Japan Bean Fund ..., 1987). The effect of soil temperature decreases and air temperature increases after germination in temperate regions, i.e. azuki yields and growing season air temperatures have shown positive correlation (Shiozaki and Tsuchiya, 1985). Azuki plant height, number of main stem nodes, number of branches and yield components (pods per plant, seeds per pod, yield per plant) are influenced more by air temperature than soil temperature two weeks before and after anthesis (Narikawa *et al.*, 1973).

Azuki is especially sensitive to cool temperatures beginning three weeks before and during anthesis, as well as during early pod filling (Nomura, 1964, 1974; Chiba, 1980), but is most sensitive during early anthesis (Matsukawa, 1973). A low temperature treatment of 15°C (day) and 12°C (night) for seven days after anthesis had minimal effects, but these low temperatures applied for 10 to 14 days after anthesis caused delayed maturation and decreased seeds per pod and average seed weight and a 13% post-treatment increase in flowers per plant (Nomura, 1969b). In a similar experiment Satoh *et al.*, 1975b found that plants were 16 to 23% shorter, had 55 to 71% fewer pods and 48 to 69% less seed yield than controls. Similar results were observed for various cultivars by other researchers (Nomura, 1965/66; Matsukawa, 1973; Satoh *et al.*, 1975a; Goto *et al.*, 1984). Normal azuki flower progression is not affected by low air temperatures during the growing season but the rate of flowering, number of flowers and the length of the flowering period decrease (Nomura, 1962; Nomura and Asanuma, 1970; Satoh, 1972, 1979).

Satoh (1972, 1979) found that in-pod embryo abortions varied with air temperature. Cold weather during the growing season resulted in a 65% whole seed rate compared to 87% for warmer conditions. He found that the rate of embryo abortions was highest at the base, and lowest at the tip of each pod, and the size of the aborted seeds was larger under low temperatures and smaller when subjected to higher temperatures (Satoh, 1979). Leaves and pods both suffer frost damage at about -3.0°C (Murata, 1983).

The effects of low temperatures on yield components varies by cultivar. For example, *Chagara wase* seed yields were reduced by fewer mature pods per plant, *Takara shouzu* had fewer seeds per pod, and *Maruba 1-5* seed had fewer mature pods per plant and seeds per pod because of low temperature. Other researchers have observed a reduction in the number of mature pods (Nomura, 1965/66; Satoh *et al.*, 1975ab; Japan Bean Fund ..., 1987). The reduction in the number of mature pods was caused by an overall reduction in azuki growth, delay in flowering and reduced pollination levels in some cultivars (Nomura, 1962, 1969a).

Low temperatures during anthesis have been shown to cause an increase in seed weight (Matsukawa, 1973; Satoh *et al.*, 1975ab; Satoh, 1979). Final seed weight is determined 40 to 50 days after anthesis and can be attributed to fewer seeds per pod (Nomura, 1965/66, 1969ab, 1974). For example, the 1000 seed weight of azuki cultivars *Hayate* and *Chagarawase* increased from 10 to 27% and the number of seeds per pod decreased approximately 25% under cool temperature conditions (Japan Bean Fund ..., 1987).

Two tests for determining cool temperature tolerance in azuki have been developed. Matsukawa (1973) developed a ratio computed from the results of two low temperature treatments on seed weight and mature pod number divided by the same values resulting from more optimum conditions. The researcher calculated ratios for *Man Azuki*, *Hanko Tsubuke-1* and *Azuki W69* that were consistently higher than for 12 other cultivars. Low temperature tolerant azuki cultivars can also be identified by subjecting early progeny to 15°C daytime and 12°C nighttime temperatures for 10 days after anthesis (Konno and Narikawa, 1978). Cultivars having earlier and profuse anthesis may be the most yield-stable under the cool growing conditions of Hokkaido Prefecture (Satoh, 1970).

High temperatures also affect components of yield. Relationships between the number of flowers, mature pods and mature pods/flowers were similar for *Hayate* and *Takara* cultivars under higher than normal temperatures. The highest number of flowers and mature pods were found at the second, third and fourth main stem nod positions in both cultivars while the highest mature pod rate occurred at nodes 5, 6 and 7. A positive correlation exists between the number of flowers produced per plant per day, and the rate of flower and pod abortions; however, there was no causal relationship between flower and pod abortion rates (Kon, 1979).

Palmer (1972) reported that the cumulative temperature, calculated from the median temperature for the growing period multiplied by the number of days for the growing period, was directly correlated with yield on Hokkaido. The cumulative heat needed to exceed 2100°C to obtain yields of one mt or more per ha. In nine plots sites in Korea, an accumulation heat range of 1161 to 1319 was needed for flower initiation (Park and Hong, 1970). A broad range of cultivars were observed in Washington State by the authors. Accumulated heat degrees to flowering ranged from 750 to 1200 and the growing season ranged from 2300 to 2800 over three years. In China and Taiwan, an accumulated temperature of 1000°C was required by azuki for development from the seedling to anthesis stage (Wu and Hsu, 1986; Duan, 1989) and 2000 to 2100°C for maturity of a 95 to 99 day cultivar (Zhang *et al.*, 1981). In a comparison of 10 white azuki cultivars, 100-seed weight was found to negatively correlate with the accumulated temperature between May and November, while branches and seeds per plant were positively correlated with hours of sunshine (Terai and Horie, 1991).

Planting time effects

The effects of different planting times on azuki has been the focus of many studies (e.g. Jin *et al.*, 1989). The response to delayed planting has been the most common topic, while some others have been on early planting effects. Wang (1990) reported notable decreases in pods per plant and seed yield when azuki was planted 5 to 15 days earlier in Hebei Province, China. McClary (1990) recorded no significant differences in components of yield when azuki was planted as much as 30 days early in the Columbia Basin of Washington State. However Yoshida and Sato (1990) found that mean seed size was positively associated with the duration of the reproductive period and negatively associated with seed size variation. Size also varied widely from grower to grower but not from year to year.

Overall vegetative growth, plant height, days to anthesis and days to maturity for six azuki cultivars decreased when planting was delayed from mid-May to mid-June in the Tohoku region of Japan (Furusawa *et al.*, 1974).

Similar results from late planting of azuki have been shown by Satoh (1974), Lawn (1979a), Matsui (1984) and Wang (1991). In Tainjin, China the number of branches decreased with delayed planting (Hu and Ma, 1985). Lawn (1979a) observed that late-planted azuki took longer to reach physiological maturity after the completion of flowering than early-planted plants; a characteristic he attributed to cooler daytime temperatures later in the growing season. In Korea, Lee *et al.* (1991) found that the number of flowers and mature pods decreased with late planting, and late planting decreased the duration of flowering. They also found that about 15 days were required for full pod elongation after flowering.

Satoh (1974) reported yield reductions for late-planted azuki due to fewer pods per plant, but also noted increases in seeds per pod. In contrast, Harada (1953) stated that a delay in planting would result in a decrease in seeds per pod. Reduced numbers of pods per plant may be the result of a decrease in the total number of flowers per plant; fewer flowers can be attributed to the production of only a single set of flowers on late-planted plants versus the normal two to three sets for the early-planted plants (Harada, 1953). There may also be a strong correlation between total dry matter per plant and seed yields (Lawn, 1979ab).

Yield reductions due to late planting were not significant in Iwate Prefecture until planting on June 20th. Azuki planted on June 20th averaged 44.1% of the seed yield of azuki planted May 20th. The yield reduction was attributed to fewer seeds per pod, and not a reduction in seed size. Furusawa *et al.* (1974) concluded that azuki has some degree of tolerance for planting up to 20 days later than recommended in Iwate Prefecture. There is a reported lack of planting time flexibility for azuki in Hokkaido where the growing season is short; a delay of 15 days in the normal planting date resulted in an 8 to 20% decrease in yields for five cultivars (Japan Bean Fund ..., 1987). In Kyoto Prefecture, the number of pods per plant and pods per m^2 decreased across six planting densities for July 20th plantings as compared to azuki planted July 1st (Matsui, 1984). Wang (1990), in contrast to other delayed planting studies with azuki, reported a large increase in the number of branches, pods per plant and seed yield of azuki planted in Hebei Province, China.

Plant density and configuration effects

Plant densities and configurations have significant effects on the vegetative and reproductive characteristics of azuki. The effects of different plant densities are partially influenced by varietal characteristics; highly branched cultivars respond poorly to increased plant density while cultivars that produce the majority of their pods on the main stem respond favorably. As expected with this type of research, results from different experiments are sometimes contradictory.

Azuki plant densities of about 667,000 to 1,333,000 plants per ha have been attempted in China, but met with disastrous results: anthesis was delayed, plants were emaciated with few or no pods and yields were low (Han *et al.*, 1987). Increasing the planting density from 100,000 to 350,000 plants per ha delayed maturity, increased internode length and lodging, but resulted in the highest yield (Adachi *et al.*, 1979). Azuki grown at 500,000 plants per ha versus a common Japanese production density of 300,000 were shorter, had more pods per m^2 and seeds per pod and had higher yields; at 148,000, plants were shorter and averaged more seeds per pod and fewer pods per m^2 than azuki grown at 300,000 (Satoh, 1974). Yoshida (1987) also reported results for a 300,000 versus 550,000 plant per ha comparison, observing that azuki plants at the higher density were taller, produced fewer pods, had more seeds per pod and higher seed yields than plants at the lower density. Pods per plant substantially decreased with increasing density while seed size and seeds per pod differences were small (Lawn, 1983). Increasing plant density from 100,000 to 350,000 plants per ha reduced mature pods per plant, seeds per pod, and 100-seed weight (Furusawa *et al.*, 1974). Azuki produced at a density of 100,000 plants per ha resulted in more flowers and seeds per plant and more total flowers and pods per m^2 than at 183,000 plants per ha (Satoh, 1979).

Furusawa *et al.* (1974) recorded more mature pods and higher seed yields for *Iwate dainagon* when planted at 360,000 per ha, but suggested that the results may have been affected by an abnormally dry growing season. These researchers normally recommended a density of 183,000 plants per ha for this cultivation since higher densities, under normal growing conditions, would result in fewer seeds meeting quality standards. Nakaseko (1983) observed that densities greater than 160,000 per ha did not change the maximum dry matter production of 620 per m^2. Increasing plant densities have been reported to cause decreases in the dry weight of primary branches and increases in CGR and LAI (Nakaseko and Gotoh, 1981; Nakaseko, 1983; Lawn, 1983). Substantial LAI increases were achieved when plant density was increased from 200,000 to 800,000 plants per ha (Lawn, 1983).

Planting configuration differences have been reported to have significant effects on azuki morphology and yield components. Planting at 10 cm between seeds within the row, instead of the standard 20 cm, resulted in an increase in pods lower than 10 cm from the ground (Nomura and Satoh, 1977). On Hokkaido (Hokkaido Prefecture ..., 1972a) an experiment with three cultivars studied the influence of the space left by missing plants on adjacent plants. In general, plants adjacent to the space of a missing plant in the same row had increased height, weight, yield and number of branches and pods. The closest plant in the adjacent row and the second plant away in the same row were not different from the control. Azuki planted in 60 cm rows, with 20 cm between seeds, compensated for 48% of the total yield losses resulting from the missing plants (Ushirogi *et al.*, 1986). Huang and Lie (1987) found that azuki compensated for 92% of the potential yield losses when 30% of all potential plants were missing due to poor cultivation practices or inclement weather.

One research team reported yield differences between broadcast versus row-planted azuki (Nagayama *et al.*, 1978). Average seed yield for three densities of broadcast-planted azuki were 3513 kg/ha as compared to 2477 kg/ha for three densities planted in 60 cm rows. The researchers attributed the lower yield of the row-planted azuki to fewer support branches and greater subsequent branch breakage (Nagayama *et al.*, 1978). Nakaseko and Gotoh (1981) determined that optimum dry matter production resulted from equidistant planting with a minimum of 160,000 plants per ha. Oka (1976) found that the number of pods per plant decreased from approximately 60 to 30 as intra-row spacing decreased from 50 cm to 10 cm.

Several researchers have reported on the effects of both planting density and configuration on azuki growth and development. Lawn (1983) determined that the highest yield (1450 kg/ha) was achieved at a row width of 50 cm and a plant density of 400,000 per ha. Lawn also recorded increases in azuki total dry matter per m^2 as row widths decreased and plant densities increased up to 800,000 per ha.

Azuki yields decreased in a planting configuration of 60 cm by 20 cm with three plants per hill (450,000 plants per ha) was used rather than two plants per hill (300,000 plants per ha). The overall decline in production was attributed to intra-plant competition (Oka, 1976).

The cultivation environment

The conversion of paddy rice fields to the production of other crops has been supported by the Japanese government as a way to reduce overproduction of rice. Azuki production research in converted rice paddy fields has been conducted at the Hokkaido National Agricultural Experiment Station (Motomiya and Ito, 1972). Four azuki cultivars grown in a converted rice paddy field had suboptimal early growth as compared to plants produced in an upland field although yields were comparable. These researchers postulated that the growth differences were the result of soil texture and temperature differences between the fields. The comparable yield in the rice field azuki was due to increased branching compensation for short overall height (Furusawa *et al.*, 1974). Yoshida *et al.* (1988) identified a correlation between seed protein content and converted paddy fields.

The pruning of azuki plants is an ancient Chinese practice; pruned parts were then eaten as a green vegetable. This practice has recently been shown to dramatically increase yields if the pruning of one-third to one-half of the branch length occurs at the full bloom stage (Wang, 1990). Most of the yield increase comes from an increase in pods per plant and some from seed weight.

4

PRODUCTION

To date, the only comprehensive publications on the production of azuki are published in Chinese and Japanese. The best of these include reports or books by the Hokkaido Agricultural Experiment Station (1977), Chen (1979a), Murata (1983b), Ushirogi *et al.* (1986) and Duan (1989).

INITIATION OF CULTIVATION

The earliest reported site of azuki use and production is in North Korea and dates to approximately 3000 BP (Anonymous, 1986; Hammer *et al.*, 1989a), but too few archaeological sites have reported azuki findings to state with confidence where azuki was domesticated. Fujita (1974) reported that archaeological evidence for azuki cultivation dates to approximately 2300 BP, but the site was not identified. Several ancient Chinese books discuss azuki. The book, *Shen Nong Shu* (神農書) (*The Book of the Heavenly Husbandman*), from about 2300 to 2500 BP, discusses the number of days until flowering (60) and maturation (110) for azuki (Li, 1957). *Huang Di Nei Jing Su Wen* (黃帝內經素問), written about 2300 BP, states that the color of azuki is red and that the flavor belongs to the sour class (Li, 1957). The book, *Qi Min Yao Shu* (齊民要術) (*Essential Techniques for the Peasantry*), written by S. Jia about AD 533 to 544, states that azuki cultivation began about 2000 to 2500 BP in China (Shih, 1974, 1979; Guo, 1980; Cao, 1981; Tong, 1981; Hoshikawa, 1985). Supporting this, carbonized azuki seeds dating to AD 22 with average dimensions of 5.54 mm by 3.74 mm, were found in a Han dynasty tomb near Changsha, Hunan Province, China (Cultural Relic ..., 1978; Duan, 1989).

Cultivation of azuki in Japan is thought to have started later than in China and Korea, but the exact period is unknown. The *Nihonshoki* and *Kojiki*, two Japanese books about mythology and history from the 8th century AD, contain references to azuki (Hoshikawa, 1985). The *Kojiki* states that when the god *Susano no Mikoto* came to earth, he killed a man and azuki sprang from the man's nose, while the *Nihonshoki* states that the god *Tsukiyomi no Mikoto* killed another god and azuki sprang from the testicles of the corpse (Igata, 1977; Maeda 1987). The remains of azuki seeds were found in Tenno historical ruins (3rd century BC) at Mitsuo-mura,

Toro historical ruins (1st century BC to 1st century AD) in Shizuoka-ken (Igata, 1977). Seeds were also found at a 9th century AD Ainu (indigenous inhabitants of Japan) archaeological site on the island of Hokkaido. Ainu called azuki *Antuki* (Fujita, 1974; Crawford and Yoshizaki, 1987).

China is the most frequently stated origin of Japanese azuki (Narikawa, 1972; Hoshikawa, 1985). Fujita (1974) believes that azuki migrated along two paths from Southeast Asia into Japan. He conducted a comparative linguistics study which indicated that one path was through China, across the Korean peninsula and the other path was via Taiwan to Okinawa. These migration paths are the same proposed for the dominate ethnic group in Japan today. Tasaki (1963) used varietal differences in the degree of plant branching to suggest that azuki migrated from central or southern China to southern Japan and then eventually up the Japanese archipelago. This migration probably occurred during the 3rd to 8th century AD (Ushirogi *et al.*, 1986). Yamaguchi (1992) reported the discovery of wild azuki in Japan, which raises the possibility of domestications as well as introductions within Japan.

PRODUCTION AREAS

Azuki is the only member of the six major species in the subgenus *Ceratotropis* that is produced in temperate regions of East Asia (Ahn *et al.*, 1981). In 1944, Allard and Zaumeyer observed that azuki adaptation extended only to 40°N latitude. Currently most commercial azuki production in Japan and China is between 35 to 47°N (Fig. 4.1). Research and commercial azuki production has been successful between 46 to 48°N in the Columbia Basin of Washington State, between 43.5 to 46°N in Minnesota State in the U.S., and also in southern Canada. Azuki is or could be grown in environments with between 530 to 1730 mm of annual precipitation and a mean annual air temperature range of 7.8 to 27.8°C (Ag. Lexicon ..., 1979; Duke, 1981). Humid, hot weather is required for most of the growing season (Piper and Morse, 1914; Japan Bean Fund ..., 1987; Duan, 1989). Azuki grows well under irrigation in the deserts of central Washington State and Ningxia Province in China although dry air during maturation and harvest exacerbates pod shatter problems (mentioned 1st century BC, Shih, 1974) and may increase the percentage of seeds with a hard seed coat.

Duan (1989) recommends sandy soil to promote early maturity in areas where the growing season is short, and clay loam or loam soils to promote high yields where the growing season is long. Azuki grows on many soil types, but soils less than pH 5.0 and higher than 7.6 hinder development (Noda, 1952; Thomas *et al.*, 1974; Hernandez *et al.*, 1978; Kay, 1979; Duan, 1989). On highly acidic soils, rice bean, called *Tsuru* azuki, *Kanime* or *Baka* azuki in Japan, has been substituted for azuki on a limited scale for hundreds of years throughout Asia (Noda, 1952; Tasaki, 1963; Hoshikawa, 1985).

Noda (1952) found that two rice bean accessions from Hyogo and Tottori Prefectures outperformed local azuki cultivars when grown on acidic soils. The Usu soil of Japan has been shown to have a growth disorder related to copper content (Mizuno *et al.*, 1971; Mizuno and Takao, 1972).

Almost all production of azuki occurs in four countries of northeast Asia, i.e. China, Japan, South Korea and Taiwan (Motomiya and Ito, 1972). Elsewhere, azuki has been or is being produced or studied in research plots in Argentina, Australia, Belgium, Brazil, Congo, India, Italy, Kenya, Malaysia, Mexico, New Zealand, North Korea, Philippines, Taiwan, Thailand, the former USSR and the United States (Sawada, 1919; Piper, 1936; Motomiya and Ito, 1972; White, 1972; Dorer, 1973; Purseglove, 1974; Arias, 1975; Sacks, 1977; Hernandez *et al.*, 1978; Kay, 1979; Duke, 1981; Foschine *et al.*, 1983; Gupta and Bahl, 1983; Hardman, 1985, 1987; Duan, 1989; Hammer *et al.*, 1989b; van Oers, 1989; van der Maesen and

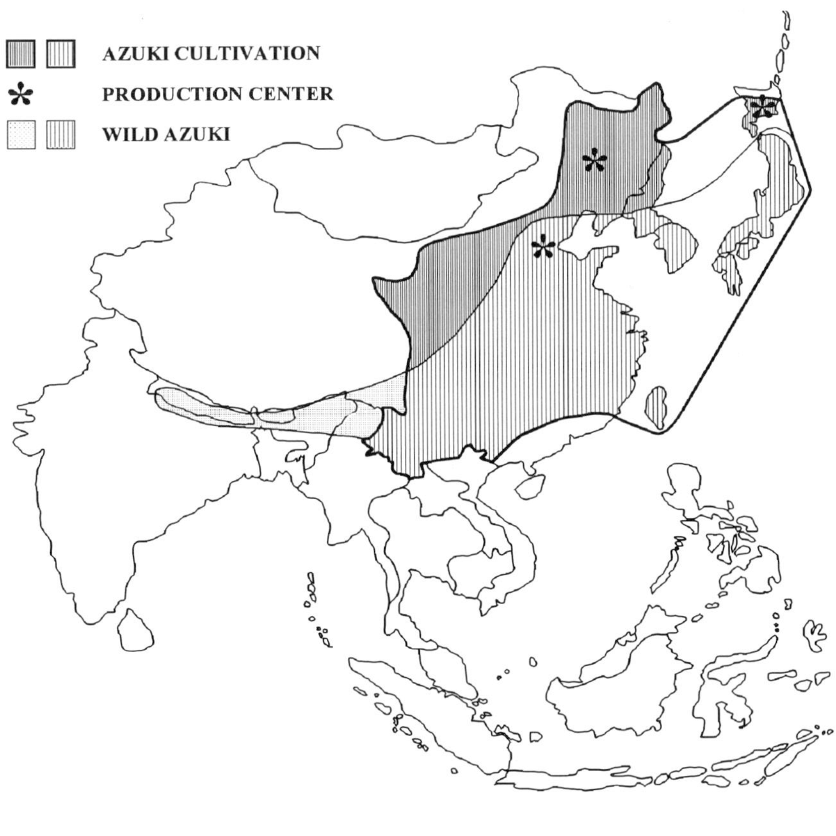

Fig. 4.1. Range of traditional azuki production in Asia (thick line), centers of concentrated production (∗) and possible range for wild azuki, *Vigna angularis var. nipponensis* (thin line).

Somaatmadja, 1989). Duan (1989) stated that azuki is produced in 24 countries. Japanese import/export data (Japan Bean Fund ..., 1987) indicates that, during the past 35 years, 15 countries exported azuki to Japan, the world's largest azuki importer. The five largest exporters of azuki to Japan in the 1980s were China, Taiwan, the United States, Thailand and Canada in descending order of quantities (Chao, 1982; Tanaka and Konno, 1982; American Embassy Office of Agricultural Affairs, 1987). Korea averaged azuki exports of about 2.2 metric tons (mt) from 1981 to 1986 (Choi and Lee, 1987; Choi, 1988) but imported approximately 5000 mt from China in 1989. Exporting countries that did not appear in the lists above include Burma, Canada, Chile, Colombia, Indonesia and Hong Kong (probably trans-shipment). It is difficult to determine whether all of these or any other countries actually grow azuki since its production is often grouped into categories such as "miscellaneous" and "specialty" crops in national and international statistical publications. Research and production in Angola and Zaire was cataloged by the F.A.O. (Plant Production ..., 1959), but has not been mentioned elsewhere.

China

Jin and Jiang (1988) and Duan (1989) reported that approximately 670,000 hectares of azuki are produced annually in China, but government statistics cannot be found to confirm this. Statistics on azuki are not reported in China's Agriculture Yearbook nor in known publications by the State Statistical Bureau. However, exports have recently been reported in the Yearbook of China Foreign Trade.

China exports azuki beans under several trade names. Tianjin Red Bean is the most famous. Other export trade names include Dahongpao, Tangshan Red Bean, Chunming Red Bean (Duan, 1989) and Jiamusi Red Bean (Du, 1986). Recent export volume and prices were: 43,932 mt at U.S. $515/mt in 1988, 36,742 mt at $720/mt in 1989, 33,468 mt at $635/mt in 1990 and 25,280 mt at $477/mt in 1991 (Yearbook Committee ..., 1992).

Most Chinese azuki is produced in the provinces of Hebei, Henan, Shandong and the northeast provinces (Tong, 1981; Duan, 1982, 1989; Liang and Zhang, 1991). The main azuki producing provinces of China during the 1980s were Hebei, Liaoning and Jilin (Japan Peas ..., 1987). Zheng and Zhang (1983b) claim that Heilongjiang Province is the main site of azuki production in China. Heilongjiang produced red, mottled and white azuki on 40,000 to 50,000 ha between 1950 and 1980, but this dropped to about 10,000 ha during the 1980s. Three thousand mt were exported from Heilongjiang in 1984 (Editorial Com. ..., 1993). Han *et al.* (1987) and Zheng (1988) maintain that the main site of Chinese production occurs in Wuqing and Jinghai Counties, Hebei Province, with about 4500 to 6500 ha produced annually. This area near the port city of Tianjin is famous for the production of high quality azuki and exports about 5000 mt annually (Zhu, 1989). Exports from Tianjin have fluctuated considerably, e.g. 27,146 mt in 1971, 17,020 mt in 1973, 5028 mt in 1975, 2874 mt in 1977, 8321 mt in 1979 and 6926 mt in 1980 (Ding and Zhang, 1984).

Other areas in China that produce azuki include Shulian and Tailai Counties, Jilin Province; the Mudan River and Three Rivers regions of Heilongjiang Province; Gansu, Guangxi, Liaoning, Shaanxi, Shanxi, Sichuan and Yunnan Provinces; and, the Inner Mongolia Autonomous Region (Guo, 1980; Duan, 1982; Dai and Zhang, 1982; Du, 1986; Hu, 1986; Hu *et al.*, 1987; Zhang and Li, 1988; Zheng, 1988; Duan, 1989). In Shaanxi Province, azuki is produced on the plateau of Weibei and plain of the Qin Mountains (Zhou, 1983). The small amount of azuki produced in Gansu Province is grown in an azuki-winter wheat-azuki crop rotation in three regions between 32 to 39°N latitude (Shi, 1988). The minor production in Inner Mongolia (Qi, 1957) averages yields of only 525 to 600 kg/ha from erect to semi-erect, small-seeded cultivars that require 120 to 136 days to reach maturity (Hu *et al.*, 1987). Lodging is one of many problems affecting azuki production in China (Dou *et al.*, 1989).

Fig. 4.2. A winter crop of azuki growing as an intercrop between palms near Tainan, Taiwan, where most of Taiwan's azuki is produced.

Azuki is an important winter crop in Taiwan (Fig. 4.2), where it is usually grown in rice paddies during the winter, often for export to Japan. Pingtung and Kaohsiung Provinces produce 98 to 99% of all Taiwanese azuki (Chao, 1982; Tsai, 1982; Chang 1988a). A detailed description of Taiwanese production practices was written by Wu and Hsu (1986). Much of the crop is affected by powdery mildew (Wu *et al.*, 1985a; Hsu *et al.*, 1989a; Hsu, 1989b). Production increased dramatically in the late 1970s in response to Japanese market demands, but by the late 1980s, low-cost mainland Chinese exports reduced profitability and the production area in Taiwan (Fig. 4.3). By 1990 mainland azuki was being smuggled into Taiwan, further undercutting profitability for Taiwanese farmers.

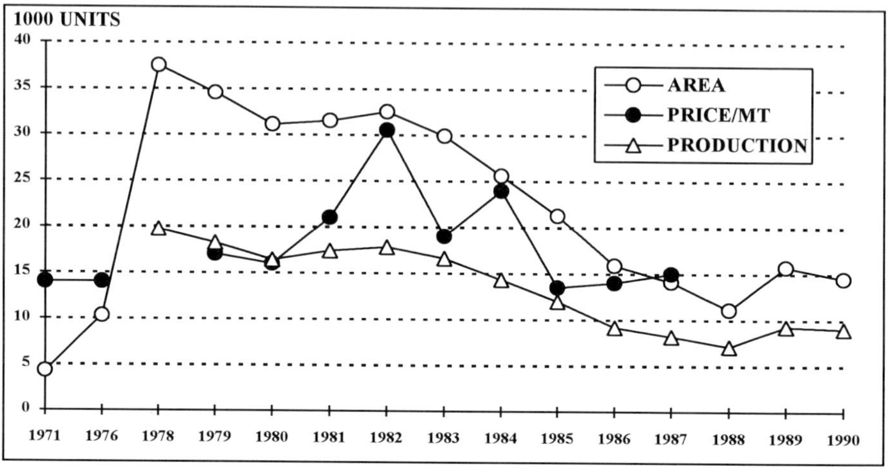

Fig. 4.3. Area (ha), production (mt) and price (NT$/mt) of azuki in Taiwan. Note missing data and years between 1971 and 1979, and data between 1987 and 1990 (Planning Division ..., 1990; Economic Daily News, 1991).

Japan

Azuki is Japan's second most important legume (Fig. 4.4), after the soybean, comprising 23% of all grain legume production. About 100,000 mt are produced annually, equaling, along with imports (Figs. 4.5 and 4.6), an annual consumption of about 1 kg per capita. The area of azuki production (Fig. 4.4) has shown a declining trend since the late 1950s, that is only partially offset by a trend of rising yields (Fig. 4.4). Imports of raw beans are supposedly used to meet shortfalls in domestic production and are strictly controlled by a quota system. However, imports do not seem to have a leveling effect on long-term trends or year to year availability (Fig. 4.5). Part of the erratic nature of imports could result from the variation of azuki stocks in China, the dominant supplier, which is subject to the same weather systems affecting Japan.

All prefectures in Japan produce azuki, but about two-thirds of commercial production occurs on the prefectural island of Hokkaido (Fig. 4.7) (Japan Bean Fund ..., 1987; Kagiwara, 1992) where weather and the erratic length of the short growing season cause annual production to be highly variable (Satoh, 1972). The climate of Hokkaido has a mean annual temperature of less than 8°C, frost-free days range from 130 to 200 and annual precipitation ranges from 800 to 1500 mm (Nuttonson, 1951; Tanaka and Konno, 1982; Hoshikawa, 1985). Home garden production of azuki began in the southwest portion of Hokkaido about 1695 and became extensive until the introduction of paddy rice production. During the 1920s and 1930s rice production shifted azuki to the northeastern districts, especially Tokachi and Kamikawa (Hokkaido Ag. Expt. Sta., 1967; Narikawa, 1978), where it has been established as an important cash crop as a result of agronomic improvements and the establishment of marketing channels (Ushirogi *et al.*, 1986). Approximately 32,000 farm households produce and market azuki beans on Hokkaido and earned from 328 to 400 yen per kg between 1984 and 1989 (Economic Bureau, 1991).

The Tokachi District is Hokkaido's main production area and averages 134 frost-free days, from May 20 to October 1, with a mean growing season air temperature of 16.7°C, and average growing season precipitation of 287 mm (Narikawa, 1972; Tanaka and Konno, 1982; Yoshida *et al.*, 1989b). Tokachi soils are derived either from Tokachi river alluvium or volcanic ash with some requiring correction for excessive acidity or moisture (Yoshikawa, 1970). Hokkaido has six secondary districts with significant production: Kamikawa, Kitami, Iburi, Shiribeshi, Sorachi and Ishikari (Yoshida and Satoh, 1988). Eight Hokkaido producing regions have shown differences in seed testa color and highly significant correlations between azuki protein content and seed weight (Yoshida *et al.*, 1988; Yoshida and Satoh, 1988; Yoshida *et al.*, 1989b).

Production constraints in Hokkaido include dust storms in May and June composed of peat soil material and volcanic ash that can damage azuki seedlings, fog along the east coast that reduces daytime temperatures, rainfall during harvest, and unseasonably cool weather that occurs every four to five years (Nuttonson, 1951; Furusawa *et al.*, 1974; Tanaka and Konno, 1982; Hoshikawa, 1985; Japan Bean Fund ..., 1987; McClary *et al.*, 1989).

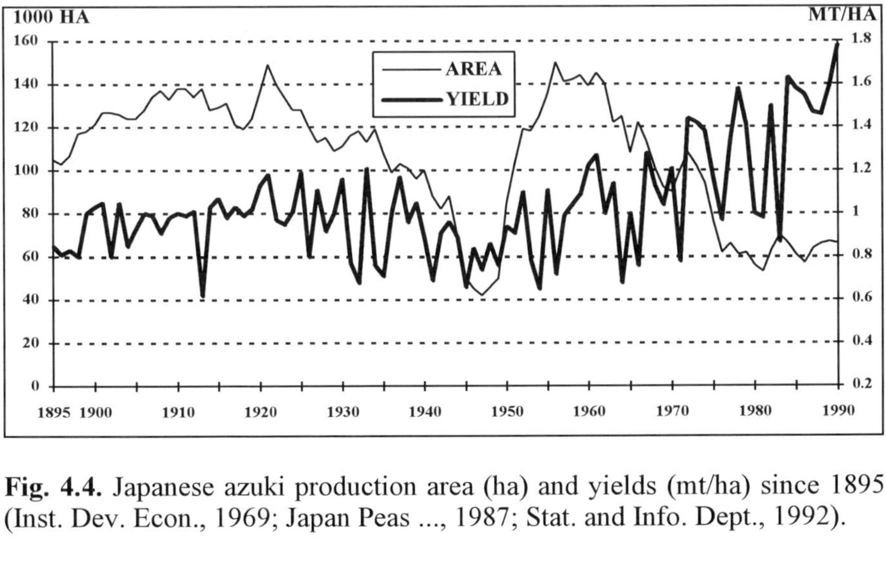

Fig. 4.4. Japanese azuki production area (ha) and yields (mt/ha) since 1895 (Inst. Dev. Econ., 1969; Japan Peas ..., 1987; Stat. and Info. Dept., 1992).

Approximately 50 to 60% of all damage on Hokkaido is attributable to cold weather damage, 10% to insects and 10 to 20% to diseases (Tanaka and Konno, 1982). These constraints cause the high variability of year to year production, frequently necessitating imports to make up the difference between domestic demand and domestic production.

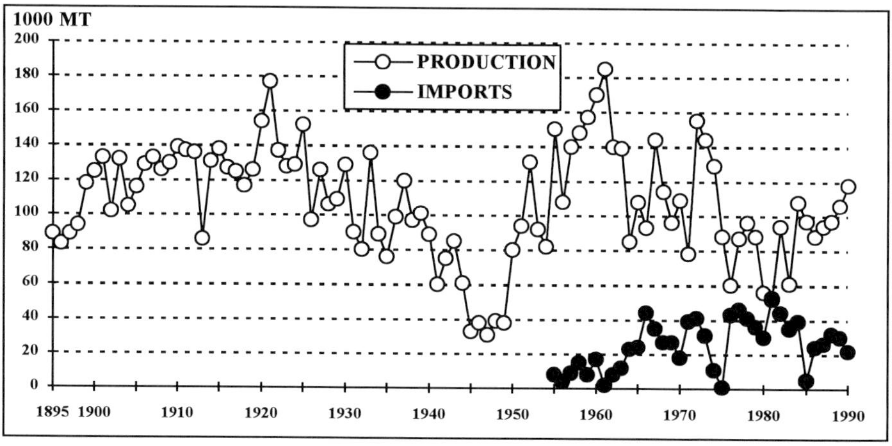

Fig. 4.5. Japanese domestic production (mt) from 1895 to 1990 and imports of azuki beans (mt) from 1955 to 1990. Statistics for imports prior to 1955 could not be located (Inst. Dev. Econ., 1969; Japan Peas ..., 1987; JETRO ..., 1991; Stat. and Info. Dept., 1992).

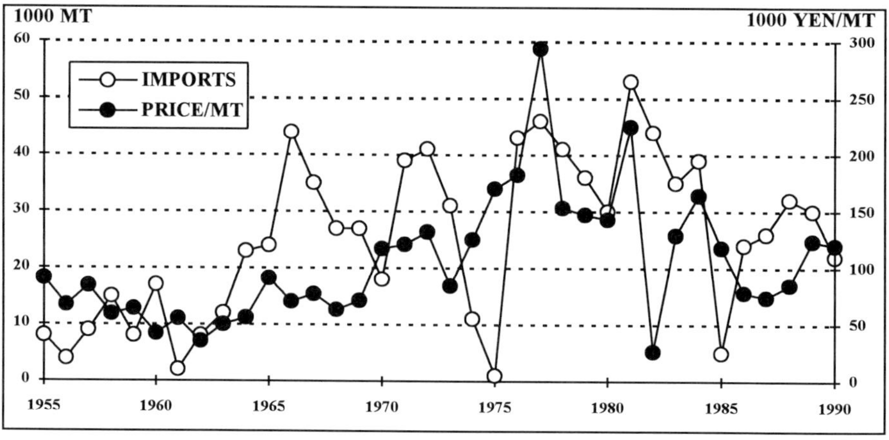

Fig. 4.6. Azuki imports into Japan (mt) and the Japanese import price per mt (1,000 yen [¥]) from 1955 to 1990 (Inst. Dev. Econ., 1969; Japan Peas ..., 1987; Stat. and Info. Dept., 1992).

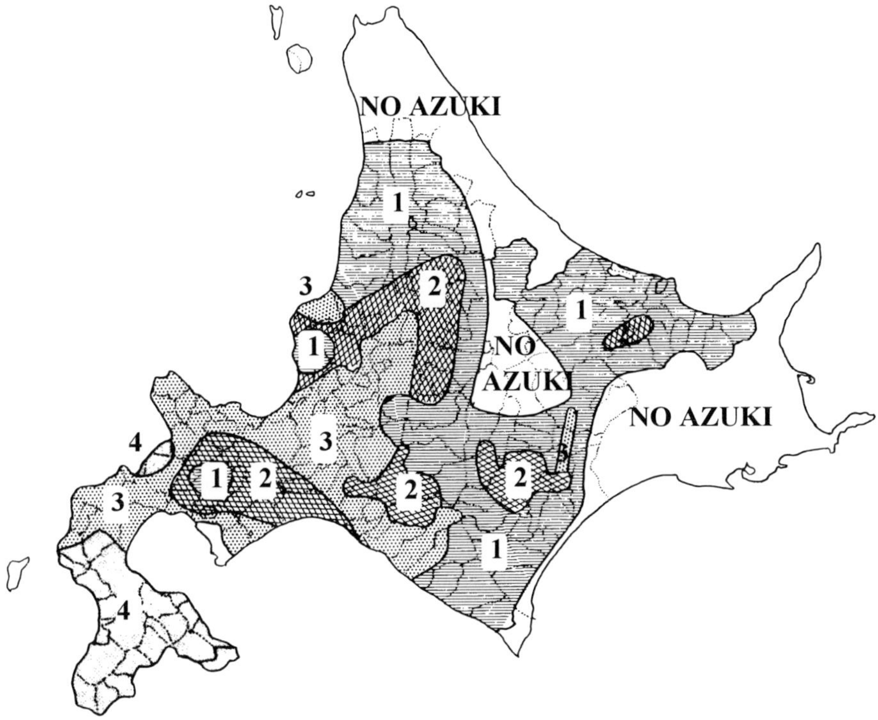

Fig. 4.7. Azuki production areas in Hokkaido Prefecture (island), Japan, correlated with four heat-unit-based production zones (adapted from Ushirogi *et al.*, 1986). Heat units are based on accumulations from May through September. Zone 1 accumulates 2400 to 2500°C heat units; *Hayate* is the most common cultivar and the first frost occurs by September 30th. Zone 2 accumulates about 2500°C; cultivars are *Erimo*, *Hatsune*, *Hayate* and *Kotobuki*, and the growing season lasts 150 days. Zone 3 accumulates about 2500 to 2600°C; cultivars are *Akane dainagon*, *Erimo*, *Hatsune* and *Kotobuki*, and the growing season lasts 150 to 200 days. Zone 4 accumulates more than 2600°C; cultivars are *Akane dainagon* and *Erimo*, and the growing season lasts more than 200 days.

The other main areas of azuki production in Japan are in the northern part of Honshu Island (the main island) and production generally diminishes to the south and southwest. The northern part of Honshu has an annual mean temperature of 8 to 13°C, frost-free period of 160 to 170 days and annual precipitation of 1200 to 1300 mm. The central and southern parts of Honshu have an annual mean temperature of 13 to 18°C, frost-free period greater than 170 days and annual precipitation of 1300 to 2500 mm (Nakayama,

1977). No other prefecture approaches the production of Hokkaido, however significant production does occur in Fukushima (2.9%), Iwate (2.7%), Aomori (2.6%), Tochigi (2.2%) and Akita (2%), and to a lesser degree in all other prefectures except that virtually none is produced in Fukui, Kanagawa, Kochi or Wakayama Prefectures (Japan Bean Fund ..., 1987; also see Piper and Morse, 1914; Hedrick, 1931; Austin, 1948; Salmon, 1948; O'Brien and Kohler, 1950; Springer, 1950; Whiteside, 1950; Nuttonson, 1951; Motomiya and Ito, 1972). Japan has three historical secondary production centers: Iwate and Kumamoto Prefectures and the Tan'ba region of Kyoto City and Hyogo Prefecture (Furusawa *et al*., 1974; Tanaka and Konno, 1982; Matsui, 1984; Hoshikawa, 1985). Iwate Prefecture has traditionally been the second largest and most stable azuki producing area (Furusawa *et al*., 1974). Cultivation in Iwate was reported in a 1716 crop survey where azuki was considered to be an essential crop since it was frequently boiled and eaten with soybeans as part of religious ceremonies (Japan ..., 1988). The Tan'ba region, to the west of the old imperial capital, Kyoto, is famous for its production of premium quality large azuki (Fig. 4.8), soybeans (Matsui, 1984) and other vegetables for the court and retainers. White azuki production is a traditional specialty in Okayama Prefecture (Murata, 1980).

Fig. 4.8. A summer crop of azuki growing on raised beds for drainage in the Tan'ba region of Japan. The highest quality *dainagon* azuki beans from this region were from pods that were individually hand harvested as they ripened on the plant.

Production constraints in Iwate Prefecture collectively result in low average yields; 20% of yield constraints are attributed to insects, 30% to diseases and the rest to drought, flood and seedling damage by birds (Satoh, 1970, 1972, 1974, 1979; Tanaka and Konno, 1982). Seed-borne viruses are the main disease problem (Furusawa *et al*., 1974). Kumamoto Prefecture has the same constraints as Iwate, except that a 5% yield reduction is caused by storms instead of floods (Tanaka and Konno, 1982). Problems in the Tan'ba region include excessive soil moisture during the rainy season and various insect pests and diseases (Matsui, 1984).

Most Japanese production comes from 17 cultivars (Mikoshiba, 1985; Japan Bean Fund ..., 1987), although almost 40% during the mid-1980s was still from landraces (Japan ..., 1988). Japanese yields have increased approximately 24% in the past 20 years (Fig. 4.4) (Tanaka and Konno, 1982; Nakaseko, 1984b) but fluctuate widely.

Korea

As in Japan, azuki is the second most important grain legume produced in South Korea in terms of planted area and production (Fig. 4.9). Azuki occupies about 10 to 15% of the grain legume area while soybean occupies about 70 to 80%. Eight of South Korea's 13 provinces have large areas of azuki production with Ch'ungbuk Province accounting for over 26% of the total area in 1986 (Choi and Lee, 1987). Other important azuki producing provinces include Chonbuk, Chonnam, Ch'ungnam, Kangwon, Kyongbuk, Kyonggi and Kyongnam. About 25,700 hectares were reported to be produced in 1985 (Ahn, 1987). National average yields during the 1980s ranged from 980 to 1240 kg/ha (National Agricultural ..., 1991), with the highest provincial yields occurring in Ch'ungbuk (Choi and Lee, 1987).

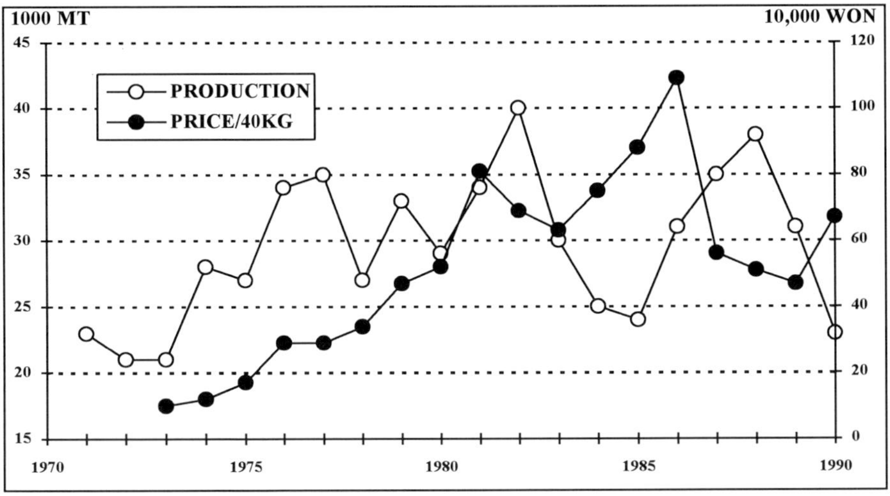

Fig. 4.9. Korean domestic production (mt) of azuki beans from 1971 to 1990 and prices (in won) received by Korean farmers for 40 kg since 1973 (National Agricultural ..., 1991).

Most South Korean farmers grow azuki for home consumption as well as for a cash crop. The average farm production area was only 0.013 ha during the late 1970s. Production is scattered throughout the country and is usually planted on hillside land in rotation with wheat or barley, or in converted paddy rice fields. *Hongcheon* was the most popular cultivar, maturing in about 105 days (Hwang, 1982; Choi and Lee, 1987; Hong *et al.*, 1987). Korea is known to import azuki but statistics have not been located.

United States

Azuki is believed to have been first introduced into the United States by the Perry expedition in 1854 (Browne, 1855; Piper and Morse, 1914; Hedrick, 1931). McCarthy and Emery (1894) mentioned testing small red Japanese beans. Galloway (1909) listed a shipment of azuki seed imported during 1908. Piper and Morse (1914) listed early introductions of azuki into the U.S., and while it has never been a major world producer of azuki, the crop has been grown experimentally and on a limited production scale in several states over the last 130 years. Early adaptation experiments were conducted in Kansas, Virginia and North Carolina (Georgeson *et al.*, 1891; Piper and Morse, 1914) and azuki has been used as a green fodder crop in some southern states (Hoshikawa, 1985).

Indirect reference is made to research and commercial production in New York, Indiana, Illinois and Alabama (Hedrick, 1931; Anonymous, 1960; Sanchez, 1987; Farr *et al.*, 1989). Some of the more recent and well-documented azuki research and commercial production attempts have been in the states of California, Colorado, Minnesota, Ohio, Oklahoma and Washington (Matlock and Oswalt, 1963; Scarbnick, 1979; Sperbeck, 1981; Hardman, 1985, 1987; Schuff, 1988; Lumpkin *et al.*, 1993ab; Nelson, 1993). Cox (1978) mentions that azuki assessment research has also occurred in Idaho, Iowa and North Dakota. Researchers in Missouri, North Carolina and Oregon have also evaluated azuki.

On-farm testing of azuki has been conducted in Ohio (Cramer, 1986), where organically grown azuki, yielding 1254 to 1906 kg/ha, was produced for the health food market in 1985 (National Research Council, 1989). Research on the adaptation of azuki to Oklahoma found that the crop proved to be relatively disease free but averaged only 302 kg/ha at six research locations from 1960 to 1962 (Matlock and Oswalt, 1963). The crop has also been tried in southwest Nebraska under dryland and irrigated conditions, the latter yielding 1852 kg/ha (Solomon, 1990).

Research on azuki began at Washington State University in the mid-1980s as part of a multi-disciplinary alternative crops program initiated by the author (Lumpkin *et al.*, 1993a), and has resulted in several reports, which are reviewed in this and other chapters (McClary *et al.*, 1989; McClary, 1990; Chilukuri and Swanson, 1991; Hsieh *et al.*, 1991, 1992; Milkov *et al.*,

1991; Tanigoshi, 1991; Engquist and Swanson, 1992; Hang *et al.*, 1993; Lumpkin *et al.*, 1993b; McClary *et al.*, 1993; Tanigoshi and Aguilar, 1993; Aguilar *et al.*, 1994ab). Production has been stimulated by proximity to Japan and bean quality but has been constrained by fluctuations in Japanese prices and demand, two-spotted spider mites and pod shatter. Cultivation area has not exceeded 500 ha and on-farm yields of grade 1 bean has been between 1300 and 2000 kg/ha.

Minnesota research literature has documented the development of azuki as a specialty crop for the state. Research was initiated in 1962 by Robinson (1964) who obtained high yields at Rosemount with a focus on developing a new livestock protein supplement source (Scarbnick, 1979; Breene and Hardman, 1987). Commercial azuki production was stimulated by the interests of a Japanese trading company in 1977 that wanted to export Minnesota-produced azuki to Japan (Cox, 1978; Sperbeck, 1981; Breene and Hardman, 1987). Production developed in the southern and central parts of the state on lighter soils and with access to irrigation (Hardman, 1985, 1987). In 1978, 100 hectares in 16 Minnesota counties produced approximately 91 mt (Scarbnick, 1979). The total production area reached approximately 400 hectares in 1980 (Sperbeck, 1981). Production and food science topics were major areas of research (Hayakawa and Breene, 1982; Drager, 1983; Robinson, 1983; Tjahjadi, 1983; Chen *et al.*, 1984).

University and on-farm testing of azuki in Colorado has focused on production practices, costs of production, and marketing opportunities for the crop (Schuff, 1988). A team of researchers at Colorado State University have studied azuki and numerous other alternative crops (Johnson *et al.*, 1989). Plots of azuki were tested in the Arkansas River Valley of Colorado and near Pueblo, Olathe and Rocky Ford. Northeastern Colorado was concluded to have better conditions for azuki. The research team analyzed production costs on two farms and found that costs were $139/acre for the organic producer and $122 for the conventional producer.

India

Azuki production has occurred in hilly regions of northern India (Piper, 1936). Seven Japanese azuki cultivars were introduced and evaluated in 1970. They were found to be free of serious pest problems and highly resistant to bacterial blight (*Xanthomonas phaseoli*), and yielded extremely well as part of a four crop rotation of wheat-azuki-sarson (*Brassica campestris* L. var. *Sarson*)-maize in the western Himalayas (Thomas *et al.*, 1974; Gill and Sharma, 1980).

New Zealand

The DSIR crop research division began research on azuki at Otara in 1969 (White, 1972) to established production for export in collaboration with Japanese trading companies. Research yields were adequate at about 1300 to 2200 kg/ha, but problems occurred with the cool maritime climate, excessive moisture at harvest, larva of a small looper moth (*Plusia chalcites*), nematodes, a wilt (*Fusarium oxysporum*) and weed control (Palmer, 1972). Growth habits and flowering under New Zealand conditions were also studied (Palmer, 1974). The research effort included the study of harvesting implements including a disc implement to cut the plants just below the ground for harvest. Palmer concluded that azuki was better adapted to a continental-type climate with higher summer temperatures than climates found in New Zealand, although good yields were possible in the South Auckland district. New Zealand has occasionally exported small quantities of azuki. In the *Vegetable Grower's Handbook*, Wallace (1987) gives general recommendations for the production of azuki in New Zealand.

PRODUCTION PRACTICES

The Chinese language has many proverbs about the production of azuki that have agronomic value and are well worth considering (Agriculture Publishing House ..., 1980; Shih, 1974). For example: "soybean or mungbean is harvested during dry years, azuki is harvested during waterlogged years"; "azuki is suitable for wet, sesame is suitable for dry"; "azuki seeds are wasted if the soil is plowed shallow"; "if one sheng (167 ml) of azuki seed is planted, 10 sheng will be harvested"; "when millet is intercropped with azuki, the millet is harvested high and the azuki is harvested low"; and, "when azuki is intercropped with kaoliang (sorghum), azuki would produce more but for the kaoliang".

Agronomic information about crop rotations, planting dates, densities and depths, use of inoculums, fertilizers and pesticides, and harvesting techniques for azuki is well documented, especially for Japan, China, Taiwan, South Korea and the U.S. In Asia, production of azuki is quite labor intensive. A large variety of small tractors and implements are used for field preparation, planting, harvest, etc. In the U.S., typical large, dry-bean production equipment is used. The small scale mechanization of azuki production in Asia can be illustrated by the data reported by Tanaka and Konno (1982) and the annual statistics report by the Japanese Ministry of Agriculture, Fisheries and Forestry (Statistics ..., 1992). For example, from 1977 to 1987 the number of labor hours to produce one ha of azuki dropped from 191 to 138. The breakdown of the 1977 labor hours was: 7 hr for tillage; 3 hr for basal dressing; 15 hr for planting; 89 hr for cultivation and weeding; 5 hr for insect and disease control; 45 hr for harvesting; and 27 hr for threshing (Tanaka and Konno, 1982).

Rotations and field selection

An old Japanese azuki proverb states, "don't make a return gift of beans". This proverb refers to recognition that continuous cropping of azuki is detrimental to yields (Japan ..., 1988). For example, the monoculture of azuki for five years resulted in reduced root nodules per plant, increases in disease and insect damage and reduced yields (Duan, 1989). After seven years of continuous culture of azuki in Korea, the linear regression coefficient (b) for yields was -21.03 (Park *et al.*, 1987). Continuous monocropping of azuki makes the root fungal flora less diverse which allows soil-borne diseases to flourish resulting in suppressed root growth and nutrient uptake. In a study of various treatments on azuki growth, the use of crop rotation was found to have the greatest impact on reducing brown stem rot. Use of farmyard manure and compost was also quite helpful (Nitta and Matsuguchi, 1988; Nitta, 1991).

In Asia azuki has been traditionally grown as a border crop on the bunds of rice fields, in vegetable gardens and in large fields (Austin, 1948; Nuttonson, 1951). In large fields most azuki farmers have recognized the problems associated with monoculture and have developed some elaborate rotations. During the 1st century BC, in China, azuki was reportedly used as a green manure in rotations with hemp or millet and intercropped with mulberry (Cao, 1981). Currently azuki is more commonly intercropped with maize, sorghum and millet, than grown in monoculture (Zhao and Tong, 1993). Chinese azuki rotations vary with the production location in the country. One-year rotations of soybean-azuki-wheat are common in southern China (Duan, 1989). In the north and northeastern regions of China, azuki is planted in rotation with wheat, along the bunds of rice paddy fields or intercropped with maize (four rows of azuki to one row of maize), sunflower, millet or kaoliang (Guo, 1980; Hu, 1986; Duan, 1989). Taiwanese farmers in the southern provinces of Pingtung and Kaohsiung use a rotation of two crops of rice grown in the late spring, summer and early autumn followed by one fall-planted crop of azuki (Chao, 1982). In recent Chinese research, azuki grown in an azuki-wheat-azuki rotation had a disease occurrence of 44% versus only 16% when grown in a sunflower-wheat-azuki rotation (Han *et al.*, 1987). In China's Jiangsu province, an annual broad bean-maize-upland rice-azuki rotation has been developed to deal with infertile soils with poor water holding capacity (Yuan, 1991).

Nuttonson (1951) described azuki rotations in Japan involving 13 crops produced over 12 years. In Iwate Prefecture, Japan, a two-year, three-crop rotation of azuki-wheat or barley-vegetables is practiced. In Kumamoto Prefecture, an azuki-buckwheat-wheat-sweet potato rotation is practiced

(Nakayama, 1977; Tanaka and Konno, 1982; Japan ..., 1988). Various three-year, three-crop rotations with forage crops, potatoes, vegetable soybeans and yams (*Dioscorea* spp.) are also practiced in Iwate and Aomori Prefectures (Japan ..., 1988). Sakurai (1986) described some complicated five- and six-year rotations used in northern Japan. On Hokkaido, the short growing season and cold winter only allows for the production of one crop per year. Hokkaido producers use azuki, sugar beets, small grains, grasses, other beans and potatoes in rotations for upland production areas (Fig. 4.10), and azuki and paddy rice for lowland production sites (Suzuki and Konno, 1982). In the Tokachi District, Hokkaido, approximately 12% of all farmland was in these azuki rotations (Statistics ..., 1992).

An intercropping study in Korea with sesame and several legumes found that a sesame-azuki combination was not economically viable (Lee *et al.*, 1992a). Azuki has been studied as a replacement for rice in Korean paddy fields if flood or drought disrupted the rice planting and was nearly as profitable as mungbean as a replacement (Park and Hong, 1970).

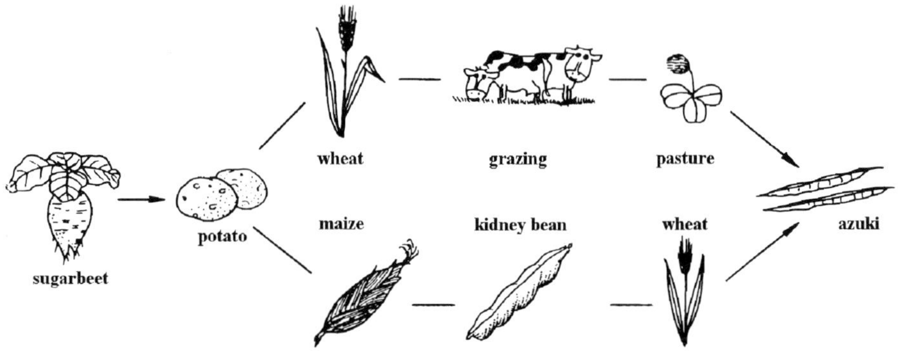

Fig. 4.10. A five- or six-year rotation recommended for Hokkaido Prefecture, Japan. Stem rot disease is one of the most important motivations for use of rotations on Hokkaido. Adapted from Murata (1983b).

In Minnesota State, U.S., a maize-grain crop-azuki-alfalfa-grain crop rotation has been recommended, with best yields occurring on sandy loam soils with a soil pH of 6.8 to 7.0 or less (Cox, 1978; Sperbeck, 1981). In the Columbia Basin of Washington State, many management problems were found to be reduced by careful selection of fields for azuki production. Good fields are located away from sources of dust, such as dirt roads, which promote the growth of two-spotted spider mites. The fields are not excessively weedy or shaded and have even slopes to reduce wet spots after irrigation. It was found that azuki should not be planted in fields that have been used for alfalfa production during the past two years or in potatoes, dry edible beans or other legumes during the previous year due to the possibility of disease inoculum and the residues of herbicides toxic to azuki. Azuki has yielded especially well after wheat or maize.

Field preparation and planting

In Japan's famous azuki production district of Tokachi on Hokkaido, standard field preparation and planting methods include a basal fertilizer application, rototilling to produce a fine-grained seedbed and initiation of planting by mid- to late May (Horiguchi, 1950) with three seeds per hill, 3 cm deep and 20 cm between hills in rows spaced 60 cm apart (Satoh, 1974; Konno and Narikawa, 1978; Tanaka and Konno, 1982; Ushirogi *et al.*, 1986). Normally two mature plants are expected to result from the three seeds planted per hill. The failure to establish three seedlings per hill is often due to damage by the insect *Delia platura* Meigen, soil infertility, soil crusting and cold-soil-related diseases (Ushirogi *et al.*, 1986). Some producers have recently begun planting azuki with 66 cm between rows and 26 cm between hills to accommodate larger tractors and planting equipment (Ushirogi *et al.*, 1986). Farmers plant either standard sized azuki (13 to 15 g/100 seeds) or the larger *dainagon* (18 to 20 g/100 seeds). Less than 10% of Hokkaido azuki production was in *dainagon* cultivars because of lower economic returns (Narikawa, 1972). *Dainagon* azuki are defined on the basis of their large size and sometimes by their unique shape called *eboshi gata* in Japanese (Narikawa, 1978). The use of row planting is challenged by the results of Nagayama *et al.* (1978) who obtained significantly higher yields by broadcast planting at all three tested planting densities when compared to row planting.

Chemical seed treatment and inoculation with *Rhizobium* at planting are not commonly practiced by Japanese farmers although they are recommended in Hokkaido (Ushirogi *et al.*, 1986; Japan Bean Fund ..., 1987). Packing the soil after planting is recommended in Hokkaido to enhance seed germination (Ushirogi *et al.*, 1986). Hokkaido farmers purchase certified azuki seed about every three years to eliminate seed-borne viruses which build up from one production year to the next in farm-produced seed stocks (Tanaka and Konno, 1982; Japan Bean Fund ..., 1987). The germination of old seed is normally not a problem since azuki which has been correctly stored for up to five years was demonstrated to have a field germination rate of 80% or greater with final yields that were comparable to those achieved with fresh seed (Satoh and Murata, 1978). Seed stored for 10 years at 6% R.H. had a moisture content of 7.9 % and germinated at 81%, while seed stored at 15% R.H. had a moisture content of 13.6% and germinated at 96% (Hokkaido Central ..., 1976).

In the Tan'ba region, near the city of Kyoto, which produces the highest quality azuki in Japan, planting occurs between July 10th and 20th at a density of 66,667 plants/ha (Matsui, 1984). A seed treatment and *Rhizobium* are applied before planting into raised beds which prevent waterlogging during heavy monsoon rains. After planting, ungerminated seed is replaced with hand-planted seedlings. A straw mulch is manually applied for weed suppression and moisture conservation.

Most Chinese azuki is produced from landraces maintained by individual farmers or state farms, although regionally developed cultivars are increasingly important (Tai, 1980; Duan, 1989). Planting density depends on seed quality, seed size, cultivar and the production location and system. The earliest mentioned planting rate, 1st century BC, was 10-12 l of seed/ha (Shih, 1974). Spring-planted azuki has optimum yields at 150,000 to 225,000 plants/ha, while early maturing cultivars planted in the summer and autumn require 300,000 to 375,000 plants/ha (Duan, 1989). Adachi *et al.* (1979) report that higher densities promoted earlier maturation. Zhang *et al.* (1981), in China's cold northern province of Heilongjiang, reported planting 45 to 53 kg seed/ha in 30 cm wide rows after the air temperature rises to 25°C. Common production steps include the replanting of areas in fields with poor stand establishment, thinning when plant densities are excessive and cultivating to control weeds two to three times before canopy closure (Duan, 1989). In northern China, azuki is normally planted in late April to mid-May and in mid- to late summer in central and southern China. Hu (1989) studied an intercrop planting of two rows of azuki between every second row of maize. The maize rows alternated between 67 cm wide and 48 cm wide. Two rows of azuki were centered between the wide maize rows with 40 cm between the azuki rows, 30 cm between hills within the row and three to four seedlings per hill. The maize was harvested first and yielded 11,250 kg/ha while the azuki yielded 1200 to 1500 kg/ha. Yao (1985) recommended a similar azuki intercropping system with maize in Jiangsu Province. In Hebei Province, Wang (1986) recommended that the azuki cultivar *Jihong* No. 1 be planted at 30 kg/ha between June and July at a depth of 3 cm in alternating 50 cm single rows of maize and azuki.

In the Taiwan, Kaoping and Tainan regions, some spring azuki is planted with two seeds per hill, 15 cm between hills within rows and 45 cm between rows. Spring-planted azuki generally produces lower yields than fall-planted azuki (Hsu *et al.*, 1989b; Lee, 1989b). The spring-planted crop produced inferior seed quality, and mechanical harvesting in the late summer was impeded by leaf retention at physiological maturity (Lee, 1989b). Until recently, most Taiwanese production relied upon the use of local landraces taking 85 to 95 days to mature. These landraces were not subject to improvement efforts (Chao, 1982; Suzuki and Konno, 1982). Taiwanese producers usually planted 70 kg/ha of home-produced seed in a square spacing pattern of 22.5 by 22.5 cm and did not use seed treatments or *Rhizobium* bacteria inoculum. With the introduction of Japanese cultivars

(*Kaohsiung* No. 1) and development of new cultivars, a fall-planting configuration of two to three seeds per hill placed 15 cm apart with 30 cm between rows was recommended (Wu *et al.*, 1985a; Hsu *et al.*, 1989ab). Hsu *et al.* (1989c) tested five ways to plant azuki after rice to determine the economic and seed yields. Conventional hand planting gave the highest seed yield followed by broadcast planting before rice harvest and covering with straw, which gave the highest economic yield.

In Korea, azuki is commonly planted during mid-June after barley, in 60 cm rows with 10 cm between hills at a density of 333,000 plants/ha, or as a full season crop planted during mid-May in 60 cm rows with 15 cm between hills at a density of 222,000 plants/ha (Hong and Park, 1982).

In India, field preparation recommendations included two to three plowings before the basal fertilizer application (Thomas *et al.*, 1974; Gill and Sharma, 1980). Both research groups recommended a planting rate of 16 to 25 kg/ha at a depth of 1.5 to 2.0 cm with a spacing of 10 to 12 cm within rows and 30 to 50 cm between rows.

In New Zealand, agronomic practices in the research fields included planting during late November at 35 kg/ha in 75 cm rows with 5 cm between seeds, and harvesting in March and April (Palmer, 1972). Pre-emergence herbicides and later cultivation were used for weed control. Nodulation was effective but a 50 kg/ha sidedressing of N at the four- to six-leaf stage was recommended.

In the Columbia Basin of Washington State, U.S., the authors have recommended the breakup of hardpan or tillage compaction layers during plowing. Under dry conditions the fields should be pre-irrigated to ensure uniform and sufficient soil moisture from planting to seedling emergence; excessive irrigation can lower soil temperatures and delay emergence of the seedlings. Irrigation after planting but before seedling emergence, especially by overhead systems on sandy loam soils, can cause surface crusting which reduces seedling emergence. Azuki is planted early since most cultivars can take up to 120 days to mature. Planting normally occurs in early to mid-May when average soil temperatures reach a minimum of 16°C, which promotes azuki germination and stand establishment. Cooler soil temperatures lengthen emergence time which increases the possibility of seed rot, reduces the ability of azuki to compete with weeds and increases the possibility of toxic effects from preplant incorporated herbicides.

Planting depth can vary because of seed size and quality, moisture conditions and soil texture. Planting depth recommendations vary in different parts of the world between 1.5 to 9 cm (Georgeson *et al.*, 1891; Thomas *et al.*, 1974; Kay, 1979; Duke, 1981; Hardman, 1985; Duan, 1989; Liang and Zhang, 1991). Researchers in the dry loess area of central China recommend a pre-treatment of sun-drying the seeds for two to three days, followed by planting at 5 to 7 cm when the soil is moist or 7 to 9 cm when the soil is dry (Liang and Zhang, 1991). Farmers in Washington State normally plant at depths of 2.5 to 4 cm in moist soil, followed by a roller/packer to insure good seed-to-soil contact. Planting is done in 56 cm rows with single seeds placed 5 to 7.5 cm apart for the cultivars *Erimo, Adzuki Express, Hatsune* or *Takara*. Densities of about 240,000 plants/ha are achieved by planting 35 to 40 kg/ha which has been found to reduce weeds populations. Azuki seed is normally inoculated with cowpea nitrogen-fixing *Rhizobium* bacteria the first time azuki is planted in a field. If a seed treatment is considered necessary, Apron (metalaxyl) has been shown to be safe for the *Rhizobium*, but seed treatments containing captan may kill the *Rhizobium*.

Fertilization

The range of fertilizer recommendations and the results of fertilizer research for the production of azuki are very diverse (Table 4.1). A 1 ha crop of azuki yielding 2160 kg/ha was reported to take up about 74 kg N, 18 kg P and 50 kg K from the soil (Duan, 1989). From another perspective (also see mineral nutrition in Chapter 3), 1 kg of stems, leaves and pods was reported to contain the equivalent of 26.7 g N, 3.5 g P, 13.5 g K, 1.9 g Ca and 3.9 g Mg in one study (Kushizaki *et al.*, 1961) and 1 kg beans contained 37 g N in another report (Osaki *et al.*, 1992). Average mineral absorption by azuki on alluvial and volcanic soils was 150.5 kg N, 17.4 kg P, 74 kg K, 38 kg Ca and 14 kg Mg/ha for a complete growing season (Japan Bean Fund ..., 1987). Nishimune (1984) reported a total nitrogen utilization range of 60 to 180 kg/ha for azuki grown on three different soil types. Azuki utilized only about 46 to 60% of applied nitrogen fertilizer and obtained the balance of its nitrogen requirements from soil and symbiotic-derived sources.

Table 4.1. Recommended or actual fertilizer rates in kg/ha for azuki in different countries. The Washington State recommendations are minimum levels combining both soil fertility and applied chemical sources.

COUNTRY	NITROGEN	PHOSPHORUS	POTASSIUM	COMPOST	OTHER	REFERENCE
China	74	18	49			Duan, 1989
India	21	26	42			Thomas *et al.*, 1974
India	40	26				Gill and Sharma, 1980
Japan	20-50	40-150	40-120	10,000	20-40 Mg	Ushirogi *et al.*, 1986
Japan	20-30	80-100	70-100	10-15,000		Japan Bean Fund ..., 1987
Korea	40	24	37			Hwang, 1982
Taiwan	40-70	16-26	25-33			Suzuki and Konno, 1982
Taiwan	63	23	50			Chao, 1982
Taiwan	40-60, split	11	23-35			Hsu, 1989a; Wu *et al.*, 1985a
USA, MN	18-27, split	like dry beans	like dry beans			Hardman, 1985
USA, WA	34 soil + 56 split	9, soil + chem.	22, soil + chem.		10 Zn/3-4 yr	author's

To meet the crop's nutritional needs, a urea topdressing of 225 kg/ha at anthesis was found to increase yields by 25% (Han *et al.*, 1987). Similar results are reported by Duan (1989) in which a topdressing of 83 kg/ha of ammonium sulfate was applied just after flowering which increased yields by almost 16%. Split applications of 40 to 60 kg N/ha, i.e. 35% before planting, 30% 20 days after planting and the remainder at anthesis, have been recommended for the new Taiwanese cultivar *Kaohsiung* No. 5 (Hsu, 1989ab). A similar fertilizer recommendation has been made for cultivar *Kaohsiung* No. 3 (Wu *et al.*, 1985a). Yields increased when a sidedressing of basal fertilizer was applied 5 cm below the seed during planting and a topdressing was applied later (Murata, 1983b). Yamanouchi (1964a) recommended an NPK topdressing 60 days after the basal application. Azuki yields responded positively to an application of 28 kg N/ha at the beginning of flowering (White, 1972; Kay, 1979). A basal application of 112 kg N/ha increased seed protein under dryland conditions but had little effect on yield under dryland or irrigated conditions (Robinson, 1983). In this same study, under irrigation, seed from fertilized plots had more K, Al, Pb and B but less Zn, Mg, N and P, compared to fertilized dryland plots. In Japan, no significant differences were recorded in plant height when fertilized with 25 kg N, 80 kg P, 50 kg K and 32 kg Mg/ha as compared to those that were fertilized with an additional 50% and 100% of these elements; azuki plants with the increased fertilizer application did have more pods per m^2, fewer seeds per pod and increased yields as compared to azuki fertilized at standard levels (Satoh, 1974). Azuki yields also responded positively to five phosphorus application rates but the response curve was bell shaped, peaking at about 80 kg P/ha (Mou and He, 1985).

Sawaguchi (1986; also see 1987) has published the most extensive report and literature review on the nutrition of azuki, especially for Hokkaido where temperatures are cool in early summer. The basic conclusion of his report is that, on Hokkaido, the growth of azuki is highly correlated with phosphorus until flowering and with nitrogen after flowering. He recommends a sidedressing of 130 kg or more of water-soluble P/ha and less than 30 to 40 kg N/ha at planting, followed by 50 to 100 kg N/ha of ammonium sulfate as a topdressing at the beginning of flowering. The topdressing could also be applied as a foliar application of a 2% urea solution split into three applications applied every other day or as a topdressed mixture of 40 kg slow-release CDU-N/ha and 30 kg fast-acting N/ha. He presumes that highest yields can be achieved by maintaining a hot-water-soluble nitrogen concentration of 10 mg/100 g soil. He also found the best correlation between phosphorus concentration and early growth by using the Truog - P_2O_5 methods of analysis.

In Washington State, the authors usually recommend a sidedressing of 30 kg N/ha at planting which helps to reduce the availability of fertilizer to weeds. A second application of 30 kg N at the time of flowering and early pod development has also been recommended. Nitrogen is not recommended at the pod development stage since excessive fertility can then delay maturity and result in additional vegetative biomass at harvest time. Some farmers have achieved high yields without the use of N fertilizer. Phosphorus and potassium requirements of azuki are considered to be similar to other dry edible beans and these fertilizers are applied when soil tests indicate deficiencies. Some soils in Washington require zinc which should be basally incorporated at 10 kg/ha every three to four years.

Allen and Allen (1981) report that the tropical species of the Asian mungbean group, including azuki, are nodulated only by a slow-growing, monotrichously flagellated *Rhizobium* of the cowpea group. Nodules, 4 to 10 mm in length, form about three weeks after germination and remain active throughout development (Tsai, 1988). Inoculation with the cowpea *Rhizobium* has been shown to significantly improve yields (Kay, 1979; Duke, 1981). Scientists in China reported that azuki fixes about 41 to 99 kg N/ha in Shaanxi Province (Liang and Zhang, 1991) and about 86 kg N/ha in other areas (Duan, 1989). In Japan, a five-year field experiment demonstrated a nitrogen fixation range of 1 to 60 kg N/ha on three different soil types (Nishimune *et al.*, 1983). Sawaguchi (1986, 1987) reported that inoculated azuki fixed 30 to 50 kg N/ha with the greatest fixation occurring at the early pod-filling stage then decreasing rapidly after that. He recommended a 30 to 40 kg N/ha basal application as optimum if azuki seeds are inoculated. Australian scientists have found that cowpea strain CB 756 was effective for three azuki cultivars tested in Queensland (Lawn, 1979ab, 1983) and all 11 cultivars in New South Wales (Brockwell *et al.*, 1991). American scientists have identified other effective strains or mixes (Dorer, 1973) and Nitragin AL inoculant (Duke, 1981; Hardman, 1985). In Minnesota, U.S., the injection of a Nitragin company inoculant into the soil at planting increased seed protein concentrations under irrigation and dryland growing environments by 0.3% and 1.9%, respectively (Robinson, 1983). Liming was found to improve rhizobial activity in soil with a pH less than 5 (Wu and Hsu, 1986).

Nishimune *et al.* (1983) determined that after reaching a certain point, yields per unit of nitrogen taken up decreased as nitrogen fixation increased. Nishimune (1984) later determined that azuki makes initial use of fertilizer sources of nitrogen early in the season while relying on soil nitrogen as it matures. He calculated a regression equation for the response of seed yield to nitrogen fixation, i.e. seed yield = 149 + 13.6 (nitrogen fixed by nodules) ($r = 0.7676$).

Azuki growth and development may be improved in low phosphorous soils when inoculated by indigenous or applied vesicular-arbuscular mycorrhizae (VAM) of the *Glomus* spp. (Wang, 1989; Tawaraya *et al.*, 1992). Three (VAM) *Glomus* spp. fungi caused inoculated azuki plants grown in sterilized soil to be significantly taller and heavier than uninoculated plants and to have improved uptake of P and Mg (Wang, 1989). However, pods per plant and dry seed weight differences between inoculated and uninoculated azuki grown in the field at three different phosphorus levels, were statistically insignificant. The mycorrhizal infection of azuki is reduced by the use of the fungicide benomyl and results in reduced P uptake by azuki under P-deficient conditions (Tawaraya *et al.*, 1992).

Weed control

In most areas of the world both mechanical cultivation and herbicides are used for weed control before and after the planting of azuki (Konno and Narikawa, 1978; Hardman and Meronuck, 1982; Robinson, 1983; Hardman, 1985; Japan Bean Fund ..., 1987; Durgan *et al.*, 1989; McClary *et al.*, 1993). Differences in weed control often reflect the level of commercialization of the crop. In Taiwan, where azuki is a cash crop, often grown for export, herbicides have been used on 95% of all production areas (Chao, 1982), while in India, where it has been viewed as a garden crop, early season weed control was accomplished by hand hoeing of fields 20 to 25 days after planting (Gill and Sharma, 1980).

Weed control is more critical for azuki than for many other dry edible bean crops, especially in temperate regions, where the thermophyllic azuki has a slow growth rate in the spring. Field sanitation is the first line of defense against weeds for all crops. Weeds in an azuki field can be controlled with cultivation after the second true leaf has developed on the seedlings, usually 20 to 30 days after planting, but mechanical cultivation should end before flowering begins, to avoid excessive pruning of azuki's roots. Matsuda *et al.* (1992) measured the shear strength of azuki and weeds at different heights in order to obtain basic information needed to develop a weeder. A linear relationship between height and shear strength was found. Azuki seedling stem shear strength was 3.3 kgf at 38 days.

Cultivation during the seedling stage is often used, but farmers who grow azuki as a cash crop depend on herbicides for the majority of weed control. Considerable research has been done on the usefulness and impact of herbicides on azuki. The results are summarized in Table 4.2.

While many herbicides are not toxic to azuki, research has shown that azuki is susceptible to damage from some herbicides commonly used for weed control in dry edible bean production. For example, EPTC (Fig. 4.11), alachlor, metolachlor, chloramben, Pursuit and Basagran have been found to be toxic to azuki and/or significantly inhibit normal development in Washington State, especially during cold, wet weather that slows germination and seedling establishment. In a greenhouse herbicide selection study EPTC (3.36 kg a.i./ha) resulted in significantly reduced plant height, leaf area index and shoot weights as compared to nontreated azuki (McClary *et al.*, 1993). Shaw and Brown (1989) reported that the preplant

Fig. 4.11. EPTC has been found to damage azuki seedlings in the Columbia Basin of Washington State. This damage was caused by preplant-incorporated EPTC at the rate of 2.24 kg/ha along with trifluralin at 1 kg/ha.

incorporation of EPTC (4.4 kg a.i./ha), EPTC + safener (4.4 kg a.i./ha), metolachlor (2.6 kg a.i./ha), CGA-180937 (2.6 kg a.i./ha), clomazone (0.75 kg a.i./ha) and the pre-emergence use of linuron (2.0 kg a.i./ha) resulted in initial azuki injury but the crop recovered. The post application of Basagran + Assist (1.1 kg a.i./ha+ 2*l*) produced minor injury while 2,4-DB (1.0 kg a.i./ha) caused considerable injury. The other 15 herbicides screened for phytotoxicity to azuki did not produce any detrimental effects (Table 4.2) (Shaw and Brown, 1989).

Table 4.2. Herbicides recommended or studied for use on azuki.

Herbicide	Ingredient/ Formulation	Application Time	Application Rate	Comment	References
Acifluorfen		POST	0.6 kg a.i./ha		Shaw and Brown, 1989.
	20.4% EC	POST	0.3 kg a.i./ha	excellent control	Hong *et al.*, 1983.
Alachlor				recommended	Hardman and Meronuck, 1982; Wu and Hsu, 1986; Durgan *et al.*, 1989.
	43.7% EC	PRE	1.31 kg a.i./ha	no phytotoxicity	Hong *et al.*, 1983.
				phytotoxic	Authors.
Alloxydim	Sodium WP	POST	1 kg a.i./ha	2-5 leaf grasses, not *Poa annua*	Takabayashi, 1978; Fukunaga, 1987.
	75% SP	POST	0.75 kg a.i./ha	less yield	Hong *et al.*, 1983.

Table 4.2. Continued.

Aresin		PRE	5-6 kg a.i./ha		White, 1972; Kay, 1979; Japan Bean Fund ..., 1987.
		PRE	1.0-1.5 kg a.i./ha		Shaw and Brown, 1989.
Asulam/Ioxynil			8-10 *l* /ha		Takabayashi, 1978.
Basagran		POST	0.84 kg a.i./ha	broadleaf weeds	Hardman and Meronuck, 1982; Hardman, 1985; Durgan *et al.*, 1989. Shaw and Brown, 1989.
		POST	1.1 kg a.i./ha	phytotoxic	Hong *et al.*, 1983.
		PRE	1.2 kg a.i./ha	less yield	McClary *et al.*, 1993.
		POST	1.1 kg a.i./ha	phytotoxic	
Bendiocarb	50% EC	PRE	0.3 kg a.i./ha	less yield	Hong *et al.*, 1983.
Butachlor	58.8% EC	PRE	1.76 kg a.i./ha	no phytotoxicity	Hong *et al.*, 1983.
CGA 180937		PPI	2.6 kg a.i./ha	phytotoxic	Shaw and Brown, 1989.
Chloramben	21.6% LQ	PRE	2.16 kg a.i./ha	no phytotoxicity	Hong *et al.*, 1983.
		PRE	10-15 ha	low toxicity	Kusanagi, 1981.
			10 kg/ha		Hwang, 1982.
	18.6%	PPI		recommended	Wu and Hsu, 1986.
		PPI	3.36 kg a.i./ha	no phytotoxicity	Shaw and Brown, 1989.
Chloramben + trifluralin		PPI	3.36 kg + 1.1 a.i./ha	no phytotoxicity	Shaw and Brown, 1989. Hardman and Meronuck, 1982; Robinson, 1983; Hardman, 1985; Durgan *et al.*, 1989.
Chloramben + trifluralin		PPI	3.36 kg a.i./ha + 1.1 kg a.i./ha		Shaw and Brown, 1989.
Clomazone		PPI	0.75 kg a.i./ha	early injury	Shaw and Brown, 1989.
DCPA	75%	PRE		recommended	Wu and Hsu, 1986.
Dinitramine	25%	PRE		recommended	Wu and Hsu, 1986.
Dinitroaniline		PPI	0.9-1.2 kg a.i./ha	residual concern	Takabayashi, 1975.
Dinoseb-amine		PRE	5-6 kg/ha		White, 1972; Kay, 1979; Japan Bean Fund ..., 1987.
Dinoseb-amine + Aresin			16.9 / 1.1 kg/ha	not *Paspalum distichum*	Palmer, 1972.
Dinoseb					Hardman and Meronuck, 1982.
Dimepiperate			10-7-10-4 M	lab seedlings	Matsumoto *et al.*, 1991
Diuron		PRE	1 kg/ha		Japan Bean Fund ..., 1987.
DNBP	49.5%	PRE	500-600 *l*	Hokkaido region	Ushirogi *et al.*, 1986.
Dowco-453	EC, Haloxyfop 10%		0.75-1 *l* in 700-1000 *l* water/ha	foliar on 3-5 leaf grasses	Kusanagi, 1986.
	24% EC	POST	0.48 kg a.i./ha	no phytotoxicity	Hong *et al.*, 1983.
EPTC					Hardman and Meronuck, 1982; Durgan *et al.*, 1989.
		PPI	4.4 kg a.i./ha	phytotoxic	Shaw and Brown, 1989.
		PRE	3.36 kg a.i./ha	stunted seedling	McClary *et al.*, 1993.
Ethalfluralin					Durgan *et al.*, 1989.
		PPI	1.1 kg a.i./ha	no phytotoxicity	Shaw and Brown, 1989.
		PPI	1.46 kg a.i./ha	no phytotoxicity	McClary *et al.*, 1993.
Ethofumesate	21%	POST		recommended	Wu and Hsu, 1986.
Fenoxaprop		POST	1-1.5 *l* ha;	5-9°C mean, 3-5 leaf stage	Kusanagi, 1986.
Fluorodifen	30%	PRE		recommended	Wu and Hsu, 1986.

Table 4.2. Continued.

Fluazifop	35% EC	POST	0.35 kg a.i./ha	no phytotoxicity	Hong *et al.*, 1983.
Fomesafen		POST	0.25 kg a.i. /ha + Agral 90 (0.1%)	no phytotoxicity	Shaw and Brown, 1989.
HOE-171 L	Fenoxaprop-ethyl 9%, EC		0.75-1 *l* in 700-1000 *l* water/ha	foliar on 3-5 leaf annual grasses	Noguchi, 1991.
HSW-802	Dinoseb acetate 40%, metola-chlor 20%; EC	PRE	3-4 *l* in 700-1000 *l* water/ha	cold region, all soils, annuals	Kusanagi, 1986.
Imazethapyr		PRE PRE POST	0.1 kg a.i./ha w/ 1 kg meto-bromuron 0.1% Agral 90 + 10-34-0 53 g a.i./ha	no phytotoxicity no phytotoxicity no phytotoxicity no phytotoxicity	Shaw and Brown, 1989. McClary *et al.*, 1993.
IPC	45.8%	PRE	200-400 *l*	Hokkaido	Ushirogi *et al.*, 1986.
IPC + prometryn	EC 45.8% WP 50%		2 *l* + 1-1.5 kg in 700-1000 *l* water	cold region, all soils	Noguchi, 1990.
Linuron Linuron + Dowco 453 Linuron + fluazufop	50% WP	PRE PRE PRE POST PRE POST	0.5 kg a.i./ha 2.0 kg a.i./ha 0.5 kg a.i./ha + 0.18 kg a.i./ha 0.5 kg a.i./ha + 0.26 kg a.i./ha	low toxicity phytotoxic excellent weed control excellent weed control	Hong *et al.*, 1983. Shaw and Brown, 1989. Hong *et al.*, 1983. Hong *et al.*, 1983.
Metobromuron		PRE	2-3 kg a.i./ha	no phytotoxicity	Shaw and Brown, 1989.
Metolachlor		 PPI PRE	 2.6 kg a.i./ha 1.96 kg a.i./ha	 phytotoxic phytotoxic	Durgan *et al.*, 1989. Shaw and Brown, 1989. McClary *et al.*, 1993.
Metribuzin	35% WP	PRE PRE	0.7 kg a.i./ha 0.75 kg a.i./ha	low toxicity no phytotoxicity	Harrison, 1988. Shaw and Brown, 1989.
NC-302 (L) (Flowable)	Quizalofop-ethyl 10%		0.75-1 *l* in 700-1000 *l* water/ha	foliar on 3-5 leaf grasses, cold	Kusanagi, 1986.
Oxyfluorfen	23.5%	POST		recommended	Wu and Hsu, 1986.
PCP (penta)		PRE	7.5-10.0 kg/ha		Takabayashi, 1980b.
Pendimethalin	 31.7% EC	 PPI PRE PPI	 0.95 kg a.i./ha 1.1 kg a.i./ha	 recommended not nightshade	Durgan *et al.*, 1989. Wu and Hsu, 1986. Hong *et al.*, 1983. McClary *et al.*, 1993.
Propham		PRE	2-4 kg/ha		Japan Bean Fund ..., 1987.
Quizalofop-ethyl		POST	0.75-1 *l* /ha	5-9°C mean	Kusanagi, 1986.
S-604 (EC)	Cethodim 25%		0.25 *l* in 700-1000 *l* water+oil	cold region all soils, grasses	Noguchi, 1991.
Sethoxydim	EC	POST	1.5-2 *l* /ha	3-5 leaf grasses, not *Poa annua*	Kusanagi, 1981. Fukunaga, 1987.
SL-236 (L)	Fluazifop-p-butyl 17.5%, EC		0.75-1 *l* in 700-1000 *l* water/ha	foliar on 3-5 leaf grasses, cold	Kusanagi, 1987; Noguchi, 1989.
SSH-55 (EC)	Trifluralin 14% prometryn 6%		9-10 *l* in 700-1000 *l* /ha water	cool region, all soils	Kusanagi, 1988.
Thiobencarb	55%	PRE		recommended	Wu and Hsu, 1986.

Table 4.2. Continued.

Trifluralin		PPI	0.84-1.12 kg/ha + Amiben		Hardman and Meronuck 1982; Robinson, 1983; Hardman, 1985; Durgan *et al.*, 1989.
	44.5% EC	PRE	2.5-3 *l* in 700-1000 *l* /ha water		Hu, 1982; Kusanagi, 1987.
	GR, 2.5% a.i.	PPI	1.1 kg a.i./ha	no phytotoxicity	Shaw and Brown, 1989.
	EC or GR	PPI	1.1 kg a.i./ha	annuals	Hong *et al.*, 1983.
	44.5% EC	PRE	0.89 kg a.i./ha	no phytotoxicity	McClary *et al.*, 1993.
		PPI	0.84 kg a.i./ha	no phytotoxicity	
2,4-DB		POST	1.0 kg a.i./ha	phytotoxic	Shaw and Brown, 1989.

FOOTNOTES:
Abbreviations used:
- PPI = preplant incorporate
- PRE = pre-emergence
- POST = post-emergence
- WP = wettable powder formulation
- EC = emulsifiable concentrate formulation
- GR = granular formulation

Common and chemical names confirmed in: Sine, C. (ed.). (1988-1991) *Farm Chemicals Handbook*. Meister Publishing Co., Willoughby, Ohio.

Herbicides that have been successfully used in Washington State, include pendimethalin which has been preplant incorporated at 1.1 kg a.i./ha for control of *Kochia scoparia, Chenopodium album, Amaranthus* spp. and annual grasses, but is not effective for control of *Solanum* spp. Trifluralin, preplant incorporated at 0.8 kg a.i./ha, controls the same weeds and has been substituted for pendimethalin when it is less costly. Ethalfluralin has been used to control *Chenopodium album* and some *Solanum* spp. under rill irrigation systems when preplant incorporated at 1.5 kg a.i./ha (McClary *et al.*, 1993). *Chenopodium album* is also a problem for azuki in the Tokachi area of Hokkaido and must be controlled before azuki flowering or yield will be reduced (Ikeda and Hirokawa, 1977). Under overhead irrigation, azuki fields treated pre-emergence with glyphosate at 0.55 kg a.i./ha + X-77 followed with an ethalfluralin treatment, had significantly higher yields. Tank mixing of ethalfluralin and pendimethalin was found to cause severe stunting of azuki (McClary *et al.*, 1993). Sethoxydim is labeled for azuki in the U.S. and can be used to control annual grasses or volunteer monocot crops such as barley, maize and wheat.

Water requirements

In East Asia's summer monsoon conditions, azuki fields are not normally irrigated. Duan (1989) reported that the growth of 1 g of dry azuki biomass requires 600 to 650 g of water, while others (Inst. Ag. Science ..., 1981) reported water requirements of 600 to 1000 g to produce 1 g of dry azuki and demonstrated the value of irrigation during the flowering (see Ag.

Lexicon ..., 1979) and pod-filling stages under rain-fed conditions. In a pot study, azuki was found to grow well for the first 40 days at soil water capacities as low as 40 to 60%, thereafter azuki required 60 to 100% and was optimum at 80 to 90% of soil water capacity (Tasaki, 1959). Hsu (1983) felt that dew might contribute a small but significant proportion of the water requirement of several crops during the winter months in southwest Taiwan. He developed instrumentation to measure the dew and found that azuki had 4.92 mm of dew on its leaves over a 46-day period that dew fell. Near Kaohsiung, Wu and Hsu (1986) recommended two to four irrigations to supplement the autumn rains. Under the dry conditions of Washington State's Columbia Basin, an azuki crop normally requires the addition of 15 cm of irrigation for high yield (Hang *et al.*, 1993). On sandy soils in Minnesota, Robinson (1983) irrigated azuki in amounts of 1.5 to 2.5 cm in June through August for a total of 24 to 30 cm in addition to the 16 to 36 cm of rainfall that occurred during this period. Yields in unfertilized plots increased from 1457 to 3260 kg/ha with irrigation, but protein percentage in the seed decreased.

Harvest

The range of azuki harvesting practices can partially be explained by the interaction of weather at the end of the growing season and azuki's generally indeterminate maturation. This interaction should be considered when determining the correct harvesting time. Good timing can optimize economic yields by reducing pod shatter and promoting the largest seed size and best seed testa luster (Ushirogi *et al.*, 1986; Japan Bean Fund ..., 1987; Duan, 1989). Arrival of cold weather from Siberia has the strongest influence on the harvest time in the largest azuki production areas of northern China and northern Japan.

Once harvest is initiated in Asia, in most locations, azuki stalks with pods are hand cut or pulled (Suzuki and Konno, 1982; Duan, 1989). Typical Japanese harvesting practices include both manual and mechanical means. On Hokkaido, harvest begins in late September and requires an average of 20 days for field drying and curing (Yoshida and Satoh, 1988). Plants are traditionally cut by hand and allowed to cure on the ground for several days before being stacked into drying piles (Fig. 4.12) which are monitored for correct ventilation to prevent molding. After stack drying to 16% seed moisture, the stalks with pods are threshed at 300 to 400 rpm in a stationary thresher (Fig. 4.13) (Ushirogi *et al.*, 1986; Japan Bean Fund ..., 1987). Some Hokkaido farmers have begun replacing hand reaping with a one- to two-row harvester. The harvester is used in the evening or early morning when azuki moisture content is highest in order to reduce pod shatter and seed cracking (Ushirogi *et al.*, 1986; Japan Bean Fund ..., 1987).

Mechanical pulling, windrowing and combining or direct combining have been used in Minnesota (Hardman, 1985) and Washington State. Farmers in Washington have reduced loses from pod shatter by dragging a heavy knife cutter under the ground to cut or pull out mature plants before windrowing. When the windrowed plants are dry, they are combined. Azuki is similar to navy and small red (*Phaseolus*) beans in their susceptibility to mechanical damage. Farmers use combine cylinder speeds between 190 and 210 rpm and carefully adjust the clearance between concave and cylinder bars to allow seed to pass through without breakage. Slower operating speeds and open concaves have been used elsewhere on combines to reduce pod shatter and prevent splitting or damaging seeds (Robinson, 1975; Hardman, 1985; Japan Bean Fund ..., 1987).

In China, Duan (1989) recommended to start the harvest when approximately two-thirds of all pods are mature or turning yellow. On Hokkaido, harvest begins when 70 to 80% of the pods are mature and average azuki seed moisture is about 30% (Yoshida and Satoh, 1988). Others report that approximately 80% of azuki pods mature at one time (Kay, 1979; Hardman, 1985). If early frost is threatening, Ushirogi *et al.* (1986) found that azuki can be harvested at the 40% mature pod stage and will still achieve 90% of optimum yield.

Fig. 4.12. Drying stacks of azuki plants in the fields of central Hokkaido. These types of stacks have been traditionally used to deal with the wet weather often associated with the harvest period, including typhoons. See Figure 4.13 for details about the construction of these stacks.

Mechanical harvesting techniques result in substantial yield losses because many azuki pods are set close to the ground (White, 1972; Nomura and Satoh, 1977). In New Zealand, White (1972) addressed this problem by using an experimental harvester that cut azuki a little below the ground with a disc. Some growers in Washington State pull a subsurface knife through the fields to cut or pull out the stalks without damaging the lower pods. Several research attempts have been made to select for a plant architecture that is suitable for direct combining. Nomura and Satoh (1977) determined that tall, late-maturing azuki cultivars had the smallest number of mature pods between ground level and 10 cm above the ground and thus were the most adapted to successful combine harvesting.

Plant moisture content is a second major problem for combine harvesting. In northern Japan and New Zealand, excessive moisture can impede the threshing operation, while dry weather, common in the U.S. during harvest, causes pod shatter which can result in substantial yield losses during the reaping operation, especially if reaping is delayed or done late in the day (Georgeson *et al.*, 1891; Piper and Morse, 1914; Plant Production ..., 1959; Matlock and Oswalt, 1963; Narikawa, 1972; Thomas *et al.*, 1974; Hernandez *et al.*, 1978; Kay, 1979; Duke, 1981; Hardman, 1985; Hoshikawa, 1985; Ushirogi *et al.*, 1986; Japan Bean Fund ..., 1987; Duan, 1989). The ground or aerial application of Spodnam (1 kg in 360 l of water/ha) has given mixed results in reducing pod shatter. This product is difficult to apply effectively since the leaf canopy prevents the spray from contacting the pods below.

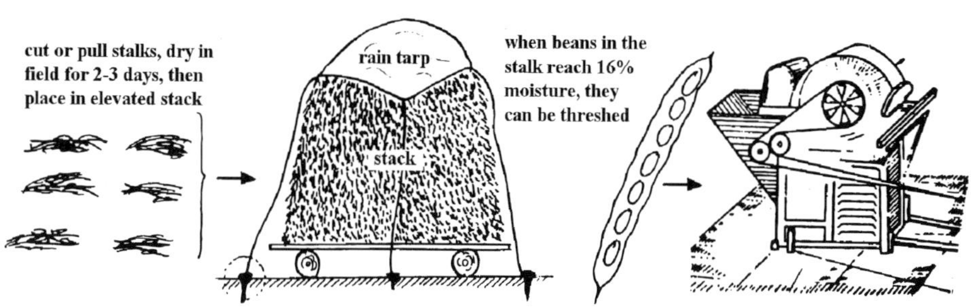

Fig. 4.13. The traditional approach to harvest of azuki on Hokkaido. This method is an adaptation to the wet, cool weather that often occurs at harvest time in northern Japan (illustration adapted from Ushirogi *et al.*, 1986).

In-field processing to improve azuki seed quality before grading has become a common practice in Japan as well as the use of on-farm drying facilities (Ushirogi *et al.*, 1986). An after-harvest winnowing method was reported by Tanda (1958) who found that specific gravity differences between immature, small, insect-damaged azuki seeds and normal seeds make it possible to separate the good seeds on the farm, using a fan that produces wind velocities of 6.3 m/sec. Duan (1989) recommended that harvested azuki seed be treated with either a boiling water bath or steamed with aluminum phosphate to kill insects before storage. If seed moisture content for storage is less than 13%, the seeds can be stored for four to five years and still have a high germination rate. Wu and Hsu (1986) suggest storing azuki in sealed basins, buckets or plastic bags with a Phostoxin pill for each 100 kg to prevent pests.

5

INSECTS AND NEMATODES

Like most crops, azuki is afflicted by a wide range of insect and nematode pests. Azuki bean weevil, *Callosobruchus chinensis* L., is generally considered to be the worst insect pest of azuki, however many other insects also cause extensive damage. Sixteen common insect species have been reported to attack azuki in Japan (Kobayashi, 1978) although many additional, less common pests have been identified. In the Hokkaido island production areas up to 70 different insect pests are found (Akai *et al.*, 1986). See Figure 6.3 for the occurrence of common azuki insect pests during the growing season on Hokkaido.

The sections on azuki in the plant protection book by Akai *et al.* (1986) provide the most comprehensive guide for insects and diseases of azuki found in Japanese literature. Insects and disease of azuki in Taiwan and their control are described by Wu and Hsu (1986). The Class Hexapoda phylogeny of Borrow *et al.* (1981) was used to organize the following sequential descriptions of insect pests of azuki and entomological research reported in the world literature; different families within orders are described in alphabetical order. A list of insect and nematode pests appears in Table 5.1. Pesticides used for the control of insects are listed in Table 5.2.

INSECTS

Thysanoptera: Thripidae

The bean flower thrips, *Megalurothrips usitatus* (Bagnall) Hood is reported to be a serious pest of azuki in Taiwan (Chen, 1979b; Chang, 1987ab). *M. usitatus* occurs in southern Taiwan from late September through early January in large numbers during late azuki anthesis (Chang, 1988ab; 1990). The bean flower thrips first migrate to Taiwanese azuki fields approximately 50 to 55 days after planting and can quickly reach populations of more than ten thrips per flower (Chang, 1989). Chang (1988b) observed that first generation bean flower thrips favored azuki fields in southern Taiwan while the second generation migrated to *edamame* or grain soybean fields after azuki anthesis was completed.

The larval dispersal pattern of *M. usitatus* on azuki flowers is best described by the negative binomial during all stages of flowering (Chang, 1992). Adults are randomly distributed at initial flowering and then follow the negative binomial after 50 days of azuki growth.

Chang (1988a) quantified the preference of bean flower thrips for oviposition sites on azuki leaves and their feeding preference for azuki leaf tissue, flowers and pollen grains over those of peanut and soybean. *M. usitatus* adults and larvae preferred azuki flowers over leaf tissue (Chang, 1988a). Bean flower thrips had much higher fecundity when allowed to complete one life cycle on azuki foliage and flowers than on identical soybean and peanut components (Chang, 1987b).

Chang (1989) recommended the application of pesticide between 11 am and 2 pm, at 40 to 45 days after planting and again at 65 days for control of thrips. Chang (1987ab) observed a wide range of effects on the number of *M. usitatus* per azuki plant one day after the application of five different insecticides. Methiocarb did not affect thrip populations; phenthoate caused only a 1% reduction. Bendiocarb resulted in a 10% reduction in thrips while Advantage (carbosulfan) caused a 28% reduction. The best of the five treatments was methamidophos which caused 60% reduction in the number of thrips per azuki plant one day after application (Chang, 1987b).

Mass trapping experiments in azuki fields using different colored water pans demonstrated that blue traps (43.86 reflectance [L]; -6.97 [a]; -52.45 [b] [Hunter spectrocolorimeter designations]) caught significantly more *M. usitatus* than yellow, green and white water pans (Chang, 1990). The author suggested that the blue trap's high percentage of reflectance in the 450 to 480 nm light wavelength may account for its attraction to thrips, but recommended that more research be conducted before colored water traps were used to control *M. usitatus* in azuki fields.

Frankliniella intonsa Trybom (*Frankliniella formosae* Moulton) is reported to infest azuki growing in Taiwan and Japan (Takahashi, 1936; Chang, 1989; Hachiya, 1990). *F. intonsa* adult and larval numbers greater than 100 per growing point or flower cluster cause a significant decrease in the number of mature pods (Hachiya, 1990). This researcher tested the effectiveness of 45% Tokuthionin (prothiophos) for the control of *F. intonsa* in Japan and reported insignificant differences in mature pods, seed number and seed weight per plant between the insecticide treatment and the nontreated control. *F. intonsa* exhibited no significant difference in color preference between blue, yellow, green or white water traps set in azuki fields in Taiwan (Chang, 1990).

Hemiptera: Miridae

Lygus spp. are serious pests of alfalfa (*Medicago sativa* L.) in the western U.S. They have not been previously reported as pests of azuki, but have been observed by the authors to cause minor damage to azuki especially in fields near alfalfa (Fig. 5.1). If frequent sweep net samples indicate the presence of lygus bugs at flowering, dimethoate at 1.7 kg a.i./ha applied by air, first at the initial bloom stage and a second application two weeks later, has given effective control.

Homoptera: Aleyrodidae

Azuki was shown by Lin (1981) to have little resistance to *Trialeurodes vaporariorum* Westwood, greenhouse whitefly. Whiteflies, *Trialeurodes* spp., can be a problem in southern Taiwan during winter azuki production; nymphal stages and adults feed on the ventral side of leaflets and can transmit plant viruses (Chang, 1989).

Homoptera: Aphidae

Aphids, called *mame abura mushi* in Japanese, can damage azuki and be vectors for different viruses. Young azuki leaves, pods, stems and flowers are subject to cowpea aphid, *Aphis craccivora* Koch, which attack in Japan's Hokkaido Prefecture by mid-June, especially if warm, dry weather promotes high populations. Cowpea aphids are also a production concern in Taiwan, especially when they quickly reach high population levels in dense plant canopies with poor air circulation (Chang, 1989). Azuki at early stages of development is frequently infested with aphids in Liaoning Province, China (Guo, 1980).

Affected azuki leaves curl, turn yellow and die while plants exhibit poor pod set and seed formation (Akai *et al.*, 1986). Viral diseases may also develop (Kobayashi *et al.*, 1976; Kobayashi, 1979; Tsuchizaki and Omura, 1987; Chang, 1989). Cowpea aphid populations decline in Hokkaido Prefecture when azuki reaches full maturity and stem chlorophyll content decreases (Akai *et al.*, 1986). However, another aphid, *Aphis medjcaginis* Koch, is a problem during azuki anthesis in China (Duan, 1989).

Sugano *et al.* (1981) observed that black and transparent silver-stripped plastic mulch diminished field levels of aphids, reduced virus infection and significantly improved azuki yields over controls. Mukoo (1975) recommended two to three applications of various insecticides every 10 days after the expansion of the first true leaf of azuki for aphid control and virus transmission. However, there are Chinese cultivar differences in the level of aphid infestation (Guo, 1980) that could be exploited in breeding for tolerance.

Coleoptera: Elateridae

Wireworm is the prominent soil pest of dryland small grain crops and some legume crops in western North America. An unidentified wireworm species has caused minor damage to azuki growing in the Columbia Basin (Fig. 5.2).

Fig. 5.1. (Left) *Lygus* spp. in the western U.S. feed on azuki leaves and developing pods and seeds. Feeding causes testa imperfections and internal destruction of the cotyledons in developing azuki seeds.

Fig. 5.2. (Right) Wireworms, the immature form of click beetles (Coleoptera: Elateridae), feed on azuki seeds and seedlings. This damaged seedling and larva were observed in the Columbia Basin of Washington State.

Coleoptera: Bruchidae

The azuki bean weevil (*Callosobruchus chinensis*), *azuki zou mushi ga* in Japanese, is the most destructive pest of stored azuki and other *Vigna* spp. in East Asia (Cheng and Qou, 1967; Lin, 1981; Shinoda and Yoshida, 1984; Fernandez and Talekar, 1990). The weevil can complete four generations a year under optimum storage conditions (Yoshida *et al.*, 1986b; Kim and Choi, 1987; Han and An, 1990). All seeds in storage can be damaged or destroyed by *C. chinensis* in three to four months if the pest is not controlled (Fernandez and Talekar, 1990). Han and An (1990) reported a base development temperature of 12.3°C and 381 days per life cycle of *C. chinensis*. One female weevil can lay about 15 eggs per day (Chun and Ryoo, 1992) and one generation can be completed in 21 days at 30°C (Yoshida *et al.*, 1986b).

Feeding trials with *C. chinensis* (Umeya and Shimizu, 1968) on members of the Fabaceae family demonstrated that azuki seed was preferred for egg laying and had the highest weevil hatching and emergence rates as compared to mungbean (*Vigna radiata* [L.] Wilcz), rice bean (*Vigna umbellata*

[Thunb.] Ohwi and Ohashi), broad bean (*Vicia faba* L.), soybean and kidney bean (*Phaseolus vulgaris* L.). Pea (*Pisum sativum* L.), black soybean and peanut (*Arachis hypogea* L.) were least preferred (Lin, 1964; Kim and Choi, 1987; Han and An, 1990). Shinoda *et al*. (1991) collected seeds of 11 wild legumes growing in Okayama Prefecture, Japan. Adults of *C. chinensis* emerged from seeds of only two of these species, i.e. wild azuki (*V. angularis* var. *nipponensis*) and *Dunbaria villosa* (Thunb.) Makino. Adults were placed with seeds of each of the 11 legume species. After incubation, adults emerged from only four species, the two mentioned above plus *Amphicarpaea edgeworthii* Benth var. *japonica* Oliver. However emergence was only 20% on the latter two species versus 85% for the former, which was the same as cultivated azuki. The weevil development period was nearly twice as long on the latter two species and adult longevity was decreased. Shinoda *et al*. (1992) continued this research by observing the life cycle of *C. chinensis* on native stands of the wild azuki and *D. villosa*. Larval survival rate was low until the first instar larvae penetrated in to seeds through the pods, where survival rate improved. This pattern is similar to cultivated azuki. Nagasawa and Yamada (1988) observed 1.4 and 8 times the number of F_1 weevils emerging from azuki as compared with peas and soybeans, respectively. It is unclear if *C. chinensis* preference for azuki is the result of ecochemicals (Honda and Ohsawa, 1990) although an azuki bean weevil oviposition stimulant, D-catechin ([2S, 3S]-3,5,7,3',4'-pentahydroxyflavan) was isolated from azuki seed coats by Ueno *et al.* (1990). Density pressure also plays a role in emergence (Murai and Fujii, 1970).

Factors affecting optimum population growth of *C. chinensis* were studied by Utida (1941ab). Research studies have determined that ambient storage temperature, degree of azuki seed conditioning, existing population levels, food characteristics and the presence of a biological conditioning substance (BCS) influence weevil populations. Kim and Choi (1987) found that *C. chinensis* had an optimum oviposition temperature of 25 to 30°C. The presence of sugar-water increased the adult longevity, oviposition length and total fecundity (Han and An, 1990). Tanaka (1993) estimated moderate or high heritability for longevity and age specific fecundities by *C. chinensis*. He also calculated a highly negative genetic correlation between longevity and early fecundity.

The young female weevils migrated to new oviposition sites if the azuki seeds contained a high number of eggs (Shinoda and Yoshida, 1984; Honda and Ohsawa, 1990). High populations of *C. chinensis* caused physical interference to copulation frequency and mechanical destruction of eggs by weevils (Utida, 1941c). The physical destruction of eggs may be increased due to the tendency of *C. chinensis* to lay a disproportionate number of eggs in the top few centimeters of stored azuki seeds in contrast to the more vertically-even distribution of eggs laid by the cowpea weevil, *Callosobruchus maculatus* Fabricius (Watanabe, 1984). Body size versus seed size plays a role in the distribution (Watanabe, 1985, 1986). Competition between these two species has been studied by Fujii (1969, 1970)

Honda and Ohsawa (1990) determined that a biological conditioning substance (BCS) released by azuki weevil had ovicidal properties when egg densities were high. Oshima *et al.* (1973) first isolated and experimented with this *C. chinensis* oviposition marker pheromone. They found that females selected untreated azuki seeds and glass balls over those prepared with the marker but would still lay eggs if all azuki seeds had been treated. Similar results were reported by Sakai *et al.* (1986) for *C. maculatus* as well as *C. chinensis*; *C. chinensis* was found to produce more of the marking pheromone than *C. maculatus*, and *C. chinensis* females released more than their male counterparts. *C. chinensis* also produces the copulation pheromone erectin (Tanaka *et al.*, 1981, 1982). Thus potential for improved biological control or bio-friendly control of *C. chinensis* has been boosted by a report of successful synthesis of the pheromone (R)-callosobruchusic acid from methyl (R)-3-carboxybutanoate by Abo and Mori (1993).

Utida (1941d) observed that females slightly outnumbered male weevils at high equilibrium population levels which the researcher attributed to differences in egg stage mortality. Utida (1943) later observed that *C. chinensis* females laid their eggs evenly on azuki seeds, a characteristic that changed at an unspecified adult weevil density. Later research determined that at three to four eggs per seed and/or a minimum of 80 μg BCS per azuki seed, females oviposited randomly rather than preferentially (Honda *et al.*, 1989). The number of eggs laid by a female of *C. chinensis* stays the same across a broad range of azuki bean densities, but the number of infested beans increases since the female tries to avoid seeds with existing eggs (Chun and Ryoo, 1992)

C. chinensis eggs hatch approximately one week after oviposition and larvae burrow through their egg shell's underside and into the seeds. The shell serves as a cover over the entrance hole; pupation occurs about three weeks after feeding with adults emerging about a week later (Utida, 1941a; Akai *et al.*, 1986). Males emerge one to two days before the females (Shinoda and Yoshida, 1984). Nagasawa and Yamada (1988) formulated an asymptotic regression equation $Y=6.540 - 5.9963\ (0.5657^{X})$ to describe the emergence of weevils in stored azuki at 28°C. The presence of azuki seeds suppressed the migration of recently emerged young *C. chinensis* adults for five days which suggested that they were reproductively active for that time period and that azuki seeds were the preferred mating and ovipositional sites (Utida, 1941a; Shinoda and Yoshida, 1984).

Azuki seeds damaged in storage by the weevil are not acceptable for human consumption or commercial seed use because of extensive internal damage, presence of an odor and low germination rate. Kim and Choi (1987) reported no germination for severely damaged azuki seeds in comparison to 90% for seeds not subjected to a *C. chinensis* infestation.

C. chinensis are bivoltine in Hokkaido Prefecture; they overwinter in the larval stage while mature females oviposit from mid-September to late October, 15 to 20 cm above the ground on mature azuki pods which have a specific curvature (Shinoda and Yoshida, 1985, 1987; Akai *et al.*, 1986; Honda and Ohsawa, 1990). Adult longevity in the field decreased with cooler weather, from late August to late October (Shinoda and Yoshida, 1987), while further south, *C. chinensis* is most abundant in the azuki fields of Okayama Prefecture in early October (Shinoda and Yoshida, 1984, 1985).

Significant levels of seed destruction can occur in azuki fields if *C. chinensis* is not controlled; in-field infestation levels of 14 to 15% are documented in the literature (Shinoda and Yoshida, 1984, 1985; Akai *et al.*, 1986). Drying harvested seed in the sun, soaking in warm water, applying insecticides, fumigation or gamma ray irradiation of packaged azuki seeds has been suggested for *C. chinensis* control (Lin, 1964; Hu *et al.*, 1985; Akai *et al.*, 1986; Duan, 1989).

Four to five hours of exposure of infested azuki seeds to a solar-generated temperature of 55°C was found to kill all *C. chinensis* life stages (Yoshida and Gichuki, 1983). Han and An (1990) reported that one to three days of seed drying at temperatures above 45°C was adequate to kill all adults and reduce the rate of damaged seeds. Internal seed temperatures of 33.6°C and water content of 2.5% after one to three days of drying did not significantly affect the survival rate of weevil larvae (Shinoda and Yoshida, 1985). A single 100 Gy dose of gamma radiation applied to bagged azuki seeds caused a significant reduction in the number of adults, eggs and damaged seeds as compared to the control (Hu *et al.*, 1985).

Fujii and Miyazaki (1987) suggested a two-step interspecific hybridization of a weevil-resistant *Vigna sublobata* race II cultivar to mungbean and then azuki to the *V. sublobata*-mungbean cross to transfer the resistance gene(s) to azuki. Fernandez and Talekar (1990) reported the presence of *C. chinensis* resistance in several black gram and mungbean accessions as well as wild *Vigna mungo* sub. sp. *silvestris* and *Vigna glabrescens*. The resistance seemed to be the result of a combination of characteristics such as general plant pubescence, seed coat thickness and presence of trypsin inhibitors. It is worth noting that the level and nature of saponins in these *C. chinensis*-resistant accessions has not been documented; pea and soybean saponins have been shown to inhibit *C. chinensis* development (Applebaum *et al.*, 1969).

Several other approaches to control have been studied. Ishimoto and Kitamura (1991) and Smitanond *et al.* (1991) report that larval growth is strongly inhibited by a kidney bean α-amylase inhibitor found in some cultivars of kidney bean, opening the opportunity for transferring this gene to azuki. A resistance to *C. chinensis* has also been found in mungbean seeds (Ishimoto and Kitamura, 1993a). Hermawan *et al.* (1993) found that extracts from *Andrographis paniculata* (Kalmegh), a medicinal plant, influence the oviposition of *C. chinensis* and the feeding of *Plutella xylostella* (L.). Also, the wasp *Dinarmus basalis* Rondani has been found to parasitize larvae or pupae of *C. chinensis* within azuki beans and shows preference for 13-day-old hosts over 12- or 17-day-old hosts (Nishimura, 1993).

Azuki is a new host for *C. maculatus* (Birch *et al.*, 1985) with large competitive differences between intraspecific strains (Messina and Mitchell, 1989; Messina,1991). Chen (1973) found that azuki seed was favored by *C. maculatus*, over seven other grain legume species (Fig. 5.3). The highest level of damage occurred at 25°C which correlated to the highest reproductive rate of the weevil. Giga and Smith (1987) reported that azuki seed was the preferred substrate for eggs laid by *C. maculatus* at 20°C as compared to five other grain legumes while preference for azuki seeds at 30°C was less apparent. Fox (1993) has also studied the oviposition preference and larval development of *C. maculatus*. Singh and Chua (1976) observed abnormalities in *C. maculatus*-infected azuki such as damage to epicotyls, roots, cotyledons, primary leaves and apical meristems. *C. maculatus* produces ether-soluble oviposition markers on azuki and oviposited on seeds marked by *Callosobruchus rhodesianus* Pic. (Giga and Smith, 1984/85).

Approximately 63% of all *C. maculatus* eggs laid on azuki seeds hatched and 34% of the eggs resulted in live larvae at 20°C; while at 30°C, 80% of oviposited eggs hatched and 71% of the larvae survived (Giga and Smith, 1987). Sano-Fujii (1984) identified azuki seed moisture content and temperature as affecting the number of developing weevils. A 16 to 22% water content range was quite favorable for *C. maculatus* development while 9 to 12% was too dry for development.

Ishimoto and Kitamura (1992, 1993b) report that the α-amylase inhibitor of *C. chinensis* discussed above also inhibits larval growth of *C. maculatus*, but not *Zabrotes subfasciatus* (Boheman) and *Acanthoscelides obtectus* (Say).

Giga and Smith (1987) reported that azuki seed was preferred over five other grain legumes by *C. rhodesianus* for egg laying at 20°C, but the preference was less obvious at 30°C. These researchers also observed that at 20°C, 65% of all *C. rhodesianus* eggs on azuki seeds hatched and 44% resulted in live larvae; at 30°C, 72% of all eggs hatched and 46% became larvae. *C. rhodesianus* produced ether-soluble oviposition markers on azuki but would not lay eggs on seeds already marked by *C. maculatus* (Giga and Smith, 1984/85).

Callosobruchus analis F. has been documented to infest azuki seed in storage. Takeuchi *et al.* (1977) hypothesized that a *C. analis* population starting from one breeding pair, would consume approximately 200 kg of seed if maintained for six months at a storage temperature of 25°C.

Seed weevils, *Zabrotes ubfasciatus* Boheman, feed on azuki but can be controlled by the parasitic wasp, *Anisopteromalus calandrae* Howard (Fujii, 1983). *A. calandrae* was also used by Shimada (1985) along with

Heterospilus prosopidis to study the population dynamics of competing species on azuki.

Coleoptera: Curculionidae

The bean narrow-mouth weevil (*mame hosoku chizou mushi* in Japanese), *Apion collare* Schilsky, is a serious pest of azuki and other legumes produced in Hokkaido Prefecture (Fig. 5.5) (Sakamoto *et al.*, 1983; Akai *et al.*, 1986). Morphological features of *A. collare* that distinguish it from four other Hokkaido Apioninae species are described by Sakamoto *et al.* (1983). Adults oviposit eggs on flower buds, stipules and apical meristems in early July; larvae begin active feeding in mid- to late August and cause internal destruction of ovaries and stigmas. General growth suppression due to meristem damage and a lack of pod set can occur from high populations of *A. collare* larvae. Reproductively inactive adults appear in late August to early September and feed on leaves before beginning hibernation under fallen azuki debris. Mechanical control of *A. collare* includes the removal of azuki field residues, the use of early maturing cultivars and autumn plowing (Akai *et al.*, 1986).

Holloway (1986) observed that rice weevil adults, *Sitophilus oryzae* L., were killed by natural seed toxins present in pelletized azuki after feeding for 14 days. The researcher recorded large genetic variability in resistance to azuki seed toxins between *S. oryzae* strains from Africa, Trinidad and India; the Indian strain showed the most resistance to azuki seed toxins. Holloway and Mackness (1988) reported that monocygenase and glutathione-S-transferase enzymes present in rice weevils are most likely involved in the process of detoxifying secondary compounds present in azuki and other grain legumes. This may explain why the researchers observed population differences between rice weevils reared 40 generations on wheat or split peas; eleven different wheat-fed *S. oryzae* populations had a mean 14-day mortality rate of almost 85% when fed pelletized azuki seed as compared to 24% for three populations first reared on split peas.

Larval and adult forms of the broad-nosed weevil, *Meotiorhynchus querendus* Sharp, attack azuki and other crops in Hokkaido Prefecture; larvae feed on the main roots and stem surfaces while adults notch leaves (Akiyama and Torikura, 1984). Larva of a Coleoptera of the Tenebrionidae, *Tribolium freemani* Hinton, were fed 23 food media and found to be unable to survive on seeds of azuki.

Lepidoptera: Pyralidae and Noctuidae

The Japanese butter bur borer, *Ostrinia scapulalis-pacifica* Mutuuta and Munroe (*O. nubilalis* Hubner), *Maruca testulalis* Geyer and *Matsumuraeses* spp. (perhaps *Matsumuraeses falcana* Walsingham [Oku *et al.*, 1983], *M. phaseoli* Matsumura [Kobayashi, 1978] or *M. azukivora* Matsumura [Suzuki, 1973]) were estimated by Katayama and Suzuki (1984) to have damaged 8 to 13% of harvested azuki seeds and caused yield reductions of 18 to 61% in infected and untreated fields of central Japan. *O. scapulalis-pacifica* proved to be the most serious pest based on stem, leaf and pod damage assessments. *O. scapulalis-pacifica* and azuki pod worm (*M. azukivora*) along with the oriental pea aphid (*Acyrthosiphon pisum*) are the major azuki pests in the famous Tan'ba region of central Japan (Matsui, 1984).

Destruction of vegetative parts and a decrease in azuki yield and quality can result from *O. scapulalis-pacifica* (in Japanese, *awa no meiga*) infestation (Yamaguchi and Yoshitada, 1959; Katayama and Suzuki, 1984; Akai *et al.*, 1986). Yamaguchi and Yoshitada (1959) observed in an insecticide selection trial for control of *O. scapulalis-pacifica*, that all azuki plants in nontreated plots had stem injury (Figs. 5.4, 5.5 and 6.3), almost 55% of all pods were damaged, more than 16% of all seed was unmarketable and total yields were reduced. A multiple planting date experiment conducted by Katayama and Suzuki (1984) demonstrated that azuki planted earlier in the growing season had larger *O. scapulalis-pacifica* populations than azuki planted at traditional times.

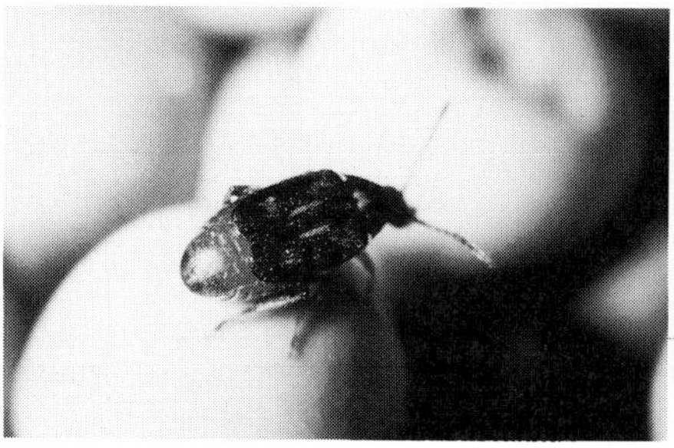

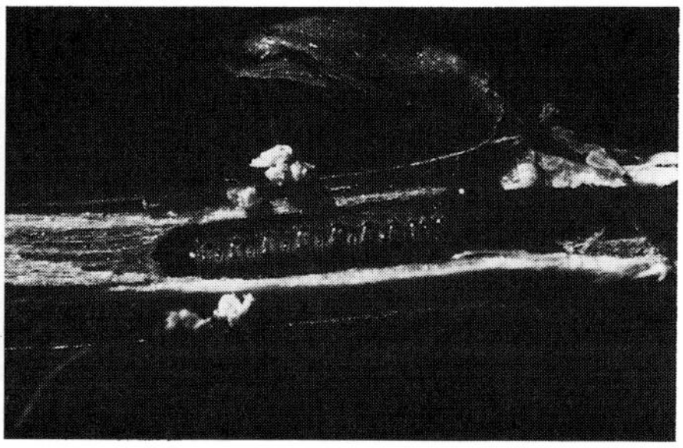

Fig. 5.3. (Left) A *Callosobruchus maculatus* adult standing near damage done to a soybean seed by a larva of the species. Damage to azuki is similar.

Fig. 5.4. (Right) Stem injury caused by *Ostrinia* spp. and *Matsumuraeses* spp. is common in Japan and Taiwan. This larva was observed in an azuki stem in the Tan'ba region of Kyoto Prefecture, Japan.

O. scapulalis-pacifica oviposit on azuki in early August; larvae bore into stems, petioles, flowers, pods and seeds (Yamaguchi and Yoshitada, 1959; Katayama and Suzuki, 1984; Akai *et al.*, 1986). The larval growth rate is related to spring and summer rainfall with quicker maturation occurring when there is adequate moisture early in the growing season. *O. scapulalis-pacifica* hibernate as mature larvae inside several host plants and in field debris (Akai *et al.*, 1986). Autumn incorporation of azuki field residues is one method of mechanical control (Akai *et al.*, 1986). Two applications of cyanophos provided effective *O. scapulalis-pacifica* control in Kyoto Prefecture, Japan (Suzuki, 1973). Endrin and EPN were found to be two of the best insecticides tested by Yamaguchi and Yoshitada (1959).

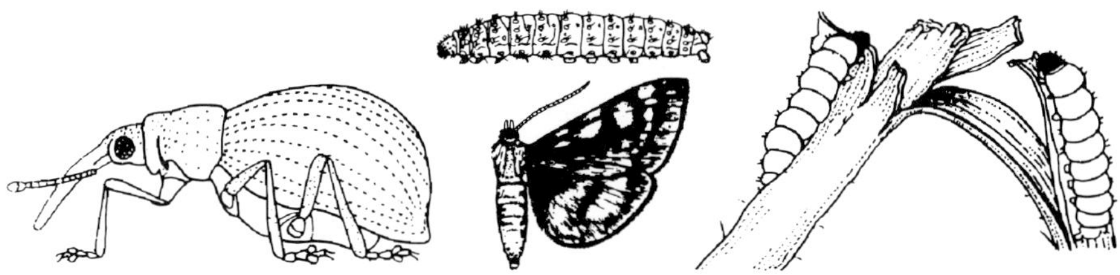

Fig. 5.5. Japanese illustrations of an adult of *Apion collare* (left), a larva and adult of *Ostrinia scapulalis-pacifica* (center) and damage to the stem by *Ostrinia* (right) (adapted from Murata, 1983b).

M. testulalis (*mame no meiga* in Japanese) populations increased the most rapidly of three species studied by Katayama and Suzuki (1984), from 0 to over 30 larvae per 10 plants in 10 days after azuki anthesis. *M. testulalis* is also reported to be a production problem in Taiwan (Cheng and Qou, 1967; Chang, 1989). Adults lay eggs on the top of immature azuki pods; emerging larvae feed on the seeds, leaves and sometimes stems (Akai *et al.*, 1986).

Matsumuraeses spp. infestations are a consistent problem in all major Japanese azuki production areas (Katayama and Suzuki, 1984; Akai *et al.*, 1986). *Matsumuraeses azukivora* Matsumura, *azuki saya mushi* in Japanese, is a pest of azuki in the field and in storage. *M. azukivora* bores into mature azuki pods when plants are at physiological maturity and larvae feed on stored azuki seeds (Suzuki, 1973). This author determined that a pesticide application in the middle of September controlled *M. azukivora* but recommended the identification of a nonpersistant insecticide that could be applied just before harvest for complete control.

The azuki podworm adult, *Matsumuraeses phaseoli*, appears in Hokkaido azuki fields by early June. Adults lay eggs on the underside of young leaves and petioles while larvae feed for about one month on the leaves and then inside stems and on pods. The second generation appears in August and the third in late September (Kobayashi, 1978; Akai *et al.*, 1986). Mean annual air temperature is positively correlated with *M. phaseoli* damage (Kobayashi, 1978). Late azuki plantings correlate positively with increasing populations during latter stages of plant growth (Katayama and Suzuki, 1984). Specific insecticidal controls have not been identified in Japan (Akai *et al.*, 1986).

Spodoptera litura F. is the most important pest of azuki in Taiwan where it causes extensive leaf, stem and pod damage (Chang, 1989). It is especially serious during autumn azuki production in southern Taiwan with high populations occurring between November and February (Lee, 1989a). Applications of pesticides or use of sex pheromones are controls recommended by Chang (1989). Fujimura (1976) reported that in the laboratory, the spotted cutworm, *Xestia c-nigrum* (L.) (*Amathes c-nigrum*), would oviposit on azuki leaves and the resulting larvae would feed and successfully pupate on the leaves as well.

Orgyia postica Walker and *Porthesia taiwana* Shiraki can cause minor damage to azuki in Taiwan, especially meristematic buds, leaves and pods (Chang, 1989). The cotton bollworm, *Heliothis zea* (Boddie), can be a production problem in China (Wang, 1986). *Acrocercops coerulea* Meyrick and *Chlorita flavescens* F. are occasional problems in Taiwan, especially during azuki pod development (Chang, 1989).

Diptera: Agromyzidae

Azuki was demonstrated by Lin (1981) to be susceptible to leaf miner, *Liriomyza* spp., damage. Foschine *et al.* (1983) studied leaf miner fly damage to azuki in Brazil. These researchers observed that adults fed and oviposited on azuki while larvae formed internal leaf galleries that reduced photosynthetic area.

Agromyza phaseoli Conquillett and *Melanogromyza sojne* Zehntner are major problems in Taiwan during early growth stages of azuki, causing stunted plants, delayed flowering and plant death (Chang, 1989). Adults oviposit on leaflets, leaf peduncles and apical meristems; larvae burrow into the vascular tissue resulting in sudden wilting. Chang (1989) recommended the application of an insecticide once a week for three weeks after seedling emergence to control *A. phaseoli* and *M. sojne*.

Azuki was determined by Lin (1981) to be highly susceptible to the bean fly, *Ophiomyia phaseoli* Tryon. The soybean leaf miner, *Ophiomyia centrosematis* De Meijere, was observed by Talekar and Lee (1988) to complete all developmental stages on azuki, although cultivar differences in soybean leaf miner susceptibility exist.

Acarina: Tetranychidae

Four different tetranychid species have been documented to cause damage to azuki. Lin (1981) documented the susceptibility of azuki to damage by *Tetranychus cinnabarinus* Boisduval, the carmine spider mite. *Tetranychus urticae* Koch and *Tetranychus kanzawai* Kishida can successfully complete their life cycle on azuki in the greenhouse (Lee *et al.*, 1990). *T. kanzawai* and *T. cinnabarinus* Bois. reach sufficient population levels in Taiwan, especially during dry weather, to cause azuki leaf damage by the late vegetative to early reproductive development stages (Chang, 1989). *Tetranychus* spp. are a production problem in China (Wang, 1986).

Tanigoshi (1991) and his assistant (Tanigoshi and Aguilar, 1993; Aguilar *et al.*, 1994ab), members of a multidisciplinary team of Washington State University agricultural researchers assessing the potential of azuki as an alternative crop in the arid, irrigated Columbia Basin of Washington, initiated acaricide screening trials in 1988 and 1989. He observed outbreak populations of *T. urticae* by early August which approached economic threshold levels of 8 to 10 females per trifoliate in control plots; the often hot, dry and dusty growing conditions, characteristic of the Columbia Basin, favor the tremendous biotic potential of the two-spotted spider mite (Fig. 5.6). Abamectin and dimethoate reduced *T. urticae* populations below economic levels 21 days after application of the acaricides. Acephate provided no acaricidal activity while bifenthrin and cyhalothrin performed only marginally as compared to abamectin and dimethoate (Tanigoshi, 1991). Once the infestation on undersurfaces of leaves becomes webbed over with their silk and dust particles, they are difficult to control.

Fig. 5.6. A two-spotted spider mite adult **(Left)**. This mite, *Tetranychus urticae* Koch, is a pest in all major azuki production areas. Under the right environmental conditions, *T. urticae* can quickly reach high populations in azuki fields. High uncontrolled two-spotted spider mite populations result in general leaf necrosis **(Center)** and eventual plant defoliation and senescence **(Right)**.

NEMATODES

Mani *et al.* (1982) compiled a list of nematodes reported in the scientific literature to infect azuki. The earliest report of a nematode infestation in azuki was by Piper and Morse (1914). Damage caused by *Heterodera* spp. is the most well documented in the literature. Ichinohe (1953) reported that the soybean cyst nematode, *Heterodera glycines* Ichinohe, was able to successfully complete its life cycle on azuki as well as soybeans, the two most susceptible plant species of 32 tested.

Sortland and MacDonald (1985, 1987) observed in greenhouse studies that the total number of soybean cyst nematode, race 5 egg cysts were higher from azuki infestations than pea, alfalfa, oat (*Avena sativa* L.), maize, sugarbeet (*Beta vulgaris* L.), sunflower and wheat. The researchers also determined that azuki was less effective in rotation than either alfalfa, maize, oat, pea or sunflower in reducing the number of *H. glycines* females infesting soybean. Azuki is an identified host of *H. glycines* race 3 in the U.S. midwest (Melton *et al.*, 1984) which also infected azuki at low levels in Valle del Cauca, Colombia (Quintero *et al.*, 1988). Races 1, 3 and 5 of the soybean cyst nematode have been identified in Japan's major azuki producing region, the Tokachi District of Hokkaido Prefecture, but neither races 2 nor 4 were found (Ichinohe, 1988). Kigasawa (1978) reported a significant negative correlation between azuki yield and the number of soybean cyst nematode cysts and north root-knot nematode cysts per 100 grams of soil.

Table 5.1. Reported phyla Nematoda, class Secernentea, and Arthropoda, class Hexapoda and Arachnida, pests of azuki.

Pests and Citations	
PHYLUM NEMATODA	
Heterodera glycine Ichinohe	Tsutsumi and Sakurai, 1966; Yoshikawa, 1970; Kobayashi *et al.*, 1972; Kay, 1979; Duke, 1981; Tanaka and Konno, 1982; Masamune *et al.*, 1982; Negishi and Kobayashi, 1984; Shimizu and Mitsui, 1987; Sortland and MacDonald, 1987; Ichinohe, 1988
Heterodera lespedezae Golden and Cobb	Edwards and Malek, 1971; Fagbenle *et al.*, 1986
Heterodera schachtii Schmidt	Duke, 1981
Meloidogyne spp.	Piper and Morse, 1914; Duke, 1981; Tanaka and Konno, 1982
Meloidogyne hapla Chitwood	Kigasawa, 1978; Ruelo, 1980
Pratylenchus penetrans (Cobb) Chitwood and Oteifa*	Duke, 1981

Table 5.1. Continued.

PHYLUM ARTHROPODA	
CLASS ARACHNIDA	
Tetranychidae spp.	Kobayashi *et al.*, 1976; Tanaka and Konno, 1982; Wang, 1986; Han *et al.*, 1987
Tetranychus cinnabarinus Boisduval	Chang, 1989
Tetranychus kanzawai Kishida*	Kobayashi *et al.*, 1972; Chang, 1989; Tanaka, 1988; Lee *et al.*, 1990
Tetranychus telarius L.	Lin, 1981
Tetranychus urticae Koch	Lee *et al.*, 1990
CLASS HEXAPODA	
Acrocercops coerulea Meyrick	Chang, 1989
*Acusta despecta**	Kobayashi *et al.*, 1972
Acyrthosiphon kondoi Shinji*	Kobayashi *et al.*, 1972; Akai *et al.*, 1986
Adoretus sinicus Burmeister	Lin, 1981
Agromyza phaseoli Conquillett	Chang, 1989
*Anacanthocoris concolaratus**	Kobayashi *et al.*, 1972
Anomala rufocuprea Motschulsky*	Kay, 1979; Tanaka and Konno, 1982
Aphididae spp.	Wang, 1986; Fukunaga, 1987
Aphis craccivora Koch	Kobayashi *et al.*, 1972, 1976; Kay, 1979; Tanaka and Konno, 1982; Matsui, 1984; Akai *et al.*, 1986; Han *et al.*, 1987; Zhang and Li, 1988: Chang, 1989
Aphis medicaginis Koch	Duan, 1989
Apion collare Schilsky	Sakamoto *et al.*, 1983; Akai *et al.*, 1986
Armandillidium vulgare Latreille*	Kobayashi *et al.*, 1972
Callosobruchus chinensis L.*	Kobayashi *et al.*,1972; Kay, 1979; Lin, 1981; Yoshida *et al.*, 1986b; Tanaka, 1991ab
Callosobruchus maculatus Fabricius	Chen, 1973; Sano-Fujii, 1984; Birch *et al.*, 1985; Sakai *et al.*, 1986
Callosobruchus rhodesianus Pic*	Giga and Smith, 1984/85
Callosobruchus analis Fabricius	Takeuchi *et al.*, 1977
Chlorita flavescens Fabricius	Chang, 1989
*Colposcelis signata**	Kobayashi *et al.*, 1972
*Cydia phaseoli**	Hwang, 1982
Delia platura Meigen	Ushirogi *et al.*, 1986
Dolycoris baccarum L.*	Kobayashi *et al.*, 1972
Empoasca solana De Long	Lin, 1981
Epilachna varivestis Mulsant	Strand, 1943; Yarnell, 1965
Etiella zinckenella Treitschke	Duan, 1989
Frankliniella formosae (Moulton) Buttle	Akai *et al.*, 1986
Frankliniella intonsa Trybom	Takahashi, 1936; Chang, 1989; Hachiya, 1990
Heliothis zea Boddie	Wang, 1986

Table 5.1. Continued.

Halyomorpha mista Uhler*	Kobayashi *et al.*, 1972
*Heliothis maritima-adaucta**	Kobayashi *et al.*, 1972; Akai *et al.*, 1986
Hylemya platura (Meigen)	Strand, 1943; Kobayashi *et al.*, 1972; Kay, 1979; Tanaka and Konno, 1982; Akai *et al.*, 1986; Fukunaga, 1987
Liriomyia spp.	Lin, 1981; Foschine *et al.*, 1983
Maruca testulalis Geger	Chang, 1989
Matsumuraeses spp.	Katayama and Suzuki, 1984
Matsumuraeses phaseoli Matsumura	Kobayashi *et al.*, 1972; Kobayashi, 1978; Kay, 1979
Matsumuraeses azukivora Matsumura	Suzuki, 1973; Matsui, 1984
Matsumuraeses falcana Walsingham	Oku *et al.*, 1983
Megalurothrips usitatus (Bugnall) Hood	Chang, 1987ab, 1988ab, 1989
Melangromyza sojne Zehntner	Chang, 1989
Meotiorhynchus querendus Sharp*	Akiyama and Torikura, 1984
*Nezara autennata**	Kobayashi *et al.*, 1972
Nezara viridula L.*	Kobayashi *et al.*, 1972
Ophiomyia phaseoli Tryon	Lin, 1981
Ophiomyia centrosematis De Meijere*	Talekar and Lee, 1988
Orgyia postica Walker	Chang, 1989
Ostrinia nubilalis Hubner	Yamaguchi and Yoshitada, 1959
Ostrinia scapulalis-pacifica Mutuura et Monroe	Akai *et al.*, 1986
Ostrinia varialis Bremer*	Kay, 1979; Tanaka and Konno, 1982; Matsui, 1984; Fukunaga, 1987
*Ostrinia zaguliaevi**	Kobayashi *et al.*, 1972
Palomena angulosa Motschulsky*	Kobayashi *et al.*, 1972
*Piezodorus hybneri**	Kobayashi *et al.*, 1972
Plusia spp.	White, 1972; Kay, 1979
Porthesia taiwana Shiraki	Chang, 1989
Prodenia litura Fabricius*	Kobayashi *et al.*, 1972
Pyrrhia umbra Hufnagel*	Kobayashi *et al.*, 1972
Riptortus clavatus Thumberg*	Kobayashi *et al.*, 1972
Spodoptera litura Fabricius*	Kay, 1979; Lee, 1989a; Chang, 1989
Thrips hawaiiensis Morgan	Chang, 1989
Trialeurodes vaporariorum Westwood	Lin, 1981
Xestia c-nigrum L.	Fujimura, 1976
Zabrotes subfasciatus Boheman	Fujii, 1983

*All names not marked * were confirmed in: Sutherland, D.W.S., Chair. (1978) *Common names of insects and related organisms* (1978 revision). Entomological Society of America, College Park, Maryland.

One possible explanation for the high susceptibility of azuki to soybean cyst nematode infestation is the apparent secretion of an egg hatching stimulator from azuki roots (Tsutsumi and Sakurai, 1966). The soybean cyst nematode also increased the severity of azuki brown stem rot disease, *Phialophora gregata* (Yoshikawa, 1970; Aota *et al.*, 1981). Various crop rotations are suggested using pea, red clover or alfalfa for cultural control of soybean cyst nematodes in azuki fields (Kobayashi *et al.*, 1972; Sortland and MacDonald, 1987). Shimizu and Mitsui (1987) reported that some azuki cultivars are resistant to soybean cyst nematode.

The lespedeza cyst nematode, *Heterodera lespedezae* Golden and Cobb, had difficulty in completing a life cycle on azuki (Fagbenle *et al.*, 1986) although Edwards and Malek (1971) found fair compatibility. While they recovered an average of 145 cysts from an original soil inoculation of 50 cysts, many were poorly formed and deeply embedded in the azuki roots (Edwards and Malek, 1971). Two distinctive nomenclatural systems for designating infection compatibility and genetic differences in either the *H. lespedezae* strain and/or the azuki cultivar used by these two research groups might account for the different findings.

Heterodera schachtii Schmidt (Duke, 1981), *Meloidogyne* spp. (Piper and Morse, 1914; Duke, 1981; Tanaka and Konno, 1982) and *Pratylenchus penetrans* (Cobb.) Chitwood and Oteifa (Duke, 1981) will infect azuki. Fujita and Miura (1934) used three azuki cultivars propagated in *H. schachtii* infected soil to confirm the earlier works of Katsufuji (1919) and Ito (1921) concerning the susceptibility of azuki to minor infestation levels. Ruelo (1980) reported that azuki grown in Taiwan was susceptible to *Meloidogyne hapla* Chitwood infestations.

The only nematocide mentioned in the literature for use on azuki is ethylene dibromide (1,2-dibromoethane) at 80 l/ha (Aota *et al.*, 1981).

Table 5.2. Insecticides used in azuki production.

Materials*	Specifics	Citations
Abamectin: (Avermectin B_1) against *Tetranychus urticae*	Good acaricidal activity	Tanigoshi, 1991
Acephate: (O, S-dimethyl acetylphosphoramidothioate)	No acaricidal activity on *Tetranychus urticae*	Tanigoshi, 1991
AI3-53060: (5-ethoxy-6-[1-(4-methoxyphenyl)ethyl]-1,3-benzodioxole)	*Callosobruchus chinensis* female, no male control	Nagasawa and Ushirokita, 1980
Aldicarb: (2-methyl-2[methylthio] propionaldehyde O-[methylcarbamoyl] oxime)	*Tetranychus urticae*	Wu and Hsu, 1986
Bifenthrin: ([2 methyl(1,1'-biphenyl)-3-yl] methyl 3-[2-chloro-3,3,3-trifluoro-1-propenyl] 2,2-dimethyl-cyclopropanecarboxylate)	Marginal effectiveness against *Tetranychus urticae*	Tanigoshi, 1991
Bromodan: (5-[bromomethyl]-1,2,3,4,7,7-hexachloro-2-norbornene) and **Alodan**	*Callosobruchus chinensis*; Seed coating at 0.4-0.5 kg	Lin, 1964
Callosobruchusic acid: ([E]-3,7-dimethyl-2-octene-1,8-dioic acid)	*Callosobruchus chinensis*; part copulation release phermone	Mori *et al.*, 1983

Table 5.2. Continued.

Camphor	*Callosobruchus chinensis*; fumigant	Abivardi and Rahimian, 1977
Carbaryl: (1-naphthyl N-methylcarbamate)	*Callosobruchus chinensis*; 50%WP	Han and An, 1990
Carbosulfan: (2,3-dihydro-2, 2-dimethyl-7-benzofuranyl [(dibutylamino)thio]methyl carbamate)	*Megalurothrips usitatus*	Chang, 1987b
Chlorbicyclen: (1,2,3,4,7,7-hexachloro-5,6-bis[chloromethyl] -2-norbornene)	*Callosobruchus chinensis*	Lin, 1964
Cyanophos: (O-4-cyanophenyl O,O-dimethyl phosphorothioate)	*Ostrinia scapulalis*; *Callosobruchus chinensis*; *Matsumuraeses azukivora*	Suzuki, 1973
DDVP: (2,2-dichlorovinyl dimethyl phosphate)	*Callosobruchus chinensis*; 50% EC	Matano and Sato, 1963; Han and An, 1990
Deltarin	*Callosobruchus chinensis*; 1% EC	Han and An, 1990
Demeton methyl: (S-[2-ethyl-thioethyl]O,O-dimethyl phosphorothioate)	Aphids: Seed treatments + spray + oat intercropped = No Control	Mukoo, 1975; Kobayashi *et al.*, 1976
Diazinon: (O,O-diethyl O-[2-isopropyl-4-methyl-6-pyrimidinyl] phosphorothiote)	Scarabs and *Hylema platura*	Matano and Sato, 1963; Fukunaga, 1987
Dicofol: (4,4-dichloro-alpha-trichloro-methylbenzhydrol)	*Tetranychus* spp.	Hachiya, 1990
Diflubenzuron: (1-[4-chlorophenyl]3-[2,6 difluorobenzoyl] urea)	*Callosobrucus chinensis*	Nagasawa *et al.*, 1978
Dimethoate: (O, O-dimethyl S-[N methylcarbamoylmethyl] phosphorodithioate)	*Aphis* spp.; *Tetranychus urticae*	Matano and Sato, 1963; Guo, 1980; Zhang *et al.*, 1981; Tanigoshi, 1991
Disulfoton: (O,O-diethyl S-[2-(ethylthio)ethyl] phosphorodithioate)	4 kg/ha; insects; mites *Aphis craccivora*; *Acysthosiphon kondoi*; *Tetranychus kanzawai*; 30 kg/ha with 60 kg/ha PSP-204; No Control- *Heterodera glycines*, *Callosobruchus chinensis*	Furusawa *et al.*, 1974; Kobayashi *et al.*, 1976; Kobayashi, 1979
Dowco-186: (triphenyl tin hydroxide)	*Callosobruchus chinensis*	Nagasawa *et al.*, 1965
Dowco-187: (allyl triphenyltin)	*Callosobruchus chinensis*	Nagasawa and Nakayama, 1968b
Endosulfan: (6, 7, 8, 9, 10, 10-hexachloro-1, 5, 5a, 6, 9, 9a-hexahydro-6, 9-methano-2, 4, 3-benzodioxathiepin-3-oxide)	*Aphis craccivora*; *Myzus persicae*	Mukoo, 1975

Table 5.2. Continued.

Endrin: ([1R, 4S, 5R, 8S]-1,2, 3,4,10,10-hexachloro-1,4,4a,5, 6,7,8,8a-octahydro-6,7-epoxy-1,4:5,8 -dimethanonaphthalene [principal constituent])	*Ostrinia scapulalis-pacifica*	Yamaguchi and Yoshitada, 1959
ENT-51007: (N, N, N', N'-tetramethyl-P-methyl-P-piperidino-phosphonic diamide)	*Callosobruchus chinensis*	Nagasawa and Nakayama, 1968b
EPN: (O-ethyl [O-4-nitrophenyl] phenylphosphonothioate)	*Matsumuraeses phaseoli*; *Ostrinia scapulalis-pacifica*	Yamaguchi and Yoshitada, 1959; Kobayashi *et al.*, 1972
Fenitrothion: (O,O-dimethyl O-[3-methyl-4-nitrophenyl] phosphorothioate) (5-ethoxy-6-[1-1-methoxyphenyl]ethyl-1,3-benzodioxole)	*Ostrinia varialis* and aphids *Callosobruchus chinensis*	Fukunaga, 1987; Nagasawa and Ushirokita, 1980
Folimat: (O,O-dimethyl-S-[2-(methylamino)-2-oxoethyl] phosphorothioiate)	Aphids and spider mites	Han *et al.*, 1987
Hempa: (hexamethyl-phosphoramide)	*Callosobruchus chinensis*	Nagasawa *et al.*, 1966; Nagasawa and Nakayama, 1968a, 1974
Hexythiazox: (trans-5-[4-chlorophenyl]-N-cyclohexyl-4-methyl-2- oxothiazolidine-3-carboxamide)	*Tetranuchus kanzawai*; WP	Tanaka, 1988; Hachiya, 1990
Lambacyhalothrin: ([1α(S*), 3a(Z)]-[±] cyano [3-phenoxy-phenyl] methyl 3-[2-chloro-3, 3,3-trifluoro-1-propenyl]-2,2-dimethylcyclopropane-carboxylate)	Marginal control of *Teranuchus urticae*	Tanigoshi, 1991
Lindane: (gamma isomer of 1, 2,3,4,5,6-hexachloro-cyclohexane)	*Callosobruchus chinensis*	Lin, 1964
Malathion: (O,O-dimethyl phosphorodithoate of diethyl mercaptosuccinate)	*Callosobruchus chinensis*	Matano and Sato, 1963; Lin, 1964
Metepa: (tris[2-methyl-1-aziridinyl] phosphine oxide)	*Callosobruchus chinensis*; egg hatching inhibitation	Nagasawa and Nakayama, 1968b, 1974
Methamidophos: (O,S-dimethyl phosphoramidothioate)	*Megalurothrips usitatus*	Chang, 1987b
Phenthoate: (O,O-dimethyl-S-[α-ethoxycarbonylbenzyl]-phosphorodithioate)	*Aphis* spp., *Myzus* sp., etc.	Mukoo, 1975; Fukunaga, 1987
Phosphorus amides and **s-triazine:** chemosterilants	*Callosobruchus chinensis*; 17 materials with 100% sterility effects	Nagasawa and Nakayama, 1974
Pipercide (III): ([2E,4E,10E]-N-isobutyl-11-[3,4-methyl-enedioxy-phenyl]-2,4,10 undecatrienamide)	From *Piper nigrum* fruit; *Callosobruchus chinensis*	Miyakado *et al.*, 1979, 1980
Piripo	*Callosobruchus chinensis*; 25% EC	Han and An, 1990

Table 5.2. Continued.

Prothiophos: (O-[2,4-dichloro-phenyl]o-ethyl S-propyl phosphorodithioate)	*Frankliniella intonsa*	Hachiya, 1990
PSP-204: (S-[(ethylsulfinyl)methyl] 0,0-bis[1-methylethyl] phosphorodithioate)	*Chlorita* spp.; 5% granules; 30-100 kg/ha	Kobayashi *et al.*, 1976
Salithion: (2-methoxy-4H-1,3,2-benzodioxaphosphorin-2-sulfide)	*Aphis craccivora* and *Myzus persicae*	Mukoo, 1975
Tedion: (2,4,5,4'-tetrachlorodiphenyl sulfone)	Aphids and spider mites	Han *et al.*, 1987
Thiohempa: (hexamethyl-phosphorothioic triamide)	*Callosobruchus chinensis*	Borkovec *et al.*, 1968
Thiourea	*Callosobruchus chinensis*	Shukla and Pandey, 1980; Pandey and Pandey, 1988
Vamidothion: (O,O-dimethyl-S-[2-([1-methylcarbamoyl-ethyl]thio) ethyl] phosphoro-dithioate)	*Aphis craccivora* and *Myzus persicae*	Mukoo, 1975

*Common and chemical names confirmed in Sine, C. (ed.). (1988-1991) and (1992). *Farm Chemicals Handbook*. Meister Publishing Co., Wiloughby, Ohio.

6

DISEASES

A large body of scientific literature addresses the numerous diseases which afflict azuki, especially the two major pathogens of azuki in Japan, i.e. brown stem rot, *Phialophora gregata* (Allington and Chamberlin) Gams and azuki mosaic virus (AzMV). Narita (1980) and Kishi (1988) have prepared very useful references for the diseases of azuki, including excellent descriptions of their history and epidemiology. As mentioned in Chapter 5, the sections on azuki in the plant protection book by Akai *et al.* (1986) provide a very comprehensive guide to the diseases of azuki found in Japan. See Figure 6.3 for an illustration of the occurrence of common diseases and insects found in Japan's largest azuki production area on Hokkaido.

FUNGAL DISEASES

The most extensively recorded fungal disease of azuki is brown stem rot caused by *Phialophora gregata* (Allington and Chamberlin) W. Gams, but many other fungal diseases are found on azuki. All fungal pathogens reported in azuki literature are listed in Table 6.1 and fungicides used on azuki are listed in Table 6.4. and some in Takabayashi, 1980a. This section is organized according to the taxonomic system used by Alexopoulos and Mim (1979) with genera arranged alphabetically within classes.

Division Mastigomycota, Subdivision Diplomastigomycotina, Class Oomycetes

Phytophthora vignae (Purss) f. sp. *adzukicola* Tsuchiya, Yanagawa and Ogoshi

Phytophthora vignae f. sp. *adzukicola*, the azuki stem rot disease, was first observed in Hokkaido Prefecture in 1967 and is now present in at least eight districts (Tsuchiya and Kodama, 1978; Narita, 1980; Akai *et al.*, 1986). The phenotypic description of the fungal pathogen which served as the basis of the identification is provided by Tsuchiya *et al.* (1978), Kitazawa and Yanagita (1978) and Akai (1988). It was first reported in 1981 infecting azuki in Korea (Han *et al.*, 1982). The fungal pathogen overwinters as oospores in the soil and initial disease transmission in the spring takes place via zoospores (Tsuchiya, 1985; Akai, 1988). *P. vignae* f. sp. *adzukicola* initially causes water-soaked spotting of azuki seedling hypocotyls near ground level; reddish-brown lesion development, stem streaking, damping-off-type symptoms and plant death can soon follow (Kitazawa *et al.*, 1978; Tsuchiya *et al.*, 1978; Han *et al.*, 1982; Akai, 1988; Tsuchiya, 1988). The development of greenish-brown lesions on azuki stems can occur later in plant development if environmental conditions are favorable; scattered spots of white powdery *P. vignae* growth soon develop and secondary infections from other fungal pathogens may also appear (Akai, 1988). Yield losses of over 60% can result when infection occurs three months after planting while plants infected earlier are frequently killed (Tsuchiya and Tanaka, 1984; Tsuchiya, 1988). Han *et al.* (1982) determined that 20 Korean landraces had a *P. vignae* f. sp. *adzukicola* infection percentage range of 20.7 to 95%.

Cultural properties, host range, taxonomy and ecology of *P. vignae* f. sp. *adzukicola* are described by Tsuchiya (1988). Tsuchiya (1985) reported differences in the base and optimum temperatures for the formation of *P. vignae* f. sp. *adzukicola* zoosporangium and zoospores. This researcher reported no zoosporangium development below 10°C and the largest zoosporangium formation at 25°C; zoospores have a base development temperature of 15°C and an optimum temperature of 27 to 28°C. Infection is greatest in the temperature range 15 to 32°C, with 25 to 30°C corresponding to the highest incidences of the disease (Kitazawa and Yanagita, 1978).

Low humidity conditions reduce the rate of disease progression after initial infection while high soil moisture levels, which can be positively correlated with periods of rainfall, enhance the development and spread of the disease (Han *et al.*, 1982; Tsuchiya and Tanaka, 1983; Akai, 1988). Tsuchiya (1985) observed that when azuki was grown in soil at 25% moisture, the incidence of *P. vignae* f. sp. *adzukicola* infection was less than 2%; the incidence of diseases increased to almost 72% when the soil was saturated. This researcher concluded that soil moisture levels at or below 45% caused motility of the zoospores to be retarded and the incidence and spread of the disease to be reduced.

High soil moisture conditions are especially prevalent in former paddy fields converted for expanded azuki production (Tsuchiya *et al.*, 1978; Tsuchiya and Kodama, 1978, 1981a; Tsuchiya, 1988; Tabiki and Tsuchiya, 1990). The spread and severity of the disease in Hokkaido Prefecture seems to be correlated with the conversion of rice paddy fields in the late 1960s and early 1970s. A two-year disease survey determined that almost 50% of all Kamikawa District production fields, a majority of which were former rice paddies, had *P. vignae* f. sp. *adzukicola* inoculum present (Tsuchiya and Kodama, 1978; Tsuchiya and Tanaka, 1983).

The *P. vignae* strain described by Kitazawa and Yanagita (1978) was different from the original Waterhouse (1963) description; it was pathogenic to azuki but not to nine other Fabaceae members and three Solanaceae members (tomato [*Lycopersicon esculentum*], eggplant [*Solanum melogena*] and pepper [*Capsicum spp.*]). Three azuki cultivars, *Takara shozu*, *Dainagon* and *Sakae shozu,* were completely susceptible to the pathogen while 14 other crop species were disease free (Tsuchiya and Tanaka, 1983; Tsuchiya, 1988). *P. vignae* f. sp. *adzukicola* isolates from cowpea (*Vigna sinensis* [L.] Endl.) and azuki showed cross-pathogenicity to azuki and cowpea, respectively (Kitazawa *et al.*, 1979ab; Tsuchiya and Kodama, 1979). Electrophoretic analysis of proteins and isoenzymes of four cowpea and three azuki isolates of the fungus, determined that they are essentially the same, being distinguished only on the basis of host range (Yanagawa *et al.*, 1985). Liew *et al.* (1991) used Restriction Fragment Length Polymorphisms (RFLP) for successful identification of groups of isolates of the fungus. The name *Phytophthora vignae* f. sp. *adzukicola* was recommended for isolates from azuki, and at least three races have been identified based on electrophoresis of proteins and isoenzymes (Tsuchiya *et al.*, 1986).

Tsuchiya and Tanaka (1983) suggested and Tsuchiya (1988) recommended an azuki seed treatment consisting of a wet or dry application of metalaxyl 35% powder or a mixture of metalaxyl, TMTDS and a third indistinguishable chemical at 0.3% by seed weight for effective early season control of *P. vignae* f. sp. *adzukicola.* Two or three foliar applications of a mixture of captan, maneb and a third unidentifiable material applied at the rate of 15.5 g/l, or captan alone, reduced the incidence and severity of the disease when sprayed on azuki plants and the ground (Tsuchiya and Tanaka, 1983; Tsuchiya, 1988). Crop rotations can also reduce the effects of *P. vignae* f. sp. *adzukicola.* Tsuchiya and Akai (1977) demonstrated that as the number of rotations between crops of azuki increased, plant height, branches per plant and seed yield per plant also increased; a six-year crop rotation between azuki plantings for example resulted in a 14-fold increase in seed yield as compared to continuous cropping of azuki in *P. vignae* f.spp. *adzukicola*-infested fields (Tsuchiya and Akai, 1977).

The azuki cultivar *Kotobuki* had fewer symptoms of *P. vignae* f. sp. *adzukicola* infection when compared to eight other cultivars that were inoculated with seven different isolates of the fungus (Kitazawa *et al.*, 1979ab; Tsuchiya and Kodama, 1979; Tsuchiya, 1988). Tsuchiya (1988) recommended the production of this relatively resistant cultivar in locations where the disease was a problem. Tabiki and Tsuchiya (1990) tested 193 azuki cultivars and identified three, *Niigata-zai Tochiohanzokin*, *Hanaadzuki* (*Kari* 154) and *Aomoria-zai* 34, with greater *P. vignae* f. sp. *adzukicola* resistance than *Kotobuki*. These researchers also determined that 9 out of 26 advanced breeding lines were resistant to *Phytophthora* stem rot infection. Other resistant cultivars include *Urasa*, *Chunagon*, *Soshokushu*, *Kensaki*, *Noto shozu* and *Oodate* 2 (Kitazawa *et al.*, 1979ab; Tsuchiya and Kodama, 1981b; Tsuchiya, 1988).

Division Amastigomycota, Subdivision Ascomycotina, Class Ascomycetes

Erysiphe polygoni D.C.

Powdery mildew caused by the fungus, *Erysiphe polygoni* (*E. pise* D.C. or *Sphaerotheca fuliginea* [Schlechtend.] Pollacci), has long been recognized as a disease of azuki in Japan (Mukoo, 1975; Akai, 1988). Symptoms of initial infection are the appearance of white hyphae and conidial masses on stems, leaves and pods; yellow to brownish-black cleistothecia are then formed; infected leaves turn yellow and infected plants become defoliated (Mukoo, 1975; Liu, 1987; Akai, 1988). The primary infection is caused by ascospores produced by cleistothecia overwintering on infected leaf debris (Akai, 1988).

Only the air-borne conidia of powdery mildew are found in Taiwan, especially on the autumn crop (Liu, 1987; Hsieh *et al.*, 1988). No single stage of azuki growth and development was identified by Hsieh *et al.* (1988) as being significantly more resistant to infection. These researchers recommended planting azuki in either late spring or early autumn in the Taiwan regions of Pingtung, Chiayih and Peikang to reduce *E. polygoni* disease levels. Hsu *et al.* (1989ab) identified 34 breeding lines out of 108 screened that had some form of powdery mildew resistance.

Mycosphaerella cruenta Latham

Mycosphaerella cruenta Latham (Anamorph: *Pseudocercospora cruenta* [Sacc.] Deighton; syn: *Cercospora cruentae* Sacc. or *Cercospora phaseolum* Cooke), is one of the principal azuki pathogens in parts of China (Duan, 1989), Japan (Ohata, 1988) and Taiwan (Cheng and Qou, 1967; Plant Protection ..., 1979). Lesion metamorphosis is from minute red-purplish leaf flecks to dark green-brown sooty lesions that can encompass entire leaves and cause premature defoliation (Akai, 1988). The method of overwintering is not clear for this fungal pathogen but is believed to be on infected plant debris left in the field (Akai, 1988).

Sclerotinia sclerotiorum (Libert) de Bary

Sclerotinia sclerotiorum can cause large yield reductions in Japan during years when temperatures remain high. The resulting disease causes damping-off-type symptoms after infection of the flowers, pods and stems near ground level (Narita, 1980; Akai *et al.*, 1986). The fungal pathogen overwinters on diseased tissue such as sclerotia. Apothecia developing from the sclerotia in turn produce aerial ascospores that cause the initial infection of azuki (Akai, 1988).

Division Amastigomycota, Subdivision Basidiomycotina, Class Basidiomycetes

Athelia rolfsii (Curzi) Tu and Kimbrough

Athelia rolfsii (anamorph: *Sclerotium rolfsii* Sacc.) is an important azuki disease in Taiwan that develops on stems just below or at ground level (Liu, 1987). *A. rolfsii* causes azuki white silk disease and is a minor source of damage in Japanese azuki producing regions (Akai, 1988). Symptoms of the disease include a mass of white mycelium and yellow-brown sclerotia produced acropetally on plants, causing a reduction in growth, yellowing, withering and eventually plant death (Liu, 1987; Akai, 1988). The severity of the disease caused by *A. rolfsii* intensifies with increasing temperatures (Liu, 1987). The fungal pathogen can survive saprophytically on dead plant material (Liu, 1987) or in the form of sclerotia (Akai, 1988).

Thanatephorus cucumeris (Frank) Donk

Thanatephorus cucumeris (anamorph: *Rhizoctonia solani* Kühn) is an important azuki disease problem in Taiwan (Liu, 1987, 1991) and Japan (Ohata, 1988). A *T. cucumeris* strain in India attacking aerial plant parts caused azuki web blight (Singh *et al.*, 1979). In Japan, a *T. cucumeris* infection rate close to 40% caused hypocotyl degeneration, first leaf spotting and occasional seedling death under experimental conditions (Naiki and Ui, 1981). Ohata (1988) reported that the fungus caused the formation of dark green, water-soaked lesions on stems and leaves that then turn gray-brown, and leaves eventually withered and fell. *T. cucumeris* can also infect azuki at later stages of development (Liu, 1987). Three different *T. cucumeris* anastomosis groups were demonstrated to infect azuki (Naiki and Ui, 1981).

Control of *T. cucumeris* has focused on identification and use of disease-resistant cultivars, crop rotations and biological control rather than the use of fungicides. Simple seed disinfection procedures seem to be ineffective; Naiki and Ui (1981) observed that disinfecting azuki seeds planted with approximately 1.5 to 2.0 fungal bodies per 100 grams of dried soil weight resulted in a 3.3% disease reduction compared to untreated azuki. Huang *et al.* (1976) found that one of four azuki cultivars tested was resistant to *Thanatephorus* spp. This cultivar achieved an 80% germination rate in field and greenhouse studies and had a lower death rate when compared to 100 Fabaceae species. A three-year crop rotation of azuki-sunflower-wheat is used for *T. cucumeris* control in China (Han *et al.*, 1987). Liu (1991) reported that azuki planted in a field with *Trichoderma koningii*, a *T. cucumeris* antagonistic and benomyl-resistant fungus, had significantly higher yields as compared to controls.

Uromyces spp.

Hirata (1952) differentiated between three *Uromyces* spp. infecting azuki, cowpea and kidney bean on the basis of host specificity, host range, size and shape of teliospores and uredospores and their first appearance on each crop. Hirata proposed the name *Uromyces azukicola* Hirata to distinguish the fungus specific to azuki.

This pathogen infects both azuki stems and leaves in Japan with the largest number of the initially white-blue and then reddish- to black-brown spots occurring on leaves (Akai, 1988; Narita, 1980). The spots soon swell and rupture, dispersing either reddish-brown uredospores (Fig. 6.1) or blackish-brown teliospores for further infections (Ushirogi *et al.*, 1986). *U. azukicola* outbreaks can begin on azuki seedlings under conductive conditions, causing the formation of reddish-brown lesions on cotyledons and hypocotyls, but the disease is more frequently a concern during anthesis and pod formation (Ushirogi *et al.*, 1986). The fungus overwinters on infected azuki tissue in the form of teliospores (Ushirogi *et al.*, 1986; Akai, 1988).

Zhao (1990) documented significant azuki yield reductions due to *U. azukicola* in the autonomous Tianjin City region of China when susceptible cultivars were grown under a monoculture system during the wet summer season. Akai (1988) earlier confirmed that high soil moisture can increase the level of infection; these conditions often exist in Hokkaido, Japan where rice paddies are converted for cultivation of azuki because of excessive rice production and declining per capita consumption (Ushirogi *et al.*, 1986).

Zhao (1990) recommended an integrated management program for effective disease control. Specific practices to reduce disease incidence and increase yields included the use of resistant cultivars such as Chinese breeding lines 82-08, 82-14, 82-15 and 85-6, intercropping with maize, planting later in the growing season, increasing soil levels of potassium and phosphorus and monitoring fields in July and August if precipitation during these months is above normal (Zhao, 1990). Cao and Van (1991) developed *U. azukicola* infection and disease indices and then screened 1000 azuki accessions from 12 Chinese provinces, identifying 22 accessions with different levels of resistance to the rust from the Tianjin region.

Four- to five-year crop rotations, removal of infected plants and chemical control (i.e. maneb, ferbam, mancozeb and ziram) are all recommended for *U. azukicola* restraint in Japan beginning from mid-July through mid-August (Mukoo, 1975; Ushirogi *et al.*, 1986).

Azuki was found to be resistant to bean rust, *Uromyces phaseoli* (Rebent.) Wint. (Hardman and Meronuck, 1982). Cummins and Ling (1950) reported that *U. phaseoli* infects azuki in Manchuria and China. Sato and Sato (1982) demonstrated that soybean rust, *Uromyces appendiculatus* (Pers.) Unger (*Phakopsora pachyrhizi*), could penetrate the leaf cuticle of azuki; they identified appressoria, transepidermal vesicles, subepidermal hyphae and haustorial infection structures in the tissue. Leaf necrosis was the first symptom to appear, about four days after inoculation.

Division Amastigomycotina, Sudivision Deuteromycotina, Form-Class Deuteromycetes

Ascochyta phaseolorum Saccardo

Ascochyta phaseolorum is an ancient disease of azuki in China (Duan, 1989), Taiwan (Cheng and Qou, 1967) and both the warm southern production regions (Narita and Ito, 1973; Mukoo, 1975) and more recently Tohoku District and Hokkaido Prefecture of Japan (Narita, 1980; Ushirogi *et al.*, 1986; Akai, 1988). The pathogen initially causes dark green and then dark reddish-brown lesions on azuki leaves, especially if the integrity of the cuticle has been broken, as well as stem and pod lesions (Mukoo, 1975; Narita and Itagaki, 1975). More mature lesions are tan, consist of concentric rings which may become brittle and have well-defined, dark brown borders (Ushirogi *et al.*, 1986; Akai, 1988). Overwintering occurs on azuki field debris or on harvested seed as pycnidia (Mukoo, 1975; Ushirogi *et al.*, 1986; Akai, 1988). The autumn incorporation of debris, use of crop rotations, disinfection of seed, removal and incineration of diseased plant material and use of fungicides are recommended for control of *A. phaseolorum* (Mukoo, 1975; Ushirogi *et al.*, 1986). The disease is called Ascochyta leaf spot (Akai *et al.*, 1986), but has other names depending upon the symptoms expressed on azuki, such as, ring spot, brown spot, brown leaf spot, etc. (Narita, 1980).

Cercospora spp.

Several Japanese azuki cultivars tested in India were found to be completely infected with *Cercospora* spp. (Thomas *et al.*, 1974). Hoshikawa (1985) reported that azuki in Japan was sensitive to *Cercospora canescens* Ellis and Martin, causal agent of azuki brown spot disease, although it is not listed as a disease of azuki in Taiwan (Plant Protection ..., 1979). Disease symptoms usually appear in Japan following floral initiation after the pathogen overwinters on azuki field debris (Mukoo, 1975) although Akai (1988) suggested that first symptoms can occur earlier. Unusually wet summer and autumn weather in Japanese azuki producing regions is correlated with *C. canescens* outbreaks (Akai, 1988).

Disease symptoms on azuki leaves, stems and pods progress in an acropetal direction with the development of yellowish-brown to reddish-brown lesions which eventually turn gray; infected leaves ultimately fall off (Mukoo, 1975; Akai, 1988). Autumn incorporation of azuki field debris into the soil to bury fungal conidia, proper field selection (those with good drainage), disinfection of seed before planting, removal of infected plants and chemical control are proposed for the management of *C. canescens* (Mukoo, 1975; Akai, 1988).

Colletotrichum phaseolorum Takimoto

Azuki anthracnose caused by *Colletotrichum phaseolorum* is reported to occur in Japan where it infects stems, leaves and pods (Narita, 1980; Akai, 1988). This fungal pathogen can cause death of infected seedlings; lesion development on more mature plants causes infected leaves to yellow, wither and drop (Mukoo, 1975). Overwintering of conidia occurs on azuki field debris and infected seed (Mukoo, 1975; Akai, 1988). Mukoo (1975) recommended using crop rotations, disinfecting seed, removing and incinerating diseased plant material, planting disease-resistant cultivars, such as *Takahashi wase* or *Wase dairyu* 1, or controlling the fungus chemically for management of *C. phaseolorum*. Several Japanese azuki cultivars tested in India were found to be completely free from anthracnose infections (Thomas *et al.*, 1974). *Colletotrichum lindemuthianum* (Sacc.) B. and C. were unable to infect azuki (Rands, 1925).

Fusarium oxysporum Schlechtend. f. sp. *adzukicola* Kitazawa and Yanagita, a new strain of *Fusarium oxysporum* pathogenic to azuki, was reported by Kitazawa and Yanagita (1984) in Hokkaido Prefecture. The suggested Japanese common name for the disease is *azuki icho byo* (Kondo and Kodama, 1989); in English it is known as fusarium wilt of azuki (Homma, 1991). The disease is generally restricted to the districts of Kamikawa, Sorachi and Ishikari (Kondo and Kodama, 1989). Kitazawa and Yanagita (1989) recommended that the strain be named *Fusarium oxysporum* Schlechtendahl f. sp. *adzukicola* based on host range test results. These researchers also identified three races of *F. oxysporum* f. sp. *adzukicola* based on the responses of azuki to infection.

Transmission of the fungus originates in the soil, where it initially infects the root hairs of developing seedlings (Akai, 1988). Incipient symptoms of *F. oxysporum* f. sp. *adzukicola* infection in azuki are vascular discoloration followed by general wilting, leaf yellowing and necrosis approximately 10 to 14 days after inoculation; advanced disease symptoms include general stunting and leaf necrosis, chlorosis, rugosity and epinasty followed by plant death (Hasegawa *et al.*, 1987; Kondo and Kodama, 1989). *F. oxysporum* disease symptoms are similar to those of azuki brown stem rot, except that the pith and vascular system of *F. oxysporum*-infected tissue is red-orange in color while *P. gregata* infections are purplish-brown in color (Akai, 1988).

This fungus was pathogenic to the six major azuki cultivars grown in Minnesota (Hardman and Meronuck, 1982). A study of the inheritance of resistance to the three races of *F. oxysporum* f. sp. *adzukicola* determined that resistance to races 1 and 2 in *Hatsune* is recessive while modifying gene(s) provide some resistance to race 3; resistance to all three *F. oxysporum* f. sp. *adzukicola* races in *Toiku* 123 involves one dominant gene (Kondo *et al.*, 1990).

The application of *Microellobosporia flavea* Y-1 when planting azuki into soil infested with *F. oxysporum* FA-3 reduced the disease index and the incidence of the disease by 9 to 76% (Hasegawa *et al.*, 1987). Hasegawa *et al.* (1991) isolated additional microorganisms from the rhizosphere of azuki that produced effective *F. oxysporum* f. sp. *adzukicola* FA-3 antigrowth agents; when the compound-producing microorganisms were field-tested, the isolated antifungal compounds hemipyocianine, chlororaphin, phenazine-1-carboxylic acid, pyrrolnitrin and the monozomycin-like antibiotic Y-1 were all effective in reducing the incidence of disease by 38 to 88%.

Phialophora gregata (Allington and Chamberlin) Gams

Phialophora gregata, causal fungus of azuki brown stem rot (BSR) (Agriculture, Forestry and Fisheries ..., 1967) was first identified in Japan in 1904 (Chiba, 1982). The literature on this disease is more commonly found under the obsolete name *Cephalosporium gregatum* Allington and Chamberlin (e.g. Narita, 1980; Akai *et al.*, 1986). BSR outbreaks were reported in the years 1938, 1942, 1955, 1969 and 1970 (Narita *et al.*, 1971ab). *P. gregata* was first identified in the Hokkaido Prefecture in 1970 and confirmed by Koch's postulates (Akai *et al.*, 1971; Narita *et al.*, 1971ab; Kobayashi and Ui, 1977; Negishi and Kobayashi, 1984). It has become the most important azuki disease in the prefecture, being especially prevalent in the Tokachi District (Narita *et al.*, 1971ab; Yamamoto *et al.*, 1990).

BSR disease symptoms in azuki are characterized by the appearance of faint yellowish streaks along leaf petioles followed by leaf discoloration, withering and finally shedding in an acropetal progression; the purplish-brown appearance of root and stem vascular bundles in cross-section is a distinctive feature of this disease (Akai, 1988). Secondary and tertiary lateral roots infected by *P. gregata* are fragile and separate easily from the rest of the root system (Tsuchiya and Akai, 1975). Late stage symptoms of the disease are the complete internal discoloration of stems and roots, defoliation, withering of flowers and pods and plant death (Narita *et al.*, 1971ab; Akai, 1988).

The main inoculum source for BSR is infected azuki stems, either left on the ground or incorporated into the soil (Kondo and Kobayashi, 1983ab; Akai, 1988). Although some conidia were found unassociated with azuki field debris, specialized survival structures such as sclerotia or chlamydospores have not been identified (Kobayashi *et al.*, 1981b; Tanaka *et al.*, 1981). Conidia survival improved when soil moisture was increased to 25% and soil temperature was decreased to a range of -5 to -25°C (Kondo and Kobayashi, 1983abc). Kobayashi *et al.* (1981b) and Tanaka *et al.* (1981) developed a selective medium for *P. gregata* isolation which recovered about 40% of the conidia present in soil suspensions. These research teams found a high correlation between the number of conidia present in infected soils and an azuki disease index using this medium. The incidence of the disease increases when monocropping azuki and results in a reduction of the fungal diversity on the roots (Nitta, 1991). The application of organic matter to the soil increases the root fungal diversity and reduces the BSR incidences (Nitta and Matsuguchi, 1988). Mengistu *et al.* (1991) have since developed an improved medium for direct isolation and quantitative estimation of propagules.

Azuki yields decreased by 30 to 60% depending upon the azuki growth stage at the time of infection (Akai *et al.*, 1971). Twenty cultivars averaged a 69% decrease in the number of mature pods per plant and a 77% reduction in seed weight per plant when grown on soils infected with *P. gregata* conidia (Chiba, 1982). The disease can also cause reductions in the number of seeds per pod and average seed weight (Narita *et al.*, 1971ab).

Factors that affect *P. gregata* disease levels include growing season temperatures, extent of azuki production in a given area, percentage of continuous azuki cropping, crop rotations used and presence of soybean cyst nematodes in the soil (Narita *et al.*, 1971ab; Aota *et al.*, 1981, 1983; Negishi and Kobayashi, 1984; Akai, 1988). The warmer the growing season is, the higher the incidence of BSR becomes (Akai, 1988).

A survey conducted in Hokkaido Prefecture between 1977 and 1981 showed that *P. gregata* conidia were present in 75% of azuki production fields (Kobayashi *et al.*, 1984). Conidia numbers generally increased in fields under continuous cultivation with the highest levels occurring during the autumn and winter (Aota *et al.*, 1983; Kondo and Kobayashi, 1983a). A field that had been in continuous azuki production for at least four years had 78,600 fungal propagules per gram of dry soil (Kobayashi *et al.*, 1984).

The number of conidia found at a depth of 40 centimeters in the soil varied depending upon the cultivation of maize, potato or sugar beet (Aota *et al.*, 1983; Kondo and Kobayashi, 1983b). Aota *et al.* (1983) found comparable yields between azuki grown in *P. gregata*-free fields and diseased fields that had been in maize production four consecutive years. A decline in *P. gregata* conidia in the soil could not be attributed to significant differences between rhizosphere and rhizoplane microflora under azuki or maize production (Souma *et al.*, 1991). BSR disease expression was severe when azuki was rotated into fields that had been previously planted in a *P gregata*-infected field four consecutive years with soybean or kidney bean (Aota *et al.*, 1983).

A brown stem rot-soybean cyst nematode disease complex may exist whereby nematodes create conditions conducive for BSR initiation (Yoshikawa, 1970; Aota *et al.*, 1981; Ichinohe, 1988). Aota *et al.* (1981) conducted an experiment consisting of three treatments - groups of 18 plants either grown with only *P. gregata*, only soybean cyst nematode, or both pathogens together - to investigate this potential interaction. The researchers observed that 2 of 18 azuki plants in the treatment with only *P. gregata* became infected and neither plant exhibited root browning. They found no *P. gregata* infection or disease expression among 18 azuki plants grown in soil with the only-nematodes treatment, but browning of root tissue and destruction of root development was recorded. When azuki was propagated in soils with both *P. gregata* conidia and *Heterodera glycines*, all 18 plants became infected and 15 showed severe disease symptoms including internal root discoloration and the destruction of secondary and tertiary roots.

Negishi and Kobayashi (1984) confirmed this preliminary research; they reported that a simultaneous inoculation of 30 second-stage nematode larvae per gram of soil with 105 to 106 fungal conidia per gram of soil resulted in discoloration of over 75% of the plants as compared to less than 10% with the fungal conidia alone. Tsuchiya and Akai (1975) had reported earlier that azuki root hairs and surrounding tissue would be infected by *P. gregata* without the benefit of soybean cyst nematodes.

Kobayashi and Ui (1980a) found that morphological and cultural characteristics, growth rates and phytotoxic compound production of *P. gregata* from Japanese azuki and from an Iowa soybean were quite similar. Negishi and Kobayashi (1985) were unable to differentiate Japanese soybean, U.S. soybean and azuki BSR isolates on the basis of culture characteristics and toxin production. These observations were originally hypothesized by Narita *et al.* (1971ab). Gray and Hepburn (1989) however, determined on the basis of RFLP analysis that soybean and azuki *P. gregata* strains they tested were distantly related. Brown stem rot strains isolated from Japanese azuki and soybeans were determined by polyacrylamide gel electrophoresis to be indistinguishable for 11 of 15 isoenzymes evaluated, with the other four isoenzyme patterns being specific to the two *formae speciales* of *P. gregata* (Yamamoto *et al.*, 1990; Kobayashi *et al.*, 1991). This lab (Yamamoto *et al.*, 1991b) later found that these strains were physiologically different based on optimum temperature for mycelial growth and colony color on potato dextrose plus V-8 juice media.

Kobayashi *et al.* (1983) tested pathogenicity differences in Japanese azuki and U.S. soybean brown stem rot isolates in their respective hosts and concluded that 12 American and Japanese soybean cultivars were resistant to the azuki strain when planted in naturally infested fields but not artificially wounded. Five Japanese azuki cultivars were resistant to the soybean strain of *P. gregata*. Host specificity is maintained even when using different strains of each BSR pathotype (Negishi and Kobayashi, 1985).

Five phytotoxic substances produced by the azuki BSR fungus were isolated by Kobayashi and Ui (1975) and later classified as gregatins (Kobayashi and Ui, 1976, 1977; Reeder *et al.*, 1986). All five were identified based on absorption spectra characteristics as stereo-isomers of aspertetronin, a metabolite of *Aspergillus rugulosus*, and three of these gregatins, designated as A, C and D, were shown to cause the characteristic disease symptoms in azuki (Kobayashi and Ui, 1976, 1977; Reeder *et al.*, 1986). These three gregatins also had antimicrobial activity against some bacteria and other fungi at 25 ppm (Kobayashi and Ui, 1975, 1977). Gregatins B and E produced no disease symptoms in azuki hosts, even at concentrations over 400 ppm, but did show significant antimicrobial activity (Kobayashi and Ui, 1975). Later, Kobayashi and Ui (1977) were unable to repeat their earlier results with gregatins B and E (Kobayashi and Ui, 1975). In the later effort, B and E caused foliar symptoms of BSR in azuki and mungbean seedlings, and were ineffective against tested microorganisms. Gregatin A is considered the most significant in causing brown stem rot symptoms in azuki. Kobayashi (1980) was unable to isolate gregatin A from plants previously injected with the phytotoxin and suggested that it was changed *in vivo*. Reeder *et al.* (1986) studied the effects of gregatin A on spinach thylakoid membranes and hypothesized that azuki and soybean leaf necrosis and the interveinal chlorosis symptoms of brown stem rot disease might be due to the inhibitory effects of this gregatin on light-driven electron transport.

A new *P. gregata* strain was identified (Kobayashi *et al.*, 1979) as type B and was shown to be less virulent on azuki than type A based on differences in disease symptomology (Kobayashi *et al.*, 1980). The type B strain grew faster, produced fewer gregatins and occasionally mutated into type A in culture resulting in production of ascochlorin instead of gregatins (Kobayashi and Ui, 1980b; Kobayashi *et al.*, 1981a; Kobayashi, 1984). Kobayashi (1983) speculated that if type B could dominate type A in azuki fields, losses from the brown stem rot disease would decrease. Inoculation with type B strain resulted in the formation of nine distinct antifungal stress metabolites in azuki roots according to Abe *et al.* (1987) who also tested the metabolites' antifungal properties on *Cladosporium herbarum*.

Chiba (1982) devised a method for determining azuki cultivar resistance to brown stem rot called the "total dry weight ratio" (TDWR). This ratio is the total dry weight of azuki cultivars grown in highly infested soils divided by their total dry weight from clean fields. He was unable to find brown stem rot immunity in 419 tested azuki cultivars although a few cultivars had higher than average TDWRs. High TDWR cultivars included *Aka mame*, *Kuro azuki*, *Maruba*, *Tan'ba*, *Azuki* 64-1 and *Shochohin*-10 which are all late-maturity types. The rate of vascular tissue browning at the base of the stem is another feature separating sensitive and resistant azuki cultivars and seems to be correlated to TDWR (Chiba, 1985). Chiba (1982) attributed year to year changes in cultivar TDWRs to air temperatures during the growing season, especially minimum air temperatures early in the season. Chiba *et al.* (1987) later determined that resistance to *P. gregata* is genotypically dominant and probably involves two principal genes.

Gray and Hepburn (1992) looked at mitochondrial DNA restriction patterns of *P. gregata* isolates from azuki. Two groups of isolates could be distinguished based on RFLP. One group was associated with isolates of this species from soybean. The other group had mitochondrial DNA patterns more similar to those from the genus *Acremonium*. The authors suggest that an *Acremonium* species responsible for vascular browning in azuki may have been misidentified as a *Phialophora* species.

Phyllosticta phaseolina Saccardo

Phyllosticta phaseolina (*Phyllosticta phaseoli* Sacc., *Phyllosticta phaseolorum* Sacc. and Speg.) is the causal fungus of azuki brown crest disease and azuki brown leaf spot in Taiwan (Plant Protection ..., 1979) and Japan (Mukoo, 1975; Akai, 1988). After overwintering on azuki field debris and seeds as pycnidia, it infects azuki plants, causing small dark brown to large light brown speckled leaf lesions, with the characteristic appearance of concentric rings (Akai, 1988). The autumn incorporation of debris, use of crop rotations, disinfection of seed before planting, removal and incineration of diseased plant material and use of fungicides are recommended control measures (Mukoo, 1975).

Epicoccum phaseoli Sawada

Epicoccum phaseoli was reported to infect the leaves of azuki growing in Taiwan (Commonwealth ..., 1958; Sawada, 1958). In Japan, azuki leaf blotch, *Epicoccum nipponica* (Hiura) Hara, overwinters on infected plant debris, and during cool moist growing seasons produces a sequence of gray expanding concentric lesions and black spore masses on ventral leaf surfaces (Narita, 1980; Akai *et al*., 1986; Akai, 1988). It was initially called *Clathrococcum nipponicum* Hiura.

Macrophomina phaseolina (Tassi) Goidánich

The azuki black rot disease caused by *Macrophomina phaseolina* infects pods and stems near ground level. Disease symptoms include the presence of charcoal-colored sclerotia embedded in stem lesions and the uniform blackening and hardening of infected pods; overwintering sclerotia can survive for more than two years in the soil and are the source of the primary infection (Akai, 1988).

Sphaeropsis azukicola Miura

Sphaeropsis azukicola, the causal fungus for azuki burnt leaf disease, causes purple-ringed lesions and withering of azuki leaves. The fungus overwinters on diseased tissue in the field which is probably the inoculum source for the next year (Ohata, 1988).

Table 6.1. Reported fungal pathogens of azuki.

Pathogens and Citations	
DIVISION MASTIGOMYCOTA	
Phytophthora vignae Purss. f.sp. *adzukicola*	Kitazawa *et al.*, 1978, 1979ab; Tsuchiya *et al.*, 1978, 1986; Tsuchiya and Kodama, 1979; Han *et al.*, 1982; Allen, 1983; Tsuchiya, 1985; Tabiki and Tsuchiya, 1990
Pythium butleri Subrammanium	Lee *et al.*, 1975
DIVISION AMASTIGOMYCOTA	
Alternaria alternata (Fr.:Fr.) Keissl.	Naiki and Ui, 1981
Aristastoma oeconomicum Tehon (Ellis and Tracy)	Duke, 1981
Ascochyta phaseolorum Sacc.*	Iida, 1972; Narita and Ito, 1973; Mukoo, 1975; Narita and Itagaki, 1975; Kay, 1979; Narita, 1980; Duke, 1981; Tanaka and Konno, 1982; Ushirogi *et al.*, 1986; Akai, 1988; Duan, 1989
Athelia rolfsii (Curzi) Tu and Kimbrough	(anamorph: *Sclerotium rolfsii* Sacc.) Kay, 1979; Duke, 1981; Liu, 1987; Akai, 1988
Botrytis cinerea Per.: Fr.	Ushirogi *et al.*, 1986; Takeuchi, 1988; Ohata, 1988
Cercospora canescens Ellis and Martin	Mukoo, 1975; Duke, 1981; Hoshikawa, 1985; Akai, 1988
*Clathrococcum nipponicum**	Iida, 1972; Kay, 1979; Duke, 1981; Tanaka and Konno, 1982
Erysiphe polygoni D.C. (syn: *Erysiphe pise* D.C. or *Sphaerotheca fuliginea* [Schlechtend.] Pollacci)	Hsieh *et al.*, 1988; Akai, 1988
Epicoccum phaseoli Sawada	Commonwealth ..., 1958; Sawada, 1958
Erysiphe nipponica (Hiura) Hara	Akai, 1988
Hypochnus sasakii Shirai	Sawada, 1919, 1931
Macrophomina phaseolina (Tassi) Goidánich	Iida, 1972; Kay, 1979; Duke,1981; Akai, 1988
Macrosporium azukiae Hara (syn: *Alternaria azukiae* (Hara) Miura)	Duke, 1981
Mucor circinelloides Tiegh	Naiki and Ui, 1981
Mycosphaerella cruenta Latham (syn: *C. cruenta* Sacc. or *C. phaseolum* Cooke)	Iida, 1972; Kay, 1979; Duke, 1981; Ohata, 1988; Duan, 1989
Penicillium janthinellum Biourge	Naiki and Ui, 1981

Table 6.1. Continued.

Penicillium paraherquei Abe and Smith (syn: *P. ochrochloron* var. *paraherquei* [(Abe)] Stolk and Samson)	Naiki and Ui, 1981
Phaeoisariopsis griseola (Sacc.) Ferraris	Akai, 1988
Phialophora gregata (Allington and Chamberlin) W. Gams	Tanaka and Konno, 1982; Gray and Hepburn, 1989; Yamamoto *et al.*, 1990
Phyllosticta phaseolina Sacc.	Iida, 1972; Kay, 1979; Plant Protection ..., 1979; Duke, 1981; Akai, 1988
Rhizoctonia spp.	Yoshikawa, 1970; Huang *et al.*, 1976; Han *et al.*, 1987
Sclerotinia sclerotiorum (Libert) de Bary	Narikawa, 1972; Kay, 1979; Narita, 1980; Duke, 1981; Tanaka and Konno, 1982; Hardman, 1985; Akai *et al.*, 1986; Fukunaga, 1987; Akai, 1988
Sclerotium bataticola Taubenhaus	Duke, 1981
Sphaeropsis azukicola Miura	Ohata, 1988
Sphaerotheca fuliginea (Schlechtend.: Fr.) Pollacci	Iida, 1972; Mukoo, 1975; Kay, 1979; Tai, 1979; Duke, 1981; Hong *et al.*, 1984, 1986; Han *et al.*, 1987; Liu, 1987
Thanatephorus cucumeris (Frank) Donk (anamorph: *Rhizoctonia solani* Kühn)	Anonymous, 1960; Singh *et al.*, 1979; Naiki and Ui, 1981; Liu, 1987; Ohata, 1988; Farr *et al.*, 1989
Uromyces azukicola Hirata	Hirata, 1952; Iida, 1972; Mukoo, 1975; Kay, 1979; Duke, 1981; Hong *et al.*, 1984, 1986; Ushirogi *et al.*, 1986; Han *et al.*, 1987; Akai, 1988; Zhao, 1990
Uromyces appendiculatus (Pers.) Unger	Anonymous, 1960; Tai, 1979; Sato and Sato, 1982; Farr *et al.*, 1989
Uromyces phaseoli (Rebent.) Wint.	Cummins and Ling, 1950; Duke, 1981; Hardman and Meronuck, 1982
DIVISION AMASTIGOMYCOTINA	
Colletotrichum lindemuthianum (Sacc.) B. and C.	Rands, 1925
Colletotrichum phaseolorum Takimoto*	Iida, 1972; Mukoo, 1975; Kay, 1979; Duke, 1981; Akai, 1988
Fusarium oxysporum Schlechtend	Piper and Morse, 1914; Yoshikawa, 1970; Kay, 1979; Duke, 1981; Naiki and Ui, 1981; Kitazawa and Yanagita, 1984; Fukunaga, 1987; Hasegawa *et al.*, 1987
Fusarium oxysporum f.sp. *adzukicola* Kitazawa and Yanagita	Akai, 1988; Kitazawa and Yanagita, 1989; Kondo and Kodama, 1989
Fusarium oxysporum f.sp. *vasinfectum* (Atk.) Synder and Hansen	Anonymous, 1960; Farr *et al.* 1989
Fusarium solani (Mart.) Sacc.	Naiki and Ui, 1981
F. s. f.sp. *phaseoli* (Burk.) Snyder and Hansen	Anonymous, 1960; Farr *et al.*, 1989

*All scientific names not marked * were confirmed in: Farr, D. F., Bill, G. F., Chamuris, G. P. and Rossman, A. Y. (1989) *Fungi on Plants and Plant Products in the United States*. The American Phytopathological Society, St. Paul, Minnesota, 1252 pp.

Phaeoisariopsis griseola (Saccardo)

The azuki angular spot disease, *Phaeoisariopsis griseola* (Saccardo) Ferraris, is a constant production challenge in Hokkaido Prefecture, producing distinctive pubescent fungal structures underneath infected leaves (Akai, 1988).

Botrytis cinerea Persoon

The gray mold disease caused by *B. cinerea* has only been a minor problem, except when unusually low temperatures and high rainfall cause widespread outbreaks such as in Hokkaido Prefecture during 1977 and 1978 (Narita, 1980; Akai *et al.*, 1986; Ushirogi *et al.*, 1986). *B. cinerea* infections usually begin on degenerating floral parts adhering to mature pod tips or floral parts that fall on stems and leaves (Fig. 6.2). The initial infection spreads and develops into water-soaked, dark green lesions that eventually turn grayish-brown in color (Ohata, 1988). Gray mold will also attack seedling stems near ground level causing damping-off-type symptoms (Ushirogi *et al.*, 1986). The fungal pathogen overwinters on diseased tissue or on harvested seeds (Ushirogi *et al.*, 1986). Thiram used as a seed disinfectant and three applications of thiophanate-methyl beginning 7 to 10 days after anthesis for approximately 10 days are recommended for *B. cinerea* control in azuki (Ushirogi *et al.*, 1986).

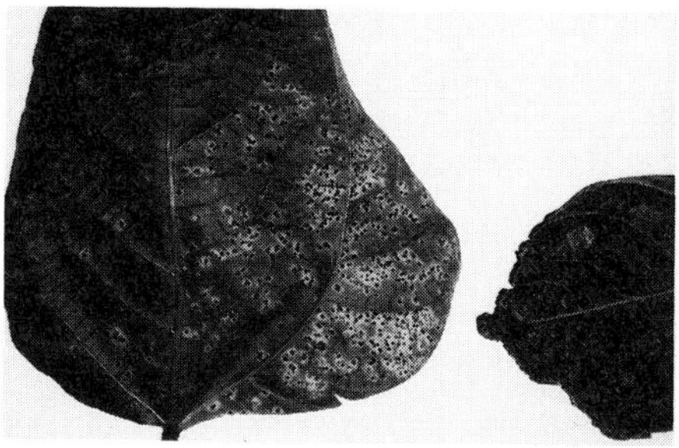

Fig. 6.1. (Left) The appearance of reddish-brown uredospores on leaves is symptomatic of an infection by azuki bean rust, *Uromyces azukicola* Hirata. This outbreak was observed in Kamikawa district, Hokkaido Prefecture, Japan.

Fig. 6.2. (Right) A gray mold, *Botrytis cinerea* Persoon, infected immature azuki pods in the Columbia Basin of Washington State. Water collected around floral keels on the tips of developing pods and prompted the infection.

BACTERIAL DISEASES

Species from three bacterial genera, *Curtobacterium*, *Pseudomonas* and *Xanthomonas*, are recorded in the scientific literature as pathogens of azuki and listed in Table 6.2. Methods for controlling bacterial diseases are listed in Table 6.4. The following section is organized alphabetically by the genus name of the bacterial pathogen as used by Bradbury (1986).

Curtobacterium spp.

Curtobacterium flaccumfaciens pv. *flaccumfaciens* is a reported pathogen of azuki (Coyne *et al.*, 1963; Bradbury, 1986). However, details are lacking.

Pseudomonas spp.

Azuki grown in Japan and in the state of Minnesota, U.S., is susceptible to bacterial stem rot, *Pseudomonas adzukicola*. This is a unique pathogenic bacteria, but the name is yet to be validated (Bradbury, 1986). The disease was first isolated in 1970 from Date-Cho, Hokkaido Prefecture, and was initially identified as *Pseudomonas syringae* (Narita, 1980; Akai *et al.*, 1986) but was later renamed *P. adzukicola* and identified as causing azuki bacterial brown stem rot disease (Tanii and Baba, 1971, 1979). Kennedy (1984ab) and Kennedy and Denny (1984) made a preliminary determination that the stem rot that devastated the Minnesota azuki industry in 1980 was quite similar to *P. adzukicola*; Breene and Hardman (1987) later confirmed *P. adzukicola* as the pathogen.

P. adzukicola can be a major disease problem when the Hokkaido Prefecture growing season is unusually wet and cool (Akai, 1988). Infected azuki seedlings exhibit small reddish-brown leaf lesions or streaks; stem girdling can result in damping-off-type symptoms. As infected plants mature, they display numerous necrotic leaf spots with outlying haloes which eventually become thin, fragile and torn; necrotic spots also develop on the surface of pods and reddish-brown streaking may occur along the placental suture (Tanii and Baba, 1971, 1979). A mid-season infection can also cause main stem node deterioration that causes azuki plants to lodge (Akai, 1988). A lack of these and other symptoms of the disease can still result in yield losses from the disease in azuki and other sensitive legumes (Kennedy, 1984b).

Azuki bacterial brown stem rot has characteristics closely resembling *Pseudomonas syringae* pv. *glycinea, Pseudomonas syringae* pv. *phaseolicola* and *Pseudomonas syringae* pv. *mori* (Tanii and Baba, 1971). The pathogen tested positive for fluorescent pigment and tobacco hypersensitivity reaction, but had negative nitrate reductase and arginine dihydrolase reactions (Tanii *et al.*, 1973). It is motile, Gram negative and aerobic. The pathogen successfully infected hyacinth bean (*Dolichos lablab* L.) and cowpea, but not pea, soybean or kidney bean (Tanii and Baba, 1971, 1974, 1979; Tanii *et al.*, 1973,). Kennedy (1984b) confirmed a hypersensitive immune reaction in soybean and pinto bean as well as navy bean. This researcher found that *Phaseolus vulgaris*, and the two main Minnesota azuki cultivars, *Takara* and *Minoka*, were susceptible to bacterial inoculation concentrations between 10^8 and 10^4 cfu/ml. The original source of Minnesota's disease is unknown, but seed-borne transmission is believed to occur in azuki (Tanii and Baba, 1979).

Tanii and Baba (1979) observed that large-seeded azuki cultivars such as *Akatsuki dainagon* and *Akane dainagon* were more susceptible to the bacterial disease than small-seeded cultivars. Apparent host range differences based on symptom expression may also result between the greenhouse and the field (Kennedy, 1984b). The number of bacteria per gram of dry infected soil can be quantified using a special culture medium developed by Tanaka *et al.* (1981).

Kennedy (1984b) and Hardman and Meronuck (1982) reported complete resistance in azuki to bean common blight and halo blight (*Pseudomonas medicaginis* f. sp. *phaseolicola*), but not *Pseudomonas syringae* pv. *syringae* under Minnesota field conditions. *P. s.* pv. *syringae* is called azuki brown mottle bacterial disease in Japan and is characterized by enlarged reddish-brown chlorotic spots on stem, petiole and leaf tissue in azuki (Akai, 1988). Transmission of the disease occurs from bacteria adhering to seeds which are then planted into infected soil (Akai, 1988).

Azuki can be successfully inoculated by artificial means with *Pseudomonas cichorii*, *Pseudomonas s.* pv. *glycinea* and *Pseudomonas viridiflava* (Bradbury, 1986).

Xanthomonas spp.

The degree of resistance in azuki to bean common blight, *Xanthomonas campestris* pv. *phaseoli*, is unclear. Coyne *et al.* (1963) determined that all USDA-ARS Plant Introduction lines were severely to very severely susceptible. Mukoo (1975) reports that *X. c.* pv. *phaseoli* is a major problem in Japan, especially during the rainy season of June and July, when it can cause leaf lesions, leaf fall and vascular browning of stems. Thomas *et al.* (1974) and Gill and Sharma (1980) reported a high degree of resistance to *X. c.* pv. *phaseoli* in the azuki cultivars they tested.

Xanthomonas campestris pv. *vignaunguiculatae* and *Xanthomonas campestris* pv. *pisi* produce an infection in azuki when artificially inoculated (Bradbury, 1986).

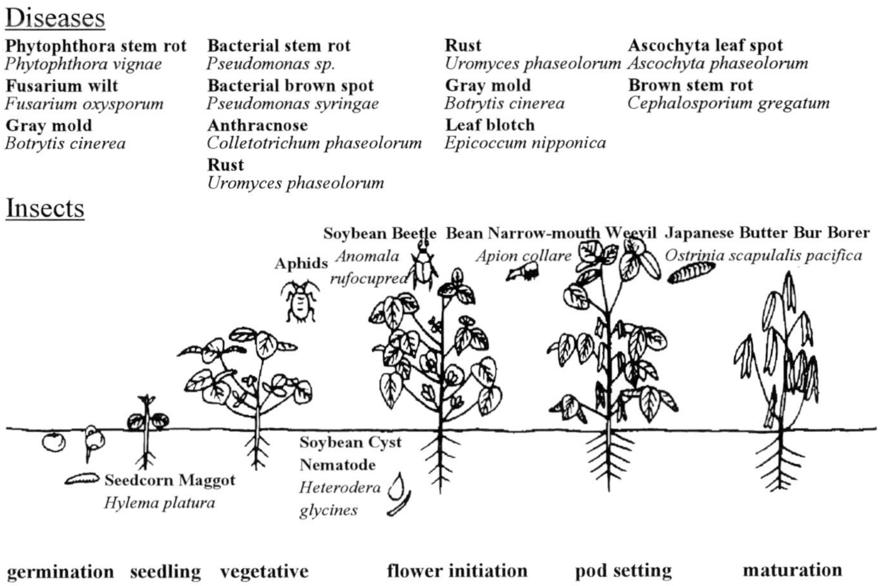

Fig. 6.3. Occurrences of insects and diseases during the development of azuki on the island prefecture of Hokkaido, Japan's largest production area. Adapted from Murata, 1983b.

Table 6.2. Reported bacterial pathogens of azuki.

Bacteria and Citations	
Curtobacterium flaccumfaciens pv. *flaccumfaciens*	Coyne *et al.*, 1963; Bradbury, 1986
Pseudomonas adzukicola	Tanii and Baba, 1971, 1974, 1979; Tanii *et al.*, 1973; Kennedy, 1984ab; Kennedy and Denny, 1984; Hardman, 1985; Breene and Hardman, 1987; Akai, 1988
Pseudomonas syringae pv. *syringae*	Hardman and Meronuck, 1982; Bradbury, 1986; Akai, 1988
Pseudomonas syringae pv. *glycinea*	Bradbury, 1986
Pseudomonas cichorii	Bradbury, 1986
Pseudomonas viridiflava	Bradbury, 1986
Xanthomonas campestris pv. *phaseoli*	Coyne *et al.*, 1963; Iida, 1972; Mukoo, 1975; Kay, 1979; Gill and Sharma, 1980; Duke, 1981; Bradbury, 1986
Xanthomonas campestris pv. *vignaunguiculatae*	Bradbury, 1986
Xanthomonas campestris pv. *pisi*	Bradbury, 1986

VIRAL DISEASES

Almost all fields of azuki are infected with viruses. Viruses reported in the literature are described in this section and listed in Table 6.3. The section is organized alphabetically by the taxonomic groups described by Francki *et al.* (1985). Some methods to control viral pests of azuki are listed in Table 6.4. A book by Kishi (1988), including color photos, is one of the best general resources for viruses and other diseases of azuki.

Alfalfa mosaic virus group

Alfalfa mosaic virus (AMV) can produce a systemic infection of azuki (Pierce, 1934). AMV is a minor viral disease of azuki in Japan, transmitted in a nonpersistent manner by aphids to azuki; it is not seed-transmitted in this host. The authors suspected the transmission of AMV from volunteer alfalfa in commercial azuki fields in the Columbia Basin of Washington State; all seedlings produced from seed of the cultivar *Erimo* that tested positive for AMV using ELISA procedures were negative for the virus (G. Mink, personal communication). Disease symptoms are characterized by bright yellow mottling of leaf tissue, leaf curling and senescence and general stunting of the plant (Iizuka, 1988). Matsumoto *et al.* (1968) observed that eight isolates of AMV caused only latent infections of azuki cultivar *Wase dainagon*, while two other isolates were expressed as leaf mosaic and mottles.

Comovirus

Horsegram yellow mosaic virus (possibly broad bean true mosaic virus [BBTMV]) is transmitted to azuki by whiteflies, *Bemisia tabaci* Genn., causing mild symptoms of infection such as nonenlarging, light yellow leaf spotting (Muniyappa *et al.*, 1977).

Cucumovirus

A virus isolated from azuki that caused leaf mosaic symptoms was identified as cucumber mosaic virus (CMV) (Lu *et al.*, 1989). Two different pathotypes of CMV, normal and P, occur throughout Japan and are transmitted nonpersistently by aphids from infected weeds and azuki plants (Iizuka, 1988). CMV can be a major disease problem during years of low rainfall because of increased populations of aphid vectors (Iizuka, 1988; Matsumoto and Inoue, 1989). The two pathotypes produce a systematic infection in azuki when sap transmitted (Inouye *et al.*, 1982; Iizuka, 1988).

Infected azuki have symptoms similar to those for azuki mosaic virus (AzMV) as well as severe leaf rugosity (Iizuka, 1988).

Iida (1972) stated that CMV is seed-borne in azuki while Ito *et al.* (1975) and Iizuka (1988) reported that the virus is not transmitted in seeds produced on infected plants. CMV was detected in the anthers and ovary but not the ovules of the azuki cultivar *Daikan 2-go* and was not seed-transmitted according to Tsuchizaki *et al.* (1970ab). Cultivar resistance to normal strains of CMV exists but not to P strains of the virus (Iizuka, 1988).

Azuki is within the host range of the peanut stunt virus (PSV) (Chang *et al.*, 1990). Iizuka (1988) reported that PSV can be spread by aphids from infected clover to azuki, causing chlorotic leaf spots.

Geminivirus

Nagaraj *et al.* (1981) reported a mungbean yellow mosaic virus (MYMV) infection rate of 100% in azuki when transmitted by whiteflies. These researchers classified the resulting infection as “mild” and found that disease expression took 25 days when graft inoculated. An MYMV azuki infection introduced by mechanical inoculation is initially characterized by vein yellowing followed by the development of yellow leaf mosaic symptoms (Honda *et al.*, 1983, 1986; Honda and Ikegami, 1986).

Bean calico mosaic virus (BCMV) isolated from *Phaseolus vulgaris* L. plants growing in Sonora, Mexico was transmitted both mechanically and by whiteflies to azuki resulting in leaf mosaic and bright calico symptoms (Brown *et al.*, 1990). Bean golden mosaic virus (BGMV) is reported by Shimizu *et al.* (1987) to cause leaf mosaic symptoms in sap-inoculated azuki.

Ilarvirus

Two distinct insect-transmitted pathotypes of tobacco streak virus (TSV) were isolated from azuki growing experimentally in eastern Washington State (Fig. 6.4). Pathotype I was successfully seed-transmitted in azuki at the rate of 29% from plants that were artificially inoculated (Kaiser *et al.*, 1991).

Nepovirus

Pierce (1934) observed a systemic infection in azuki without visible necrotic lesions after mechanical inoculation with tobacco ringspot virus (TRSV).

Potyvirus

Mosaic viruses of the potyvirus group are the most frequently reported viral pathogens of azuki. Mosaic virus diseases are one of the most important production constraints of azuki in Japan (Iizuka, 1988) and parts of China (Jiang and Chen, 1984). Mukoo (1975) stated that yield losses of 30 to 40% can occur in southern Japanese production areas from different mosaic viruses. Specific mosaic viruses of azuki in Japan include azuki mosaic virus (AzMV), bean common mosaic virus (BCMV), bean yellow mosaic virus (BYMV), blackeye cowpea mosaic virus (BlCMV) and cowpea aphid-borne mosaic virus (CABMV) (Shikata, 1986; Iizuka, 1988).

AzMV is known to infect azuki, soybean, mungbean and peanut. Azuki has shown some cultivar differences to AzMV infection (Green *et al.*, 1988; Iizuka, 1988). AzMV is nonpersistently transmitted to azuki by aphid species such as *Aphis craccivora* Koch. *(Mame abura mushi)* and *Myzus persicae* Sulver. *(Momo aka abura mushi)* (Murayama, 1941; Tsuchizaki and Omura, 1987; Iizuka, 1988; Choi and Lee, 1989). Growing seasons characterized by little rain, especially soon after germination and emergence, and high temperature and radiance levels, enhance the initiation and spread of AzMV (phenotypic symptoms of infection may be suppressed). This disease can be a major production constraint in the producing regions of southern Japan (Mukoo, 1975; Hoshikawa, 1985; Iizuka, 1988) and has been reported in northern Japan (Yamashita, 1991).

AzMV-infected azuki shed young leaves and flowers, have curled, mottled leaves that sometimes have vein clearing and puckered dark green sections and are stunted or wither and yield poorly (Murayama, 1941; Mukoo, 1975; Kobayashi, 1979; Hoshikawa, 1985; Iizuka, 1988; Choi and Lee, 1989). Leaf thickness, internal leaf characteristics and starch and sugar levels were determined to be different between the dark green and light green regions of AzMV-infected leaves (Matsumoto, 1922). A seed-borne infection is first expressed in developing seedlings as faint mosaic patterns on the primary leaves (Iizuka, 1988).

Phenotypic features of the AzMV are 750 nm long filamentous particles and pinwheel-shaped inclusion bodies in the cytoplasm (Iizuka, 1988; Choi and Lee, 1989). Serological characteristics include a crude sap thermal inactivation point after 10 minutes at 55 to 60°C, a 10^{-3} to 10^{-4} dilution end point and maintenance limit of five to seven days at 20°C (Iizuka, 1988; Choi and Lee, 1989).

Seed transmission rates of AzMV vary depending upon the cultivar and the time of infection. Murayama (1941) documented a seed-borne infection rate range of 5.1 to 10.9% using 21 azuki cultivars. Choi and Lee (1989) reported an average of 2.2 based on the screening of almost 550 plants from four regions of Korea while Iizuka (1988) stated that a 5% transmission rate is normal. AzMV seed transmission was 4.6% for the azuki cultivar *Daikan 2-go*, and 14.7% (Tsuchizaki *et al.*, 1970ab) or 11.4% (Tsuchizaki and

Omura, 1987) for *Wase dairyu*. Iida (1972) later confirmed AzMV seed transmission in azuki. AzMV transmission between generations does not occur if the time of infection is during or after anthesis (Iizuka, 1988).

The percentage of AzMV seed transmission in azuki is not affected by the time of seed formation, seed location in pods, four years of storage or by 50 to 75°C hot water or 40°C hot air seed treatments, but transmission is reduced by the presence of another virus in host plants (Murayama, 1941; Tsuchizaki *et al.*, 1970abc).

Tsuchizaki *et al.* (1971) reported that mature embryos were more than twice as likely to be infected as immature embryos, after testas had been removed and embryo surfaces had been inoculated with cucumber mosaic virus (CMV), AzMV and cowpea aphid-borne mosaic virus (CABMV). A strong relationship was identified between a high virus infection level in immature azuki embryos and eventual seed transmission. Li and Ding (1986) used host range, physical properties, method of transmission, UV-spectrum characteristics, particle morphology and serological reaction results to establish that a virus in Anhui Province, China, was AzMV.

Duke (1981) reported that azuki is a compatible host for asparagus bean mosaic virus (AbMV). AbMV is successfully sap-inoculated to azuki, causing severe leaf mottling and deformation, but is not seed-transmissible (Hino, 1960). A distinct strain of AbMV isolated in China was reported later by Wei and Chiu (1964) to cause stunting and removal of apical dominance in infected azuki plants; the virus was transmitted by aphids (*Myzus persicae*, *Rhopalosiphum pseudobrassicae* Davis, *Aphis gossypii* Glover and *Macrosiphum pisi* Kalt.) and had a seed transmission rate of 1.6%.

An unidentified seed-borne virus was isolated in China and reported to cause leaf mosaic symptoms on 19% of all azuki inoculated (Cheo and Tsai, 1959). Duan (1989) describes an unidentified potyvirus that is a major problem in China. It could be AzMV, AbMV or some strain of CMV based on what has been recently reported by Chinese researchers (Lu *et al.*, 1989). An unidentified virus of azuki in Taiwan causes disease symptoms very similar to AzMV (Liu, 1987).

CABMV-1 was detected in the anthers and ovary but not the ovules of the azuki cultivar *Daikan 2-go* and was not seed-transmitted (Tsuchizaki *et al.*, 1970ab). Tsuchizaki and Hibino (1971) found seed transmission of AzMV in azuki but no transmission of CABMV, partially on the basis of the detected distance of virus inclusions and particles from the end of azuki flower bud apical meristems. AzMV particles were found within 0.1 to 0.2 mm of apical meristems while CABMV viral inclusions and particles were 0.2 to 0.3 mm from the apical tissues. AzMV and CABMV are probably closely related based on the protection AzMV-infected plants have from subsequent cowpea aphid-borne virus infection (Tsuchizaki *et al.*, 1970a). Shukla *et al.* (1991) found evidence to suggest that CABMV is a distinct potyvirus.

The susceptibility of azuki to BCMV is unclear. Pierce (1934) found azuki not to be susceptible to mechanical inoculation with BCMV. Ito *et al.* (1975) isolated BCMV from diseased azuki plants growing in Hokkaido Prefecture; not one of eight azuki cultivars inoculated with BCMV produced disease symptoms and the virus was not found to be seed-borne in a naturally infected population. Tsuchizaki and Omura (1987) found that two strains of BCMV were not seed-borne in three azuki cultivars - *Wase dairyu, Kyoto dainagon*, and *Akatsuka dainagon* - but did cause latent systemic infections, vein banding or mosaic reactions. Aphids transfer BCMV in a nonpersistent fashion from infected bean plants to azuki; this virus is not a persistent problem in the major Japanese azuki producing areas of Hokkaido Prefecture and Tohoku District (Iizuka, 1988). This author also confirmed the usual absence of morphological symptoms in BCMV-infected azuki.

One AzMV, two bean common mosaic viruses (BCMV) and blackeye cowpea mosaic virus (BlCMV) strains were determined by a double immunodiffusion experiment to be antigenically related (Tsuchizaki and Omura, 1987). They also found these viruses to have identical protein coat molecular weights (32.8 x 10^3 kD) in SDS-PAGE, similar virus coat peptide maps using *Staphylococcus* V8 protease, and to be serologically identical. BlCMV is nonpersistently transmitted by aphids and is not seed-borne. It causes vein clearing and distinct mosaic symptoms on infected azuki leaves (Iizuka, 1988). Other researchers have also found that AzMV is related to BCMV and B1CMV, as well as peanut stripe potyvirus (PStV), based on serology (Vetten *et al.*, 1992). The PStV and AzMV appear to be proteolytically undegraded and to have nearly identical molecular weights of 35 kDa. AzMV was found to react moderately to antisera of peanut green mosaic potyvirus (Naidu *et al.*, 1991).

McKern *et al.* (1992b) used HPLC to make peptide profiles from the coat proteins of a large number of potyviruses infecting legumes. Among the tested viruses, AzMV, the Type and W strains of BlCMV, three isolates of a soybean potyvirus, and two strains of PStV seem to share the known sequence of the PStV Stripe coat protein. They (McKern *et al.*, 1992a) continued this research with 22 isolates of BCMV, and the type strains of B1CMV and PStV. As a result of this study the authors suggested splitting BCMV in to two distinct potyviruses, one with the original name of BCMV, the other with the new name of ‘bean necrosis mosaic virus’. Their results also suggest that AzMV, B1CMV, PStV and three isolates from soybean are all strains of the second potyvirus, BCMV.

Wase dairyu, Kyoto dainagon and *Akatsuka dainagon* also reacted to inoculation by two soybean mosaic virus (SMV) strains and two BlCMV isolates, but neither was seed-transmitted (Tsuchizaki and Omura, 1987). Cross-protection testing showed that AzMV in azuki provided complete protection from BlCMV infection while BlCMV-infected azuki was completely protected from two BCMV isolates (Tsuchizaki and Omura, 1987; Iizuka, 1988). These researchers believed that AzMV, BCMV and

BlCMV were not distinct viruses but viral strains. Montenegro and Galindo (1974) reported that a BCMV strain in Guanajuato state, Mexico, infected azuki but no phenotypic symptoms were expressed.

An unknown strain of bean yellow mosaic virus (BYMV) did not produce an infection when azuki was mechanically inoculated (Pierce, 1934). Two different strains of BYMV have been documented to successfully transfer and infect azuki but seed transmission does not occur (Ito *et al.*, 1975; Sugano *et al.*, 1981; Uyeda *et al.*, 1982; Sato *et al.*, 1987; Iizuka, 1988). Two other BYMV strains produced no symptoms in azuki (Kume *et al.*, 1969; Sato *et al.*, 1987). Infected azuki plants can develop chlorotic spots, mottling and deformation of leaves or maintain only a latent infection. This virus can be a production problem in southern Japan, but occurs infrequently in the Tohoku District and Hokkaido Prefecture of northern Japan (Sugano *et al.*, 1981; Uyeda *et al.*, 1982; Sato *et al.*, 1987; Iizuka, 1988).

Choopanya and Kiratiya-angul (1980) reported that peanut mottle virus (PeMoV) systemically infected azuki. An unknown virus similar to peanut mottle virus that could be nonpersistently spread by *Aphis* spp. was isolated, purified and reported by Chang (1980ab) to cause severe dwarfing and reduced yields of azuki grown in the Taiwanese province of Pingtung. *Aphis glycines* was observed to successfully transmit soybean mosaic virus to azuki and a systemic infection developed (Iwaki *et al.*, 1980).

Table 6.3. Reported viral pathogens of azuki.

Viruses and Citations	
Azuki mosaic virus (AzMV)	Matsumoto, 1922; Tsuchizaki *et al.*, 1970a; Iida, 1972; Kay, 1979; Duke, 1981; Hwang, 1982; Hong *et al.*, 1984, 1986; Tsuchizaki and Omura, 1987; Green *et al.*, 1988; Iizuka, 1988; Choi and Lee, 1989
Alfalfa mosaic virus (AMV)	Iizuka, 1988
Asparagus bean mosaic virus*	Hino, 1960; Duke, 1981
Bean calico mosaic geminivirus	Brown *et al.*, 1990
Bean common mosaic virus (BCMV)	Ito *et al.*, 1975; Meiners *et al.*, 1978; Tsuchizaki and Omura, 1987
Bean golden mosaic virus (BGMV)	Shimizu *et al.*, 1987

*Official abbreviation not found in: Francki, R. I. B., Milne, R. G. and Hatta, T. (1985) *Atlas of Plant Viruses Volume II.* CRC Press, Inc., Boca Raton, Florida, 222 pp.

Yu (1946) determined that a mosaic virus disease of cowpea could be mechanically or insect-transmitted to azuki and would cause light green blotches, marginal indentations and wrinkling of azuki leaflets. An unidentified potyvirus was isolated from soybeans growing in Taiwan and determined to be serologically similar to BCMV and AzMV. It produced systemic infection in six azuki cultivars, but was not seed-transmitted (Green *et al.*, 1986).

Tobamovirus and tobacco necrosis virus group

Tobacco mosaic virus (TMV) and tobacco necrosis virus (TNV) caused necrotic leaf lesions in azuki three to four days after inoculation (Horvath and Beezner, 1982).

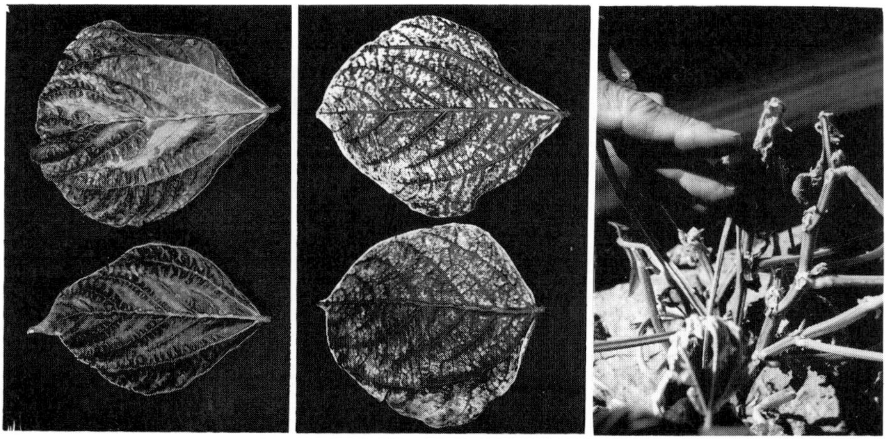

Fig. 6.4. (Left) Azuki mosaic virus (AzMV) is the most significant virus disease of azuki in all major production regions. Morphological symptoms of the disease include the development of mosaic patterns and rugose regions on leaves, leaf curling, an overall reduction in plant growth and senescence. **(Center)** Alfalfa mosaic virus (AMV) is a minor virus disease in Japan and the Columbia Basin of Washington State. Disease symptoms include bright yellow mottling and/or curling of leaves and plant stunting. **(Right)** Tobacco streak virus (TSV) causes a reddening of azuki nodes, leaf and meristem necrosis and plant death. Pathotype I is seed-transmitted at the rate of 29%.

Table 6.4. Pesticides used for bacteria, fungi and virus control in azuki production.

Materials	Specifics	Citations
	BACTERIA	
Program: foliar applied copper fungicides, Kasugamycin antibiotic (KSM) seed treatment, clean seed	*Pseudomonas adzukicola*	Tanii and Baba, 1979
	FUNGI	
Bitertanol: (β-[(+1,1'-biphenyl)-4yloxy]-α [1,1 dimethyl-ethyl]1H-1,2,4-triazole-1-ethanol)	*Ascochyta phaseolorum*; *Uromyces azukicola* or *U. phaseoli*	Takeuchi, 1986
Benomyl: (methyl 1-[butyl-carbamoyl]-2-benzimi-dazolecarbamate)	Used with antagonistic, resistant fungi, *Trichodermkoni* for *Rhizoctonia solani* control	Liu, 1991
Captan: (cis-N-trichloromethylthio-4-cyclohexene-1,2-dicarboximide)	*Phytophthora vignae*; other fungi; 5% treatment + 1200 l/ha	Robinson, 1983; Takeuchi, 1984
Captan, maneb: (coordination product of zinc, iron, manganese ethylene bisdithio-carbamate) and third unidentified material	*Phytophthora vignae*; 15.5 g/l applied to foliage	Tsuchiya and Tanaka, 1983
Chloropicrin: (trichloro-nitromethane)	*Corticium rolfsii*; *Fusarium* spp.; 300-500 l/ha	Fukunaga, 1987
Ferbam: (ferric dimethyldithio carbamate)	*Uromyces azukicola*	Mukoo, 1975
Mancozeb: (coordination product of zinc ion, manganese ethylene bisdithiocarbamate [80% a.i.])	*Uromyces azukicola*	Mukoo, 1975
Maneb: (manganese ethylenebisdithiocar-bamate)	*Uromyces azukicola*	Mukoo, 1975
Metalaxyl: (N-[2,6-dimethyl-phenyl]-N-[methoxyacetyl]-alanine methyl ester)	*Phytophthora vignae*; 35% powder	Tsuchiya and Tanaka, 1983
Metalaxyl, TMTDS: (bis[dismethylthio-carbamoyl]disulfide) and third unidentifiable chemical	*Phytophthora vignae*	Tsuchiya and Tanaka, 1983

Table 6.4. Continued.

Numbered Compound: KUF-6201 WP	*Botrytis cinerea*	Takeuchi, 1988
Procymidone: (N-[3,5-dichlorophenyl]-1, 2 dimethylcyclopropane-1,2-dicarboximide)	*Sclerotinia* spp. and *Botrytis cinerea*; WP	Fukunaga, 1987
Ridomil DT: (HSC-881) mixture of diazinon 20%+TMTD 30%+metalaxyl 15%	*Phytophthora vignae* 0.5% seed coating	Kiso,1991
Smiblend: procymidone 37.5%+diethofencarb 12.5%	*Botrytis cinerea*	Kiso,1991
Thiophanate-methyl: (dimethyl [(1,2-phenylene)bis-(iminocarbonothioyl)] biscarbamate)	*Botrytis cinerea* and *Sclerotini sclerotiorium* WP70%	Ushirogi *et al.*, 1986; Fukunaga, 1987
Thiram: (bis[dimethylthio-carbamoyl]disulfide)	*Botrytis cinerea*; seed disinfectant	Ushirogi *et al.*, 1986
Vinclozolin: (3-[3,5-dichlorophenyl]-5-ethenyl-5-methyl-2,4-oxazolidinedione)	*Sclerotinia* spp.	Takeuchi, 1980
Ziram: (zinc dimethyldithio-carbamate)	*Uromyces azukicola*	Mukoo, 1975
	VIRUSES	
Program: Insecticides to control aphids (i.e. 2 appl. 30 kg/ha **disulfuton** or **malathion** granules), 10 days apart; mulch after second application; use virus-resistant cultivars; virus-free seed; remove diseased plants from field 2 weeks after germination	Azuki mosaic virus (AzMV)	Mukoo, 1975; Kobayashi *et al.*, 1976, 1979
Program: Use of black and transparent silver-stripped plastic mulch to control aphids	Viruses	Sugano *et al.*, 1981

Common and chemical names confirmed in: Sine, C. (ed.). (1988-1991) and (1992) *Farm Chemicals Handbook.* Meister Publishing Co., Willoughby, Ohio.

MYCOPLASMA DISEASES

A mycoplasma-like organism (MLO) causing the "witches broom" disease of sweet potato was successfully transmitted by the leafhopper, *Nesophrosyne orientalis*, to azuki at the rate of 1.4%; disease symptoms consisted of dense growth of abnormally small leaves from axillary meristems (Murayama, 1966). Azuki was one of many different species screened by Murayama *et al.* (1967) for susceptibility to the MLO that causes purple top of potato; the authors did not state whether azuki is susceptible to this MLO or not.

7

BREEDING AND GENETICS

This chapter is not intended to be a reference of principles and procedures for the systematic hybridization of azuki, but instead reviews literature specific to azuki breeding and genetics and draws some conclusions. Breeding principles and procedures useful for azuki can be found in Takahashi's (1917) classic monograph on azuki breeding (in Japanese) or from generalized books such as *Breeding Field Crops* by Poehlman (1987) and *Hybridization of Crop Plants* edited by Fehr and Hadley (1980). The best sources in English may be the chapter on the Genetics of *Vigna*, including gene nomenclature, by Fery (1980) and the well-illustrated Chapter 21 in *Hybridization of Crop Plants*, by Blackhurst and Miller (1980), about the closely related cowpea (*Vigna unguiculata*). Poehlman's (1991) book about the mungbean (*Vigna radiata*) is also an excellent analogical resource.

CENTERS OF ORIGIN AND DIVERSITY

The center of origin for azuki and its boundary have not been clearly determined. The primary limiting factor in this determination is the lack of scientific consensus on the identification of azuki's wild progenitor(s) and the necessary exploration in herbariums and nature to study its range. Many researchers *think* that azuki originated in the Indo-Burma-China Center of Diversity with central/southern China, Japan, Korea and/or India as possible regions of domestication (Leese, 1958; Tasaki, 1963; Motomiya and Ito, 1972; Fujita, 1974; Ag. Lexicon ..., 1979; Kay, 1979; Jain and Mehra, 1980; Northwest ..., 1981; Dong, 1982; Smartt, 1984, 1985; Hoshikawa, 1985; Smartt and Hymowitz, 1985; Ushirogi *et al.*, 1986; Chinese Ag. Resources ..., 1987; Hymowitz, 1987; Duan, 1989; van der Maesen and Somaatmadja, 1989; Dana and Karmakar, 1990). Hoshikawa (1987) claims northeastern China as the center of domestication. Duke (1981) includes the Indochina-Indonesian Center as a center of diversity for azuki. Central China has been suggested as a possible center of diversity with Japan serving as a secondary center (Zeven and Zhukovsky, 1975). Anishetty and Moss (1988) have placed high priority on the future collection of azuki germplasm on the

Korean peninsula and in China, suggesting that these areas are either centers of origin or diversity.

Information about the location, domestication sequences, type of selection pressures, response of azuki to these pressures and subsequent genetic differentiation in the species is specifically lacking (Smartt, 1978) even in the hypothetical sense discussed for *Lens* by Blumler (1991). However, there are two hypotheses about the origin of the species: it was either domesticated from wild forms or it evolved from the rice bean (*Vigna umbellata* [Thunb.] Ohwi and Ohashi). Many researchers report that wild forms of azuki have not been discovered or do not exist (Piper and Morse, 1914; Zhukovsky, 1950; Yarnell, 1965; Sacks, 1977; Hoshikawa, 1985; Ushirogi *et al.*, 1986; van der Maesen and Somaatmadja, 1989), but Maréchal *et al.* (1978), Smartt (1978, 1985), Yamaguchi (1992) and Maeda (1993) state that wild azuki is found in Japan and several Chinese regions and call it *Vigna angularis* var. *nipponensis*. Hu (1984) reported that wild azuki had been collected in Tibet, Yunnan and Shandong. Others report that wild to semi-wild azuki are found in the Himalaya Mountains (Piper, 1936; Yi, 1961; Chinese Ag. Resources ..., 1987; Duan, 1989), for example Tateishi (1983) collected it in the Arun Valley of Nepal. However, Katayama (1979) was unable to find any wild or semi-wild *Vigna angularis* accessions during a three-month germplasm collecting trip on the Indian sub-continent and Arora (1985) reported that the wild species does not exist there. An IBPGR (International Board for Plant Genetic Resources) mission in northern Thailand found only one azuki accession (International Board ..., 1988).

Vigna angularis var. *nipponensis* (Ohwi) Ohwi and Ohashi, found in North Burma, China, Japan, Korea, Nepal and Taiwan, has been described as the wild progenitor of azuki (Smartt, 1985, 1990; Bhadra *et al.*, 1989; Yamaguchi, 1989, 1992), since there are many similarities in pollen morphology (Ohashi and Takahashi, 1981). Yamaguchi documented the basic phenotypic characteristics of two azuki weed races from the Kinki District of Japan (Yamaguchi, 1989) and wild, weed and cultivated forms from three southern main islands of Japan, plus Korea, Taiwan, China and Nepal (Yamaguchi, 1992). He described the wild type as being an annual with slender plant stature, straggling and climbing architecture, racemous inflorescence with small flowers above the canopy, rapid dehiscence of black pods and black mottled seeds. He also described (Yamaguchi, 1992) the weed races as: (1) exact progenitors of the cultigen; (2) escapees from old cultivars; or, (3) derivatives from a hybrid between wild and cultivated races. The description of the wild, weed and cultivated forms is supported by SDS-page profiles (Yamaguchi and Kosuge, 1991). This discovery of wild azuki forms could allow for the quantification of changes proposed by Smartt (1976, 1980) e.g. growth habit and maturation, number and length of internodes, increased seed size, dormancy, pod lignification dehiscence, etc., that occurred during domestication as well as an elucidation of their genetic control. The wild types of azuki are never used as food in Japan, but the weed types are occasionally used and are called either *nora-azuki, nouraku-azuki, no-azuki, norakko, yama-azuki, kuro-azuki, kitsune-azuki, taito-azuki, ishimame, bundo* or *masara* in Japanese (also see Egawa *et al.*, 1990b).

Lawn (1979a) believes that the wild azuki progenitor is the *V. umbellata.* Morphological and biochemical similarities and successful interspecific hybridization between azuki and *V. umbellata* could support the hypothesis that azuki is a derivative of *V. umbellata.* The two species are reported to have similar seed cell structure (Nguyen, 1923) and protein bands (Danimihardja and Lester, 1974). Dana (1980) placed azuki and *V. umbellata* in the same genomic group based on chromosome pairing, interspecific hybridization, hybrid fertility and similar centers of origin and diversity. Chandel (1984) prepared a diagram illustrating the close relationship to *V. umbellata,* however Chandel (1982), and later Chandel *et al.* (1988), used leaf flavonoids, isozymes and protein bands for taxonomic studies and concluded that the two species are only distantly related. Azuki pollen was used in mixture with *V. umbellata* pollen to significantly increase the production of F_1 *V. radiata* x *V. umbellata* seed (Chen *et al.*, 1978). Chandel and Pant (1982) have discussed the distribution of *V. umbellata* in India, and Jain and Mehra (1980) reported that its wild forms are distributed throughout India, and onward into central China through Malaysia. Cultivation occurs in Bangladesh, China, Fiji, Indonesia, Japan, Malaysia, Mauritius, Myanmar (Burma), Philippines and Sri Lanka. Grierson and Long (1987) found it in Bhutan and Sikkim.

INTERSPECIFIC HYBRIDIZATION

Numerous researchers have discussed the needs and problems associated with interspecific hybridization in beans (Al-Yasiri and Coyne, 1966; Smartt, 1971). Extensive reproductive isolation and compatibility studies have been conducted with azuki and other *Ceratotropis* members (Gupta and Bahl, 1983; Kim *et al.*, 1993). The results of interspecific hybridization studies between azuki, *V. umbellata, Vigna radiata* (mungbean or green gram) and *Vigna mungo* (blackgram or urd) suggest that *V. radiata* and *V. umbellata* developed independently from a shared progenitor, while *V. mungo* and azuki appeared as later derivatives (Ahn and Hartmann, 1978a; also see Egawa, 1988). Another hypothesis is that two races of *Vigna sublobata* may have been co-progenitors of these four species. Race I first served as the progenitor of *V. mungo* and then later for *V. umbellata,* while Race II gave rise to *V. radiata* and later to azuki (Jain and Mehra, 1980).

Egawa *et al.* (1990a) were able to set pods with crosses between azuki and *V. mungo, V. radiata* and *V. umbellata.* Only seeds from the crosses with *V. mungo* and *V. umbellata* germinated, but none survived to the seedling stage. They (Egawa *et al.*, 1990a) also made crosses between azuki and *Vigna glabrescens* and used embryo rescue techniques. All embryos produced seedlings and 50% developed into plants. Crosses have been made between wild *Ceratotropis* species in Japan and azuki and *V. umbellata* (Siriwardhane *et al.*, 1991). *Vigna nakashimae* was successfully crossed in both directions and produced seeds with azuki but not *V. umbellata. Vigna riukiuensis* was successful, but only as a male parent, and produced seed with both species. *Vigna reflexopilosa* was not successful with either species in both directions. These researchers suggest that both *V. nakashimae* and *V. riukiuensis* could be used as gene sources for azuki breeding and that *V. riukiuensis* could be used as a bridge to move genes from *V. umbellata* to azuki.

A bridge species could be very useful for crosses between azuki and *V. umbellata* since these crosses are reported to be unsuccessful due to cytoplasmic differentiation, however, embryo rescue techniques can be used to produce successful hybrids (Ahn and Hartmann, 1978c; Chen *et al.*, 1983). Such crosses might do better on inferior soils (Hayashi and Noda, 1961). Rashid *et al.* (1988) achieved a 3.3% successful embryo rescue rate with the *V. umbellata* x azuki cross; but embryos did not develop into plants when azuki was the female parent. Nayar *et al.* (1983) listed reasons for incompatibility with cultivated *V. umbellata* as pollen fertility, pod formation, seed setting, seed germination and hybrid survival.

Other researchers found that azuki x *V. umbellata* crosses were successful an average of 72% while the reciprocal cross averaged 44%. The different success rates were attributed to limited pollen production by azuki (Ahn *et al.*, 1986; Ahn, 1988). These researchers successfully rescued 34% of the embryos from the *V. umbellata* x azuki crosses. F_1 plants have a slight reduction in fertility as characterized by pollen grain viability and seeds per pod, but the F_1 can successfully cross with different cultivars of the parental lines (Ahn and Hartmann, 1978b; Chen *et al.*, 1983; Ahn *et al.*, 1986; Ahn, 1988). Embryo abortion appears to be the major genetic barrier between azuki and *V. umbellata,* and is stronger when *V. umbellata* is used as the female parent (Ahn *et al.*, 1986). Al-Yasiri and Coyne (1966) were unsuccessful with azuki x *V. umbellata* (= *Phaseolus calcaratus*) crosses. Pods stopped developing 12 days after pollination and no pods formed from the reciprocal cross. They believed that the high mean temperature of 25.5°C and a low relative humidity mean of 43% during pollination affected the final results by reducing pollen viability for both species when they were used as male parents. Evans (1976a) was able to produce pods and seed. Egawa *et al.* (1991) have found that the wild species *V. riukiuensis* is cross compatible with both azuki and *V. umbellata* when used as the pollen parent and could act as bridge species between the two.

Interspecific hybrids between azuki and *V. radiata* have been the most comprehensively researched of all species within *Ceratotropis*. Sawa (1973) successfully crossed *V. radiata* with a semi-wild *V. umbellata* when *V. radiata* served as the female parent and embryo rescue was practiced. F_1

pods were sterile, but colchicine was successfully used to produce fertile C_1 offspring from F_1 plants (Sawa, 1974). Such crosses with azuki have also only been successful when embryo culture techniques were used (Ahn and Hartmann, 1978a; Chen *et al.*, 1983; Egawa *et al.*, 1986). The pod-set rate is significantly affected by the choice of female parent. When *V. radiata* was used as the female parent, pod-set was 77%, whereas the percentage for azuki was only 47%. Pod-set was also influenced by the use of intraspecific hybrids. An 85% pod-set rate was achieved with *V. radiata* hybrids versus 58% for azuki hybrids. Pods from all of these crosses abscised between one and three weeks after cross-pollination and no reciprocal crosses were successful (Chen *et al.*, 1983). However, results reported by Rashid *et al.* (1988) were opposite, i.e. when azuki was used as the female parent in an interspecific cross with *V. radiata*, a higher pod-set resulted compared with using *V. radiata* as the female. The researchers suggested that some parental combinations may result in mature pods with viable embryos. Yan (1988) also successfully used azuki as the female parent. Four azuki x *V. radiata* combinations using Chinese accessions produced a small percentage of viable embryos in mature pods (Yan, 1988). Ahn and Hartmann (1978b) and Chen *et al.* (1983) concluded from their studies - based on the low incidences of chromosomal pairing, the occurrence of chromosomal irregularities and hybrid sterility - that azuki and *V. radiata* are genetically isolated. However Chen *et al.* (1978) had earlier found that azuki pollen could serve as a "mentor" pollen to increase the viability of *V. radiata* and *V. umbellata* crosses.

Indian scientists are interested in the azuki x *V. radiata* interspecific cross because of the overall resistance of azuki to many common pathogens of *V. mungo* in Himachal Pradesh (Kaushal, 1985). Hybrid seeds have been obtained from this cross and its reciprocal (Chen, 1980; Chen *et al.*, 1983). Pod-set improved when *V. mungo* or its intraspecific hybrids were used as the female parent in the interspecific cross, but interspecific F_1 embryo culture produced only a few normal seedlings which died a few weeks after transplanting due to poor root development (Ahn and Hartmann, 1978a; Chen *et al.*, 1983). Kaushal and Singh (1988) were unsuccessful in obtaining interspecific cross progeny when using either species as intraspecific hybrid parents, UV-irradiated pollen or the spraying of immuno-suppresents on female parents. Immature seed produced from F_1 intraspecific crosses produced calli in tissue culture but the calli failed to differentiate. Ahn and Hartmann (1978a) concluded from this line of research that azuki and *V. mungo* are more distantly related than azuki and *V. radiata*.

Interspecific hybridization studies with *Vigna* and *Phaseolus* species resulted in the identification of two partially compatible crosses between azuki and *Phaseolus vulgaris* (Dana, 1976), and azuki and *Phaseolus acutifolius*. These crosses initially formed pods, but the pods ceased development 15 days after pollination (Al-Yasiri and Coyne, 1966). Wei (1979) pollinated 436 azuki flowers with *P. vulgaris* which resulted in three viable seeds in the F_1 and 193 seeds in the F_2. F_2 seed produced 71 F_3 plants that were morphologically indistinguishable from azuki and 122 remaining F_3 plants that exhibited mixed characteristics of both parents. Wei identified four distinct lines from among the F_3 hybrids that had an erect growth habit.

Evolving cytogenetic distinctions can play a fundamental role in distinguishing azuki from its possible progenitors. Some researchers have looked for gene and chromosomal mutations that are manifested in interspecific hybrids as various multivalent configurations and chromosome inversion loops (Jain and Mehra, 1980). Egawa *et al.* (1988a, 1990b) hypothesized that one of the *V. glabrescens* genomes was from azuki (or *V. umbellata*) (see Egawa, 1991) based on the chromosomal and morphological characteristics of hybrids generated from the interspecific hybridization between *V. umbellata* x *V. glabrescens*, (a natural and wild tetraploid [2n=44] species). Tomooka *et al.* (1991) propose that *V. glabrescens* and *V. reflexo-pilosa* are the same species since the F_1 hybrid between these two easily produce seed, and no abnormalities were observed in F_1 and F_2 plants.

CHROMOSOMES

Azuki has a somatic diploid chromosome number of 22 (Kawakami, 1930; Darlington and Wylie, 1956; Sinha and Roy, 1979; Fery, 1980; Hoshikawa, 1985; Duan, 1989; Dana and Karmakar, 1990). Joseph and Bouwkamp (1978) prepared a pro-metaphase idiogram of azuki and determined that chromosome length ranged from 1.19 to 2.06 μm. Pushpa *et al.* (1979) reported chromosome lengths of 0.82 to 2.42 μm and described the azuki idiogram as being asymmetrical with no subterminal primary constrictions. Other researchers contend that the azuki metaphase karyotype is highly symmetrical (Sinha and Roy, 1979; Dana and Karmakar, 1990). Sinha and Roy (1979) found mostly median centromere positions for the chromosomes and reported that the azuki karyotype had lower F% range values, high TF% values and overall symmetry. They also described azuki as having one of the more primitive of 14 karyotypes despite the relatively short chromatin lengths (total chromatin length of 30.82 μm) and low DNA content. Guo and Jin (1988) identified a secondary restriction on chromosome pair number one from the Chinese cultivar *Jingnongji* No. 126. Lin and Dai (1985) reported the presence of a secondary restriction of chromosomes one and two for another, unnamed azuki cultivar. Ayonoadu (1974) compared the nuclear DNA of azuki with seven *Phaseolus* spp. and found that the isolated 4C nuclei in azuki had a mean total dry mass of 25.9 x 10^{-12} g, nucleolus dry mass of 2.1, nucleus volume of 125.1 μm^3 and nucleolus volume of 12.6. Bennett and Smith (1976) used Ayonoadu's results as part of a comparison of DNA in hundreds of different species. Zheng *et al.* (1991) compared the karyotypes and C-banding patterns of six bean species. They found centromeric bands on all chromosomes of their *Vigna* group including azuki.

Kaga *et al.* (1993) used random amplified polymorphic DNA (RAPD) analysis in an attempt to differentiate 12 cultivars of azuki. Ten amplified DNA products were found to be polymorphic from 184 produced. The 12 cultivars were separated in to eight groups by cluster analysis, however detection frequency was low, and the research team felt that RAPDs were not very useful for this intra-specific segregating population. Chloroplast DNA has also been studied to understand the diversity in azuki and its relatives (Mikami *et al.*, 1991).

GERMPLASM RESOURCES

Azuki has been collected and maintained at many locations, especially in national collections of East Asian countries (Table 7.1). Azuki germplasm resources can be classified in various ways. Smartt and Hymowitz (1985) proposed a modified form of Harlan and DeWet's (1971) hierarchical gene-pool system with five germplasm levels for azuki. Level one includes landraces, cultivars and azuki breeding lines; level two indicates wild azuki forms; level three signifies other species that successfully cross with azuki; level four contains species that produce viable but sterile progeny when crossed with azuki; and level five indicates species that produce inviable hybrids when crossed with azuki.

Table 7.1. Azuki germplasm collections.

Country	Address, Size, Citations
Institutional collections:	
BELGIUM	Faculte des Sciences Agronomiques de l'Etat, B5800, Gembloux. 8 cultivars. SEE: Ayad and Anishetty, 1980; Bettencourt *et al.*, 1989.
GERMANY	Zentralinstitut fur Genetik und Kulturpflanzenforschung, Akademie der Wissenschaften Corrensstrasse, Gatersleben, 4325. 21 azuki cultivars/landraces. SEE: Hammer and Mun-Chan, 1986; Hammer *et al.*, 1986, 1988, 1989a.
JAPAN	Iwate Agricultural Experiment Station, Morioka-shi, Iwate-ken. 214 landraces. SEE: Kiuchi *et al.*, 1987; Personal observation, 1989.
JAPAN	Tokachi Agricultural Experiment Station, Tokachi, Hokkaido-ken, Japan. Approximately 2500 accessions. SEE: Personal observation, 1989.
PHILIPPINES	Institute of Plant Breeding, College of Agriculture, University of the Philippines at Los Banos College, Laguna. 23 azuki cultivars. SEE: Ramirez, 1978.

Table 7.1. Continued.

TAIWAN	The Asian Vegetable Research and Development Centre. P.O. Box 42, Shanhua, Tainan, 741. 125-140 accessions predominately from Taiwan. SEE. Ayad and Anishetty, 1980; Ahn *et al.*, 1981; Anishetty and Moss, 1988; Bettencourt *et al.*, 1989.
National collections:	
AUSTRALIA	Division of Plant Industry, CSIRO, P.O. Box 1600 Canberra City, 2601. SEE: Ayad and Anishetty, 1980.
CHINA	Institute of Crop Germplasm Resources, Chinese Academy of Agricultural Sciences, 30 Bai Shi Qiao Road, Beijing, 100081. >3000 Chinese accessions. SEE: Jiang, 1982; Ding and Zhang, 1984; Hu, 1984.
COSTA RICA	Unidad de Recursos Geneticos, Centro Agronomico Tropical de Investigacion y Ensenanza (CATIE), Apartado 77, Turrialba. 4 Central American landraces. SEE: Bettencourt *et al.*, 1989.
INDIA	National Bureau of Plant Genetic Res., IARI Campus, New Delhi, 110012. 1200 accessions. SEE: Ayad and Anishetty, 1980; Ahn *et al.*, 1981.
JAPAN	Germplasm Storage Centre, National Institute of Agrobiological Resources, Ministry of Agriculture, Forestry and Fisheries, Kannondai, Tsukuba, Ibaraki, 305. 142 accessions. SEE: Ayad and Anishetty, 1980; Bettencourt *et al.*, 1989; Hino, 1990.
KOREA	Plant Genetic Resources Research Program, Republic of Korea. 1212 accessions. SEE: Ahn, 1987.
NETHERLANDS	Foundation for Agricultural Plant Breeding, Institute De Haaf, P.O. Box 117, 6700 AC Wageningen. SEE: Ayad and Anishetty, 1980.
USA	USDA Southern Regional Plant Introduction Station, Experiment, Georgia, 30312. 42 accessions. SEE: Anishetty and Moss, 1988; Bettencourt *et al.*, 1989.
USA	National Seed Storage Laboratory, Colorado State University, Fort Collins, Colorado, 80523. 9 Korean, Japanese and Brazilian cultivars. SEE: Bettencourt *et al.*, 1989.

Another system, called azuki gene-pool classification, consists of four levels. These are: primary (cultivar collections), secondary (*V. angularis* var. *nipponensis*), tertiary (the subgenus *Ceratotropis*) and quaternary (other species in *Vigna*). This system will be used here because it is more widely supported in the literature (Smartt, 1980, 1981; Bhadra *et al.*, 1989; Dana and Karmakar, 1990). Dana and Karmakar (1990) included landraces and wild forms of *V. umbellata* as well as cultivars as part of the primary azuki gene pool.

The number of azuki cultivars in the primary gene pool has been discussed to some extent by many authors (Tanaka and Ohno, 1895; Piper and Morse, 1914; Hedrick, 1931; Tasaki, 1963; Hernandez *et al.*, 1978; Hoshikawa, 1985; van der Maesen and Somaatmadja, 1989). Duke (1981) reported the existence of at least 60 cultivars while Anishetty and Moss

(1988) identified 167 in two collections. In Japan, about 350 cultivars were reported during 1955 (Hoshikawa, 1985) and over 300 cultivars were collected during the 1970s (Konno and Narikawa, 1978). The National Institute of Agrobiological Resources at Tsukuba in Japan maintains more than 143 azuki accessions (Watanabe *et al.*, 1986; Hino, 1990).

Two Japanese prefectural experiment stations maintain large collections of cultivars and landraces. Their existence at the prefectural versus the national level constrains their availability at the international level. A large portion of the 2500 plus accessions in the Tokachi azuki collection in Hokkaido Prefecture consists of accessions obtained from farm gardens in Hokkaido Prefecture during the early 1960s (Narikawa, 1978). The Iwate Prefecture Agricultural Experiment Station maintains a sizable collection from the Tohoku region of northern Honshu Island. This region has historically cultivated numerous azuki landraces, e.g. 52 different landraces (12 early, 15 intermediate and 25 late maturing) were recorded during a crop survey conducted between 1716 and 1741 (Japan ..., 1988). A recent collecting trip to Iwate Prefecture resulted in the acquisition of 214 landraces which were subsequently evaluated for basic morphological and physiological features (Kiuchi *et al.*, 1987). In the mid- to late 1980s, almost 40% of azuki production in the Iwate Prefecture was still landraces (Japan ..., 1988).

Large collections of azuki cultivars and landraces also exist in China. Chinese landraces are most frequently named by seed testa color or for the area of production such as a city, district or region (Duan, 1982; Zheng and Zhang, 1983ab). Over 2500-3000 different cultivars and landraces have been collected by the Institute of Crop Germplasm Resources, Chinese Academy of Agricultural Sciences in Beijing (Jiapeng Hu, personal communication, 1992: Zhao and Tong, 1993). Hu (1981) characterized basic morphological traits and classified maturity for 1040 azuki landraces collected primarily from 10 Chinese provinces and regions including Inner Mongolia. He classified these accessions by seed color, components of yield, maturity, height, collection location and their interactions. In Hebei Province, the largest azuki producing area, 254 azuki landraces were collected during the early 1980s and two of these accessions demonstrated virus resistance after three years of field and laboratory observations (Liu, 1984). Azuki germplasm from northern China has been characterized for growth and quantitative traits near Nanjing, central China (Jin, 1989ab; Jin and Jiang, 1989b). Between 1987 and 1989 about 1000 accessions were collected in Heilongjiang Province and screened for disease resistance (Chen and Wei, 1991). Accessions A-0012, A-0017, A-0024, A-0046, E0944, E0961 and E0976 were resistant, 124 accessions were mildly resistant, 316 were mildly sensitive, 376 were infected and 132 were highly infected. The long history of azuki cultivation in the Wuqing area of Hebei Province makes it an important site for azuki germplasm collection (Zheng, 1988). The features of collections from other provinces and some in Japan are summarized in Table 7.2.

Azuki is distributed throughout the Korean peninsula (Mun-Chan *et al.*, 1986), and it is cultivated by Korean people in North and South Korea and China, but not currently grown in the Primor'ye region or Sakhalin Island of far eastern Russia. Several missions for germplasm resources have visited the Democratic People's Republic of Korea (North Korea), in areas south of Pyongyang and collected 21 azuki landraces (Hammer and Mun-Chan, 1986; Hammer *et al.*, 1988, 1989a). Fifty Korean landraces were collected and classified on the basis of seven maturity groups. One cultivar reached physiological maturity between 101 and 105 days; the remaining 49 accessions took from 116 to 145 days (Chang and Han, 1965). Hong *et al.* (1987) found that genetic drift has occurred in germplasm that was originally collected and characterized in 1926, including large changes in color, seed weight, etc. The largest collection of azuki in Korea seems to be held by the Seed Management Division of the Rural Development Administration in Suwon, South Korea, with 1212 accessions of azuki reported (Ahn, 1987).

Tertiary and quaternary collections are considered adequate for anticipated interspecific hybridization, but secondary germplasm resources are quite small. However, the *V. umbellata* is another possible secondary source of genetic diversity for azuki (Tanaka and Ohno, 1895). Current efforts to increase azuki germplasm resources include an IBPGR-sponsored herbarium survey to understand the distribution and habitat patterns of wild types, clarify centers of origin and diversity and identify areas of diversity for future collection (Anishetty and Moss, 1988).

BREEDING PROGRAMS

Azuki cultivars have been developed in East Asian countries which have a long history of cultivation, especially Japan, China and Korea, as well as other countries considering azuki as an alternative crop (Koyama *et al.*, 1972; Ushirogi *et al.*, 1972; Robinson, 1980; Hong *et al.*, 1984, 1986; Murata *et al.*, 1985; Zhang and Li, 1988). Azuki breeding was initiated in Japan during 1894 (Konno and Narikawa, 1978) and in Hokkaido Prefecture by Yoshisugu Takahashi during 1908 and 1909. Takahashi wrote a monograph on azuki breeding (Takahashi, 1917) and used Mendelian principles to develop the hybrid cultivar, *Takahashi wase* (Hokkaido Agricultural Experiment Station, 1967; Narikawa, 1978). Cultivar development by hybridization continued in Japan from 1930 until the early 1940s and recommenced in 1954.

The Tokachi Agricultural Experiment Station in Hokkaido Prefecture started a selection breeding program in 1921 which resulted in a few cultivar releases. A small hybridization program began at the station in 1954 with parents selected from a collection of local cultivars (Sunada *et al.*, 1975).

Early Japanese azuki breeding efforts also occurred at the Ishioka Experiment Station, Ibaraki Prefecture (1935-1945) and the Odate Experiment Station, Akita Prefecture (1936-1950) (Narikawa, 1978).

In Taiwan, azuki breeding began in 1968 at the Kaohsiung District Agricultural Experiment Improvement Station. Pure line selection methods were used first, resulting in the identification of *Kaohsiung* 1 from among a collection of Taiwanese landraces. A hybridization program was initiated later and resulted in the development and release of *Kaohsiung* 2 and *Kaohsiung* 3 (Tsai, 1988).

In China, an azuki breeding program exists at Tianjin Agricultural College, (Zhu, 1989) and in Korea at the Suweon Crop Experiment Station, South Korea (Hong *et al.*, 1984, 1986). Farmers in Heilongjiang Province, China, were encouraged to maintain the genetic purity of their azuki cultivars for future production by making single-plant field selections and then increasing this seed for replanting (Tai, 1980). Wang and Zhao (1984) recommended that selection be based on the following characteristics: excellent adaptation to the local growing environment, bush-type growth habit, lodging resistance, high number of pods per plant and seeds per pod and disease and insect resistance. Unfortunately, Chinese and Korean azuki breeding programs are poorly documented in distributed literature.

BREEDING OBJECTIVES

Azuki is primarily self-pollinating, but has a significant percentage of cross-pollination. Bagging of flowers is necessary to eliminate natural hybrids in breeding for objectives (Piper and Morse, 1914; Kakizaki, 1923; Yarnell, 1965; Duke, 1981; Wang and Zhao, 1984; van der Maesen and Somaatmadja, 1989). In addition to taxonomic studies, azuki breeding objectives usually attempt to improve production and processing characteristics and overcome constraints in the major areas of production. Components of yield are often targets for improvement. For azuki, yields = plants per area x pod per plant x seeds per pod x seed weight (Wu and Hsu, 1986). Smartt (1985) hypothesized that the universal goal of higher and more stable yields might be achieved in the Asiatic members of *Ceratotropis* by increasing the leaf area of individual plants, improving harvest indexes, and selecting for more effective *Rhizobium* strains. Pods per plant and plants per area are the two most important factors influencing azuki seed yield, while branches per plant and pods per branch were observed to have the most significant effect on the number of pods per plant (Tsai, 1988).

In 1973, the Tokachi Agricultural Experiment Station in Japan was selected to be the nation's centralized azuki breeding center with the objectives of developing early-maturing, high-yielding azuki cultivars with good seed quality and resistance to cool weather and *Phialophora gregata* (= *Cephalosporium gregatum*) infection (Hokkaido Agricultural Experiment

Station, 1967; Sunada *et al.*, 1975; Murata, 1976; Konno and Narikawa, 1978; Murata and Narugawa, 1978; Chiba, 1982, 1985; Hoshikawa, 1985; Chiba *et al.*, 1987). Other breeding objectives include the development of new early-maturing dainagon cultivars with more seeds per pod, higher pod-set for improved mechanical harvest, increased resistance to lodging and increased determinacy (Sunada *et al.*, 1975; Suzuki and Konno, 1982; Hoshikawa, 1985). Researchers at the Kitami and Kamikawa experiment stations on Hokkaido have been successfully screening azuki accessions for *Phytophthora vignae* resistance in converted paddy fields and have identified nine superior lines (Tabiki and Tsuchiya, 1990). Nomura and Satoh (1977) conducted a detailed analysis of 26 cultivars for the higher pod-set needed for mechanical harvesting. They found that higher pod-set was positively correlated to plant height, number of branches, flowering date and late maturity dates. Plant height has been found to be negatively correlated with pods per plant, seeds per plant and yield (Jin, 1989a). Yoshida (1987) looked at the correlation of seed size with plant morphological features and found a positive correlation of seed size with the number of nodes, pods and seeds on branches per plant. No correlation was found with plant height, growing period or number of pods per plant. The Iwate Agricultural Experiment Station began azuki breeding work in 1966 with the objectives of improving quality and yield as well as conducting cultivar testing for virus resistance (Tanaka and Konno, 1982). Promising new Japanese accessions are commonly tested for their responses to cool temperatures, fertilizer levels and planting dates and densities (Koyama *et al.*, 1972; Ushirogi *et al.*, 1972, 1986; Satoh *et al.*, 1975ab).

In China, azuki breeding objectives at Tianjin Agricultural College include the development of late-maturity cultivars with high yield potential (over 5250 kg/ha), very early-maturity cultivars for use as a catch crop and medium-maturity cultivars with excellent processing qualities and virus resistance (Zhu, 1989). Eight azuki cultivars were evaluated in Shandong Province for high yield potential, early maturity and resistance to drought, flooding and virus diseases (Rong and Ma, 1986). Breeders in China's northeastern provinces have sought early maturity and high quality with some success (Du, 1986: Du and Zhang, 1987). In Taiwan, the azuki breeding objectives at the Kaohsiung District Agricultural Experiment Improvement Station include improved yields, seed quality, pest resistance and more uniform maturity (Wu *et al.*, 1985b, 1986; Hsu *et al.*, 1989ab). Fernandez and Talekar (1990) at the Asian Vegetable Research and Development Center in Taiwan have selected and tried to breed for resistance to bruchids, but this became complicated because the resistance is controlled by the maternal genome and recessive genes. Korean breeding objectives for azuki include higher and more stable yields (Han and Choe, 1977). They used days to flowering, plant height, node number, 100-seed weight, pods per plant and seed yield to select the 27 cultivars tested that had the highest and most stable yield over time.

Other breeding objectives could capitalize on the demonstrated resistance of some azuki germplasm to high temperatures, shade, drought, frost, saline soils, viruses and the effects of slope (Plant Production ..., 1959; Dong and Liang, 1980; Duke, 1981; Wang and Tian, 1990). Many Chinese accessions from Hebei, Shanxi, Henan and Shaanxi provinces are highly tolerant of saline soils (Wang and Tian, 1990).

HERITABILITY

According to Fery (1980), Matsuura (1933) was able to identify 11 specific genes for six phenotypic characters of azuki. Matsuura isolated an inhibitor gene for red seed testa color, an intensifier gene for purple stem color and a pleiotropic gene for stem and seed testa color and seed testa pattern. Matsuura was also able to identify three separate two-gene linkage groups and complementary epistasis action for purple stem color, buff seed testa color and black mottled seed testa pattern.

Kakizaki (1923) stated that blackish-brown mature pod color was dominant over medium-brown mature pods; the presence of an epicotyl intensity gene was closely linked to the gene that darkens mature pod color. Other single gene qualitative characters include a dominant gene for leaf shape (narrow leaf dominant over round leaf), red seed testa color (red dominant over white) and solid seed testa color (solid dominant over patterned) (Matsuura, 1933; Fery, 1980). Partial dominance has been described for black spotting of seed testa (Kakizaki, 1923) and trichome shape (Matsuura, 1933; Fery, 1980). Male sterility in azuki was discovered in 1974 and determined to be controlled by one recessive gene suggesting the potential for hybrid seed production (Nakashima *et al.*, 1980).

Diallel crosses have been used extensively to clarify various mechanisms of heritability in azuki. Diallel crosses of five azuki cultivars demonstrated that seed weight per plant, plant dry weight, stem diameter and flowering to physiological maturity all exhibited overdominance, while partial dominance was shown for 100-seed weight, pods per plant, node number per plant, plant height, days to physiological maturity and days from planting to flowering (Han *et al.*, 1979, 1984). A comparison with earlier results (Han *et al.*, 1979) showed that node number per plant and plant height were overdominant rather than partially dominant, and seed weight per plant was partially dominant instead of overdominant in nature (Han, 1983; Han *et al.*, 1984). They also reported that the degree of dominance may change based on the particular filial generation, for example the quantitative trait, time from planting to flowering, was completely dominant rather than partially dominant in an F_2 population. Complete dominance was also observed for the F_1 traits of seed weight per plant (rather than partial dominance) and plant height (instead of overdominance) (Han *et al.*, 1984).

Heritability values for physiological and morphological traits of azuki have been determined by many researchers (Table 7.3). Tian (1990) calculated a genetic advance range of 2 to 23% for the heritabilities of eight morphological characters and an average of 14% for seed yield per plant. Total plant weight and the number of pods per plant had the highest genetic advances of seven quantitative characteristics analyzed for seed yield (kg/ha) at an undetermined selection intensity (Chang, 1969b). Chang also recorded a slight genetic advance decline for seed yield if the number of branches per plant increased between generations. Jin (1990ab) found that the plurality hereditary capacity of 92 cultivars was 73 to 87% for groups of agronomic properties.

Nomura (1968) found highly significant and positive correlations between seed yield and plant height, longest branch length, branches per plant and plant weight. Chang (1969a) and Han *et al.* (1970) used Path Coefficient Analysis to determine that pods per plant had the strongest influence upon yield with plant height, stem diameter and 100-seed weight also having some positive effect. Han and Choe (1975) later described the importance of pods per plant for affecting positive change in seed yield. The relationship between these two genotypic features was highly correlated and the relationship remained stable between production years. This is consistent with earlier findings by Chang *et al.* (1968c) and Chang *et al.* (1969) who also included seeds per plant as a selection characteristic for analyzing potential parental lines and breeding lines. Han and Choe (1975) observed that the genotypic correlation between seed yield and 100-seed weight, plant height, total dry matter and time to flowering were highly negative one year and highly positive in another production year. A highly significant negative relationship has been reported between 100-seed weight and seed yield (Chang and Han, 1965; Nomura and Satoh, 1968). Weight of 100 seeds is also reported to be negatively correlated with pods per plant and flowering duration (Chang *et al.*, 1968c; Han and Choe, 1975). Days to anthesis, days to physiological maturity and total dry matter were positively and highly correlated with 100-seed weight (Chang and Han, 1965; Han and Choe, 1975; Han and Chang, 1976).

MUTATION AND TISSUE CULTURE TECHNIQUES

Various laboratory techniques such as tissue culture (Phillips *et al.*, 1992), electrophoresis and radiation-induced mutation have been used for selection and development of new azuki cultivars (Sunada *et al.*, 1975). Resistant and sensitive sub-sets of three azuki cultivar populations were identified after being subjected to five different irradiation treatment levels. Resistant and nonresistant plants were distinguished on the basis of leaf spotting, leaf area index (LAI) and stem height differences. Muszynski *et al.* (1980) suggested that either LAI or leaf spotting were suitable for quantifying irradiation damage since significant differences appear at low levels and measurements are easy to record. A 10 kR irradiation dose (Co^{60}) was applied to the seeds of the azuki cultivar, *Monbetsu* 26, which resulted in the eventual selection of the shorter and earlier maturing cultivar, *Beni-Nanbu* (Satoh *et al.*, 1984). Milkov *et al.* (1992) used matriconditioning to attempt to reduce redrying damage to mutagenized legume seeds but obtained mixed results with azuki.

Tissue culture may help to expedite development of new azuki cultivars. Callus culture has been used to obtain normal azuki plantlets from inoculated calli (Bhadra *et al.*, 1989). Azuki plantlets were successfully regenerated from epicotyl-derived callus on both Murashige Skoog (MS) and B5 media supplemented with auxin and cytokinin (Xu *et al.*, 1984). Similar results were achieved with a procedure consisting of an inducing MS medium supplemented with 1.0 mg/l zeatin + 500 mg/l luteining hormone followed by the use of a differentiation medium consisting of 2.0 mg/l 6-benzyl adienine (BA) + 1.0 mg/l kinetin + 0.5 mg/l naphaline acetic acid (NAA) + 500 mg/l luteining hormone (Lu *et al.*, 1985). Epicotyl explants produced shoots on an MS medium containing 0.1 to 1.0 mg/l BA (Ozaki, 1985). He also found that compact calli, not soft calli, would produce shoots if first induced on MS medium containing any of the following combinations: 0.05 mg/l NAA + 4.0 mg/l BA, 0.5 mg/l NAA + 10.0 mg/l BA, 0.1 mg/l NAA + 2.0 mg/l BA or 0.5 mg/l NAA + 1.0 mg/l BA, and then transferred onto MS medium containing 0.2 mg/l BA. Jin and Futsuhara (1993) found that primary leaf tissue obtained two days after budding was the best for tissue culture.

Ahn (1988) was successful in obtaining multiplication of F_1 hybrids from *V. umbellata* x *V. angularis* crosses by shoot tip and node culture using MS medium with 3% sucrose and 8% agar, at pH 5.8, 500 lux and maintained at 26 $\pm$ 1°C. *In vitro* formation of two different types of calli from immature azuki leaflets has been reported but no plants were regenerated; compact calli from three-day old seedlings developed into plantlets using a sequence of two MS regeneration media, the first containing 0.5 mg/l NAA + 2.0 mg/l BA for shoot development and the second 0.2 mg/l NAA for root growth (Ozaki, 1986). Other researchers (Sato *et al.* 1993) have been successful at producing whole azuki plants from protoplasts. They added 1 mg per liter each of 2,4-D and 6-benzylaminopurine (BAP) to MS medium to increase division frequency. Callus colonies were further cultured on solid MS medium with 0.5 and 1.0 mg/l, respectively, to produce calli of 2 mm diameter. These were transferred to solid MS medium containing BAP, kinetin, trans-zeatin and 3-indoleacetic acid to produce shoots, and finally these shoots were transferred to normal MS medium to produce plants.

CULTIVAR RELEASES

Azuki landraces have been and still are quite widespread and popular in Asia. Takahashi (1917) listed 57. A sizable percentage of Japanese azuki production was from landraces until the early 1960s. Some of the more important landraces were *Maruba* (round leaf), *Chunagon* (medium-sized seed), *Dainagon* (large-sized seed), *Natsu* (summer) and *Zairai* (native species) (Tasaki, 1963, 1965; Narikawa, 1978; Hoshikawa, 1985). Japanese landraces and cultivars have been classified into 19 groups according to two general physiological stages, i.e. emergence to anthesis and anthesis to physiological maturity. The largest single group based on the classification of 190 accessions took 75 to 89 days to reach anthesis and 40 to 49 days from anthesis to harvest (Kawahara, 1962; Hoshikawa, 1985).

Since the advent of modern techniques, numerous improved cultivars with superior qualities have been developed and released by breeders in Japan, China, Taiwan, South Korea and the United States. A few of the cultivars reported in the literature are discussed in Table 7.4. The most popular cultivar in Japan, *Erimo*, was registered by the Japanese Ministry of Agriculture, Forestry and Fisheries in 1981. This cultivar was selected from a *Kotobuki* x *Toiku* No. 77 cross. *Erimo* takes fewer days to reach physiological maturity and produces a slightly larger and better quality seed for *an* production when compared to the old cultivar, *Takara*. It also achieves higher yields, especially under cool growing conditions, and is more resistant to lodging (Satoh, 1982; Murata *et al.*, 1985). A new *dainagon* cultivar, *Tokachi* 125, was released for production in Hokkaido in 1990 but has not become widely popular.

Kotobuki is an F_5 selection from a *Noto* x *Wase dairyu* 1 cross that was registered in 1971. This cultivar has taller stems, 20% larger seed, earlier maturity and is 3 to 13% higher than *Takara* (Koyama *et al.*, 1972). The azuki cultivar, *Akatsuki dainagon*, was also the product of a cross between *Noto* and *Wase dairyu* 1. This cultivar was registered by 1970 and has some tolerance to ascochyta leaf spot (Ushirogi *et al.*, 1972; Satoh *et al.*, 1975a). *Akane dainagon* was released in 1974 for production in the central and southern portions of Hokkaido Prefecture. It is slightly shorter and produces fewer branches and smaller seeds than *Akatsuki dainagon*, but demonstrates greater cool temperature tolerance during anthesis (Satoh *et al.*, 1975a).

The Tokachi Agricultural Experiment Station released the white-seeded azuki cultivar, *Hokkai Shiro*, in 1980 to replace the natural white-seeded mutation *Shiro* (Kawanishi) (Adachi *et al.*, 1980; Murata, 1980). *Hayate* (line #5082 and *Toiku 85 go*) was developed from a *Takara* x *Buchishoryukei-1* cross and registered in 1976 by this station (Murata, 1976). *Hayate* produced 22% higher yields than the traditional early maturing azuki cultivar, *Chagarawase*, during four years of yield testing.

Table 7.2. Ranges for traits of some azuki landraces collected and characterized in Chinese provinces and in Japan.

Province	Provincial features	Landraces collected	Days to maturity	Height in cm	Branches per plant	Pods per plant	Seeds per pod	Grams per 100 seeds	Grams seed per plant	References
All of China	Beijing growout	869	92-161	21-169	0.8-11.8	3.6-194.7	4.1-10.2	3.3-15.89	2.4-77.5	Hu, 1981
Anhui	East Central	99	80-110	-	5-30	-	6-10	5-16	-	Zhou *et al.*, 1964
Gansu	Arid Central	60	-	-	-	16-44	7-11	-	10-25	Shi, 1988
Heilongjiang	Cold N.E.	163	135-152	84-107	1.7-2.2	12-22	5-5.4	10.1-12.1	7-11	Zheng and Zhang, 1983ab
Shaanxi	Central	320	100-131	27-129	1.6-6.3	5-43	3.5-12.5	3.1-15.9	2-30	Zhou, 1983; also see Dai and Zhang, 1982
Shandong	Center E. Coast	202	100-120	58-82	6.1-6.5	12-22	6.9-7.4	-	16.5-25.2	Jiang and Chen, 1984
Hokkaido, Japan	North Island	256	-	28-108	1.3-5.9	-	4.3-8.9	6.9-19.5	6.5-20.9	Chiba, 1980

Table 7.3. Heritability values for morphological and physiological traits of azuki.

Number of cultivars	Sense of heritability	Days to flowering	Plant height	Stem nodes	100-seed weight	Pods per plant	Stem diameter	Dry matter	Seed yield	Branches per plant	Flowering duration	Days to maturity	Seeds per plant	References
many	-	89-96	49-84	66	84-99	52-60	45-51	61-74	37-50	-	-	-	-	Han and Choe, 1975, 1977; Han and Chang, 1976
15, $100F_{6-7}$	broad	-	49, 72, 86	-	88, 96	17, 28	5, 12	24, 55	-	44, 60	-	-	-	Nomura, 1968; Nomura and Satoh, 1968, 1977
30	broad	99	-	-	97	87	-	-	42	-	97	99	99	Chang *et al.*, 1968ab
$40F_4$	broad	84	26	-	70	37	34	81	31	-	-	-	-	Chang *et al.*, 1968ab
-	narrow	-	41	-	76	9	-	-	34	69	-	-	9	Tian, 1990

Hayate is responsive to high population densities and fertilizer levels and is tolerant of cool growing conditions (Murata, 1976; Murata *et al.*, 1977). It is highly susceptible to azuki phytophthora stem rot caused by *Phytophthora vignae* (Purss) f. spp. *adzukicola* (Tsuchiya and Kodama, 1979). Hayate was recommended as a replacement for *Chagarawase* in marginal azuki-producing eastern and northern regions of Hokkaido Prefecture (Murata, 1976).

Sakae, the product of a cross between *Wase dairyu* 1 and 3407 (*Chagarawase* x *Wase dairyu* 1 [F_6]), was released in 1973. The cultivar has medium-maturity, high-yield and good quality seed for production in the central portions of Tokachi District and the central and southern parts of Kamikawa District in Hokkaido Prefecture and is characterized by higher yields and less sensitivity to cool temperature at anthesis than *Takara*, and is responsive to high fertilizer levels (Satoh *et al.*, 1975b).

Beni-Nanbu was released by the Iwate Agricultural Experiment Station and has earlier maturity, shorter height and produces slightly smaller seed than its parental line, *Monbetsu 26*. It is also determinant and moderately resistant to some viruses (Satoh *et al.*, 1984).

In Taiwan, *Kaohsiung* No. 3 (same as breeding line KS 352) was produced from a *Kaochiaotsaosheng* x *Danayan* cross that was released in 1983. The cultivar averaged yields of 2619 kg/ha, maturity in 90 days, a height of 37.9 cm, a 100-seed weight of 15.3 g and large seeds with bright red testa color during the seven years of trials. While this cultivar has medium susceptibility to powdery mildew infection from *Erysiphe* spp., it is fast growing and develops a large leaf area index for excellent early season competition against weeds. This cultivar is also lodging resistant and has indehiscent pods positioned at least 20 cm above the ground so it can be direct-harvested with a combine (Wu *et al.*, 1985a).

In China, the azuki cultivar *Baihong* No. 1 was released by the Jilin Agricultural Crops Breeding Examination and Approval Committee in 1987, and is suitable for the western portion of Jilin Province and Tai Lai County of Heilongjiang Province. It is characterized by a resistance to preharvest pod shatter, lodging, quick seedling growth and yields of 1500 to 2250 kg/ha depending on soil fertility (Du and Zhang, 1987; Zhang and Li, 1988). The cultivars *Longxiaodou* 1 and *Longxiaodou* 2 were also recently developed for Heilongjiang Province and mature in about 100 to 115 days with a yield of about 3.7 mt (Du, 1986).

In Korea, two azuki cultivars were released by the Crop Experiment Station at Suweon in the early 1980s. *Chungwonpat* was developed from indigenous cultivars and is a dark gray-seeded cultivar that outperformed the standard cultivar *Hongchonjokdu* by 15% during three years of yield trials. This new cultivar shows some resistance to *Uromyces azukicola*, *Sphaerotheca fuliginea*, powdery mildew and azuki mosaic virus, as well as yield tolerance to late planting (Hong *et al.*, 1984). Another cultivar,

Chungjupat, is red-seeded, yields 10% more than *Hongchonjokdu* and shows greater resistance to powdery mildew and rust (Hong *et al.*, 1986).

In the U.S., the Agricultural Experiment Station of Minnesota released *Minoka* azuki in 1980. *Minoka* was selected from azuki seed obtained from Oklahoma (Breene and Hardman, 1987). *Minoka* matured in 115 days and had comparable or greater yields than kidney bean during three years of testing at four of five testing sites, as well as a 29% greater seed weight than that of available Japanese cultivars (Robinson, 1980). The cultivar is not commercially available due to complications in the Minnesota Crop Improvement Association seed increase program (Breene and Hardman, 1987).

Table 7.4. Notes and references for azuki cultivars, landraces and some breeding lines. All are Japanese unless otherwise noted. Names in parentheses are usually Japanese prefectures, sometimes Chinese or Korean provinces identified by country or Chinese and Japanese regions. Some azuki breeding lines developed at the Kaohsiung District Agricultural Experimental Improvement Station (DAIS), Taiwan are listed in this table and are coded, e.g. for line **KA 71-03-92, KA** = Kaohsiung (DAIS), **71** = Taiwanese calendar year of hybridization (add 1911 years for A.D.), **03-92** = hybridization number codes.

Azuki Names (Origin) Notes and References.

73-2 (Jilin, China)**:** Large-seeded cultivar developed by hybridization. Grown in Shulian County mostly for export to Japan. SEE: Zheng, 1988.

80 Shengwai 5 (Hebei, China)**:** Two-year, three-site growout in Shanxi Province averaged 121 days to maturity, 60 cm tall, 4.4 branches, 21 pods, 5.5 seeds/pod, 13.6 g/100 seeds, 2043 kg/ha yield. SEE: Song *et al.*, 1989.

80-131(Beijing, China)**:** Advanced breeding line. Shandong Province growout averaged 102 days to maturity, 42.2 cm tall, 3.6 branches, 14.2 pods/plant, 7.1 seeds/pod, 10 g/100 seeds, 554 kg/ha yield, bright red seed testa. Beijing growout averaged 90 cm tall, 4.5 branches/plant, 12 g/100 seeds, 2438 kg/ha yield and showed disease resistance. SEE: Hu, 1986; Rong and Ma, 1986.

80-2 (Jiangsu, China): Average yield 1224 kg/ha, 19.5g/100 seeds, semi-prostrate, intercropped with maize, 52,500-60,000 plants/ha. SEE: Ye *et al.*, 1993.

81-95 (China)**:** Advanced breeding line with early maturity (90 days), 2715 kg/ha yield, good for wheat-bean rotations. SEE: Hu, 1986.

82-08, 82-14, 82-15, 85-6 (Tianjin, China)**:** Four cultivars among 11 tested that were resistant to *Uromyces azukicola*. SEE: Zhao, 1990.

83-5 (Jiangsu, China): Average yield 1224 kg/ha, 19.5g/100 seeds, semi-prostrate, intercropped with maize, 52,500-60,000 plants/ha. SEE: Ye *et al.*, 1993.

194 (Hebei, China)**:** Shandong Province growout averaged 101 days to maturity, 49.2 cm tall, 5 branches/plant, 23 pods/plant, 6.9 seeds/pod, 9.1 g/100 seeds, 704 kg/ha yield, bright red seed testa. SEE: Rong and Ma, 1986.

78065 (Hebei, China)**:** Advanced breeding line. Two-year, three-site growout in Shanxi Province averaged 123 days to maturity, 46.6 cm tall, 3.6 branches/plant, 19.4 pods/plant, 5.3 seeds/pod, 12.3 g/100 seeds, 1755 kg/ha yield. SEE: Song *et al.*, 1989; Tian, 1990.

80155 (Hebei, China)**:** Advanced breeding line. Shandong Province growout averaged 102 days to maturity, 67.6 cm erect and semi-prostrate, 3.4 branches/plant, 20.4 pods/plant, 8 seeds/pod, 9.5 g/100 seeds, 711 kg/ha yield, bright red seed testa. Two-year, three-site growout in Shanxi Province averaged 123 days to maturity, 70.3 cm tall, 4.6 branches/plant, 18.7 pods/plant, 5.5 seeds/pod, 11.1 g/100 seeds, 1508 kg/ha yield. SEE: Rong and Ma, 1986; Song *et al.*, 1989; Tian, 1990.

80166 (Hebei, China)**:** Advanced breeding line. Two-year, three-site growout in Shanxi Province averaged 127 days to maturity, 71.3 cm tall, 17.5 pods/plant, 5.2 seeds/pod, 13.4 g/100 seeds, 1357 kg/ha yield. SEE: Song *et al.*, 1989; Tian, 1990.

820039 (Hebei, China)**:** Growout in Shandong Province averaged 101 days to maturity, 48.6 cm tall, 5 branches/plant, 18 pods/plant, 8.3 seeds/pod, 9.9 g/100 seeds, 632 kg/ha yield, bright red seed testa. SEE: Rong and Ma, 1986.

Abe: SEE: Nomura, 1968; Satoh, 1974; Nomura and Satoh, 1977; Yoshida, 1987.

Acc 68: Resistant to races 1, 2 and 3 of *Fusarium oxysporum* f.sp. *adzukicola*. SEE: Kondo and Kodama, 1989.

Aka (Fukushima)**:** Late maturity, bushy, few branches, autumn planted, very daylength sensitive, little temperature sensitivity, some resistance to common virus diseases, 4% AzMV infection rate; also grown in Korea. SEE: Matsumoto, 1922; Tasaki, 1963, 1965; Tasaki and Honma, 1965; Chiba, 1982; Japan Bean Fund ..., 1987.

Akamame: Korean cultivar, male parent for Japanese cultivar Hatsune shouzu, source of recessive gene that provides resistance to *Fusarium oxysporum* f.sp. *adzukicola* races 1 and 2. SEE: Kondo and Kodama, 1989; Kondo *et al.*, 1990.

Akane dainagon: Noto shouzu x Wase dairyu 1, released in 1974, medium maturity, semi-determinate, highly branched (3.8/plant), bean size uniformity, good seed quality, medium

Table 7.4. Continued.

height (41-50 cm), light gray short pods, dark red seed testa, large seed size (15.2-19 g/100 seeds), 22% protein, 9.01% seed testa/seed weight, 47 pods/plant, 4.07 seeds/pod, harvest index 61%, 2930 kg/ha yield, susceptible to lodging and *Phialophora gregata* infections; also identified in the literature as either Toiku 69 or No-Rin 1 Go. SEE: Satoh *et al.*, 1975a; Sunada *et al.*, 1975; Murata and Narugawa, 1978; Chiba, 1980; Masaki *et al.*, 1986; Ushirogi *et al.*, 1986; Hokkaido Ag. Affairs Bureau, 1987; Japan Bean Fund ..., 1987; Yoshida, 1987; Yoshida *et al.*, 1988.

Akatsuki dainagon (Hokkaido)**:** Noto shouzu x Wase dairyu 1, released in 1970, medium plant height (45-72.2 cm), highly branched (3.4-4.6/plant), buff-colored pods, dark red seeds, large seed size (19.1-20 g/100 seeds), medium maturity (126 days), 36.9-46 pods/plant, 4.29 seeds/pod, 8.94% seed testa/seed weight, harvest index 43.5-62%, 2520-2960 kg/ha yield, very responsive to high planting densities (550,000 plants/ha), some *Ascochyta phaseolorum* and *Cercospora canescens* resistance but susceptible to viruses such as BCMV, CMV and BYMV. SEE: Nomura, 1968; Narikawa, 1972; Ushirogi *et al.*, 1972; Matsukawa, 1973; Satoh, 1974; Satoh *et al.*, 1975a; Nomura and Satoh, 1977; Konno and Narikawa, 1978; Chiba, 1980; Suzuki and Konno, 1982; Tanaka and Konno, 1982; Hoshikawa, 1985; Yoshida, 1987.

Aomori-zai 34: *Phytophthora vignae* resistance. SEE: Tabiki and Tsuchiya, 1990.

Anci zhusha hong (Hebei, China)**:** Selected from among 254 collected landraces and sold by producer, high yields, red in color, 2733 kg/ha yield over three to four years. SEE: Liu, 1984; Duan, 1989.

Anego (India)**:** Short height (44 cm), 37 pods/plant, 2400 kg/ha yield; one of the two most successful Japanese cultivars introduced into India in 1970. SEE: Thomas *et al.*, 1974.

Anegokei 2: Late maturity. SEE: Nomura and Satoh, 1977.

Annam (Chungbug, Korea)**:** Landrace, black seed, 51.1% *Phytophthora vignae* f.spp. *adzukicola* infection rate. SEE: Han *et al.*, 1982.

Asangcheon (Chungbug, Korea)**:** Landrace, red seed, 68.9% *Phytophthora vignae* f.spp. *adzukicola* infection rate. SEE: Han *et al.*, 1982.

Aso Nos. 1, 5, 6 and 13 (Kumamoto)**:** Late-maturity, autumn-planted cultivars, very daylength sensitive; no. 1 is viney, prostrate and highly branched. SEE: Tasaki, 1963, 1965.

Azuki No. 1 (Shimane)**:** Early maturity, bushy, stiff stem with thick base, few branches. SEE: Tasaki, 1963.

Baek doo (Korea)**:** semi-prostrate, viney, highly branched. SEE: Tasaki, 1963.

Baicheng 153 (China)**:** Pure line selection promoted for production. SEE: Duan, 1989.

Baihong No. 1 (**Baicheng 732**) (Heilongjiang, China)**:** Selected from landrace Ha No. 3, quick seedling growth, matures in 98 to 113 days, bushy, 60-75 cm tall, 13-37 pods/plant, 4-8 seeds/pod, 7.8-14 g/100 seeds, 23.4% seed protein content, 1500-2250 kg/ha yield, pod shatter- and lodging-resistant. SEE: Du and Zhang, 1987; Zhang, 1987; Zhang and Li, 1988.

Bakgak (Chungbug, Korea)**:** Landrace, red seed, 84.8% *Phytophthora vignae* f.spp. *adzukicola* infection rate. SEE: Han *et al.*, 1982.

Ban: Some resistance to common virus diseases. SEE: Mukoo, 1975.

Bangfu (Anhui, China): early maturity, shatter-resistant, 5.84 g/100 seeds, Chinese medicine. SEE: Zhou *et al.*, 1964.

Baoqing (Heilongjiang, China): Top quality trade name for Heilongjiang azuki. SEE: Editorial Com. ..., 1993.

Baoxi 77824 (China)**:** Pure line selection, promoted for production. SEE: Duan, 1989.

Beni: Highly resistant to common virus diseases, N.E. China cultivar used in Japan. SEE: Mukoo, 1975.

Beni dainagon: Hayate x Kiyohara azuki, dark red to medium red testa color, low testa/seed weight ratio, 16.8 g/100 seeds, used for superior ama natto processing, 1870-2560 kg/ha yield, very resistant to virus infections, susceptible to low temperatures and *Phialophora gregata* infections, Japanese M.A.F.F. registered in 1985. SEE: Masaki *et al.*, 1986; Japan Bean Fund ..., 1987; Japan ..., 1988.

Beni M 14: Early maturity. SEE: Nomura and Satoh, 1977.

Beni-nambu (Iwate)**:** Developed by irradiation of Monbetsu 26-go, lodging-resistant, high yields, good bean quality, highly susceptible to virus infections, 1978 release by Prefectural Experiment Station. SEE: Satoh, 1979; Suzuki and Konno, 1982; Masaki *et al.*, 1986; Japan Bean Fund ..., 1987; Japan ..., 1988.

Benizu: No symptoms of AzMV infection. SEE: Matsumoto, 1922.

Bin Fu (Heilongjiang, China): Produced by radiation mutation. SEE: Editorial Com. ..., 1993

Table 7.4. Continued.

Bin Xiaodou (Heilongjiang, China): selected by Bin County Ag. Inst., 95 days to maturity, 54-58 cm height, white seed, 25-30 pods/plant, 10.8 g/100 seeds and shatter resistant. SEE: Editorial Com. ..., 1993

Bloodwood: Australian registered cultivar. SEE: Strang, 1980.

Bon: Susceptible to common virus diseases, e.g. 24% AzMV infection rate. SEE: Matsumoto, 1922; Mukoo, 1975.

Bosodu (Korea)**:** SEE: Ahn *et al.*, 1986.

Buchi ryuu kei 1 (Hokkaido)**:** Cultivar maintained at the Tokachi Agricultural Experiment Station, very low seed weight (0.069 g/seed). SEE: Chiba, 1980.

Buchi shouryuu kei 1 (also called **Buchishoryu 1**)**:** Early maturity, medium height, cold weather-tolerant, susceptible to all three *Fusarium oxysporum* f.sp. *adzukicola* races, very small seed, poor quality. SEE: Narikawa, 1972; Matsukawa, 1973; Konno and Narikawa, 1978; Chiba, 1980; Suzuki and Konno, 1982; Yoshida, 1987; Kondo *et al.*, 1990.

Buchi shouryuu kei 2: Early maturity, *Phialophora gregata* susceptible. SEE: Chiba, 1982.

Chadol (Korea)**:** Symptoms of seed-transmitted AzMV are upper leaf vein clearing. SEE: Choi and Lee, 1989.

Chagarawase (Hokkaido)**:** Native cultivar selection or a pure line Wase Maruba selection (originally from the Kamikawa District) made at the Hokkaido Agricultural Experiment Station, released in 1914-15, very early maturity, bushy, few branches, little daylength sensitivity, very temperature sensitive, low yields, short height, light brown pods, small seeds, specific for cold northern regions, susceptible to races 1, 2 and 3 of *Fusarium oxysporum* f.sp. *adzukicola*. SEE: Tasaki, 1963, 1965; Tasaki and Honma, 1965; Hokkaido Agricultural Experiment Station, 1967; Nomura, 1968; Narikawa, 1972; Satoh, 1974; Nomura and Satoh, 1977; Chiba, 1980; Tsuchiya *et al.*, 1986; Japan Bean Fund ..., 1987; Yoshida, 1987; Kondo and Kodama, 1989.

Chazaya (Yamaguchi)**:** Intermediate maturity, bushy, moderate branching, some daylength sensitivity. SEE: Tasaki, 1963, 1965.

Cheongwon cheon (Chungbug, Korea)**:** Landrace, black seed, 40.4% *Phytophthora vignae* f.spp. *adzukicola* infection rate. SEE: Han *et al.*, 1982.

Cheongwon jeogdu (Korea)**:** SEE: Han and Choe, 1977.

Chichi no mine (Tohoku region, Japan)**:** Late maturity, very daylength sensitive. SEE: Kakuta and Satoh, 1953b.

Chien Shien (Taiwan)**:** Japanese (AVRDC accession number 5124 with Taiwanese name), used in interspecific crosses. SEE: Chen, 1980; Chen *et al.*, 1983.

Chihokushu (Akita)**:** Cultivar maintained at the Tokachi Agricultural Experiment Station on Hokkaido, high total dry matter. SEE: Chiba, 1980.

Chin en: N.E. China cultivar used in Japan, highly resistant to common virus diseases, e.g. 0% AzMV infection rate. SEE: Matsumoto, 1922; Mukoo, 1975.

Chu iku 1 (Hokkaido)**:** SEE: Ito *et al.*, 1975.

Chu sha (China)**:** Susceptible to a mosaic virus disease of cowpea. SEE: Yu, 1946.

Chunagon: Some resistance to common virus diseases. SEE: Mukoo, 1975.

Chunagon (Chiba)**:** Bushy, moderate branching. SEE: Tasaki, 1963.

Chunagon (Hokkaido)**:** Resistant to *Phytophthora vignae* f.sp. *adzukicola*. SEE: Tsuchiya, 1988.

Chunagon (Ibaraki)**:** Intermediate maturity, bushy, moderate branching, little daylength sensitivity, some temperature sensitivity. SEE: Tasaki, 1963, 1965; Tasaki and Honma, 1965.

Chunagon (Kagawa)**:** Late maturity, autumn planted, viney, prostrate, highly branched and daylength sensitive. SEE: Tasaki, 1963, 1965.

Chunagon (Kumamoto)**:** Grown in the 1970s, promoted in Nagano Prefecture SEE: Tanaka and Konno, 1982; Japan Bean Fund ..., 1987.

Chunagon (Nagano)**:** Late maturity, autumn planted, bushy, few branches, very daylength sensitive, little temperature sensitivity. SEE: Tasaki, 1963, 1965; Tasaki and Honma, 1965.

Chunagon (Oita)**:** Late-maturity, autumn-planted cultivar, very daylength sensitive. SEE: Tasaki, 1965.

Chunagon (Tokushima)**:** Late maturity, autumn planted, viney, prostrate, highly branched, very daylength sensitive. SEE: Tasaki, 1963, 1965.

Chunagon (Tottori)**:** 450 kg/ha yield (in Tottori) on acidic soils. SEE: Noda, 1952.

Chungjapat (Chungju Chungbuk, Korea)**:** 1984 release, *Uromyces azukicola* and *Sphaerotheda fuliginea* resistance. SEE: Hwang, 1982; Han *et al.*, 1984; Hong *et al.*, 1984, 1986.

Chungju (Korea)**:** SEE: Ahn *et al.*, 1986.

Table 7.4. Continued.

Chungwonpat (Korea): Released in 1983 from pure line selection, dark gray seeds, some *Uromyces azukicola*, *Sphaerotheca fuliginea* and AzMV resistance. SEE: Hwang, 1982; Han *et al.*, 1984; Hong *et al.*, 1984, 1986.
Daenabeon (Korea): SEE: Ahn *et al.*, 1986.
Daenam sum (Korea): Symptoms of seed-transmitted AzMV are mosaic patterns on upper leaves. SEE: Choi and Lee, 1989.
Daengang (Chungbug, Korea): Gold-colored landrace, 95% *Phytophthora vignae* f.spp. *adzukicola* infection rate. SEE: Han *et al.*, 1982.
Dahongpao (China): 90 cm tall, 7-9 seeds/pod, 13-15 g/100 seeds. SEE: Zhou, 1982.
Daikoku (Shimane): Early maturity, bushy, stiff stem with thick base, few branches. SEE: Tasaki, 1963.
Daikoku (Yamagata): Intermediate maturity, bushy, few branches, daylength sensitivity, little temperature sensitivity. SEE: Tasaki, 1963, 1965; Tasaki and Honma, 1965.
Dainagon: Name refers to large-seeded cultivars, some resistance to common virus diseases. SEE: Mukoo, 1975.
Dainagon (Aichi): Large seeded, intermediate maturity, bushy, moderate branching, daylength sensitivity, some temperature sensitivity. SEE: Tasaki, 1963, 1965; Tasaki and Honma, 1965.
Dainagon (Aomori): Large seeded, bushy, stiff stem with thick base, few branches, late maturity, autumn planted, daylength sensitive. SEE: Tasaki, 1963, 1965; Japan ..., 1988.
Dainagon (Fukushima): Large seeded, intermediate maturity, bushy, few branches, some daylength sensitivity. SEE: Tasaki, 1963, 1965.
Dainagon (Gunma): Large seeded, intermediate maturity, daylength sensitivity. SEE: Tasaki, 1965.
Dainagon (Hyogo): Large seeded, late maturity, autumn planted, viney, prostrate, highly branched, very daylength sensitive. SEE: Tasaki, 1963, 1965.
Dainagon (Ibaraki): Large seeded, intermediate maturity, bushy, moderate branching, daylength sensitivity, some temperature sensitivity. SEE: Tasaki, 1963, 1965; Tasaki and Honma, 1965.
Dainagon (Iwate): Large seeded, intermediate maturity, bushy, few branches, daylength sensitivity. SEE: Tasaki, 1963, 1965.
Dainagon (Kagawa): Large seeded, viney, prostrate, late maturity, highly branched. SEE: Tasaki, 1963.
Dainagon (Kochi): Large seeded, late maturity, autumn planted, viney, prostrate, highly branched, very daylength sensitive. SEE: Tasaki, 1963, 1965.
Dainagon (Miyagi): Large seeded, intermediate maturity, bushy, few branches, daylength sensitive. SEE: Tasaki, 1963, 1965.
Dainagon (Nagasaki): Large seeded, late maturity, autumn planted, viney, prostrate, highly branched, very daylength sensitive. SEE: Tasaki, 1963, 1965.
Dainagon (Oita): Large seeded, late maturity, autumn planted, viney, prostrate, highly branched, very daylength sensitive. SEE: Tasaki, 1963, 1965.
Dainagon (Shimane): Large seeded, late maturity, autumn planted, bushy, few branches, daylength sensitive, little temperature sensitivity. SEE: Tasaki, 1963, 1965; Tasaki and Honma, 1965.
Dainagon (Tohoku region, Japan): Numerous regional cultivars, especially in Aomori and Fukushima Prefectures, large seeded, intermediate to very late maturity, some daylength sensitivity, one has an 8% AzMV infection rate. SEE: Kakuta and Satoh, 1953b; Hoshikawa, 1985.
Dainagon (Tokushima): Large seeded, late maturity, autumn planted, viney, prostrate, highly branched, very daylength sensitive. SEE: Tasaki, 1963, 1965.
Dainagon (Tottori): Large seeded, very sensitive to acidic soils with low phosphorus availability. SEE: Noda, 1952.
Dainagon (Toyama): Large seeded, late maturity, autumn planted, bushy, few branches, very daylength sensitive. SEE: Tasaki, 1963, 1965.
Dainagon No. 1 (Kumamoto): Large seeded, late maturity, autumn planted, very daylength sensitive. SEE: Tasaki, 1965.
Dainagon No. 5 (Akita): Large seeded, intermediate maturity, bushy, few branches, daylength sensitivity. SEE: Tasaki, 1963, 1965.
Dezhou (Shandong, China): Results from a one-year growout: 101 days to maturity, 48.4 cm tall, 4.4 branches/plant, 10.2 pods/plant, 6.1 seeds/pod, 10.6 g/100 seeds, 500 kg/ha yield, bright red seed testa. SEE: Rong and Ma, 1986.

Table 7.4. Continued.

Dongyang (Chungbug, Korea): Red-seeded landrace, 89.9% *Phytophthora vignae* f.spp. *adzukicola* infection rate. SEE: Han *et al.*, 1982.
Donsu (Kanto region, Japan): Black-spotted red seeds, blackish-brown pods, reddish-purple stem. SEE: Kakizaki, 1923.
Doyo No. 1 (Toyama): Late maturity, autumn planted, bushy, few branches, very daylength sensitive. SEE: Tasaki, 1963, 1965.
Duksan (Chungbug, Korea): Red-seeded landrace, 20.7% *Phytophthora vignae* f.spp. *adzukicola* infection rate. SEE: Han *et al.*, 1982.
E0977 (Hebei, China): Breeding line. Two-year, three-site growout in Shanxi Province averaged 128 days to maturity, 47.2 cm tall, 4 branches/plant, 18 pods/plant, 4.9 seeds/pod, 12.8 g/100 seeds, 1266 kg/ha yield. SEE: Song *et al.*, 1989.
E1037-1 (Hebei, China): Breeding line. Two-year, three-site growout in Shanxi Province averaged 122 days to maturity, 59.1 cm tall, 3 branches/plant, 20.5 pods/plant, 5.2 seeds/pod, 13.1 g/100 seeds, 2045 kg/ha yield. SEE: Song *et al.*, 1989.
E1038 (Hebei, China): Breeding line. Two-year, three-site growout in Shanxi Province averaged 128 days to maturity, 61 cm tall, 3.1 branches/plant, 21.3 pods/plant, 4.8 seeds/pod, 11.9 g/100 seeds, 2005 kg/ha yield. SEE: Song *et al.*, 1989.
EC 15217, 87896, 87897, 87900: Cultivars express mild horsegram yellow mosaic virus (HYMV) symptoms when inoculated. SEE: Muniyappa *et al.*, 1977.
Erimo: Kotobuki azuki x Juiku 77 go, medium maturity, medium height (47-52 cm), sparsely branched, brown pods (34 pods/plant), small bright red seeds (11.7-13g/100 seeds), 22.4% protein, lodging resistant, cold tolerant, susceptible to *Phialophora gregata*, all 3 *Fusarium oxysporum* f.sp. *adzukicola* races, *Pseudomonas adzukicola* and virus infections, M.A.F.F. registration 1981. SEE: Satoh, 1982; Masaki *et al.*, 1986; Ushirogi *et al.*, 1986; Hokkaido Ag. Affairs Bureau, 1987; Japan Bean Fund ..., 1987; Yoshida *et al.*, 1988; Kondo and Kodama, 1989; Kondo *et al.*, 1990.
Eweol (Chungbug, Korea): Red-seeded landrace, 53% *Phytophthora vignae* f.spp. *adzukicola* infection rate. SEE: Han *et al.*, 1982.
Fang shan (China): Cultivar developed and sold by producer. SEE: Duan, 1989.
Fujimine (Ehime): Late maturity, autumn planted, viney, prostrate, highly branched, very daylength sensitive, little temperature sensitivity. SEE: Tasaki, 1963, 1965; Tasaki and Honma, 1965.
Fuji wase (Mie): Intermediate maturity, bushy, moderate branching, little daylength sensitivity. SEE: Tasaki, 1963, 1965.
Gaegoori patt (Korea): Semi-prostrate, viney, highly branched. SEE: Tasaki, 1963.
Gagoek (Chungbug, Korea): Black-seeded landrace, 27.2% *Phytophthora vignae* f.spp. *adzukicola* infection rate. SEE: Han *et al.*, 1982.
Galgaeguri (Korea): Cultivar. SEE: Han and Choe, 1977.
Gammul (Chungbug, Korea): Red-seeded landrace, 21% *Phytophthora vignae* f.spp. *adzukicola* infection rate. SEE: Han *et al.*, 1982.
Gaoxuan 1 (Taiwan): Cultivar, highly susceptible to powdery mildew infections. SEE: Wu *et al.*, 1985a.
Gengmi (Liaoning, China): Cultivar, prostrate, light green seed testa. SEE: Guo, 1980.
Gon no sho: Cultivar, AzMV susceptible. SEE: Murayama, 1941.
Gongju baegdu (Korea): Cultivar. SEE: Hwang, 1982; Han *et al.*, 1984; Hong *et al.*, 1984, 1986.
Gong xiao 12 (China): Cultivar developed and sold by producer. SEE: Duan, 1989.
Gong xiao 46 (China): Cultivar developed and sold by producer. SEE: Duan, 1989.
Gunseo (Chungbug, Korea): Red-seeded landrace, 35.4% *Phytophthora vignae* f.spp. *adzukicola* infection rate. SEE: Han *et al.*, 1982.
Gwanju jeogdu (Korea): Cultivar. SEE: Han and Choe, 1977.
Gyenggi daenabeon (Korea): Cultivar. SEE: Han and Choe, 1977.
Gyeonggi gaeguri (Korea): Cultivar. SEE: Han and Choe, 1977; Hwang, 1982; Han *et al.*, 1984; Hong *et al.*, 1984, 1986.
Hanaadzuki (Kari 154): *Phytophthora vignae* resistance. SEE: Tabiki and Tsuchiya, 1990.
Hatsune: Hayate shouzu x Akamame, resistant to *Pseudomonas adzukicola* and *Phialophora gregata*, resistant to races 1 and 2 of *Fusarium oxysporum* f.sp. *adzukicola*, but not race 3, some resistance to virus infections and lodging, medium maturity, medium plant height (48 cm), sparsely branched, brown pods, 33 pods/plant, dark red seeds, 11-12 g/100 seeds, high testa/seed weight ratio, 23.4% protein, 1720 kg/ha yield, M.A.F.F. registered. SEE:

Table 7.4. Continued.

Masaki *et al.*, 1986; Ushirogi *et al.*, 1986; Hokkaido Ag. Affairs Bureau, 1987; Japan Bean Fund ..., 1987; Yoshida *et al.*, 1988; Kondo and Kodama, 1989; Kondo *et al.*, 1990.

Hatta: Intermediate maturity. SEE: Nomura, 1968; Satoh, 1974; Nomura and Satoh, 1977; Yoshida, 1987.

Hayate: Takara shouzu x Hanshoryuke 1, released in 1976, early maturity, medium height (38-49 cm), sparsely branched, brown pods with curved tips (38-48 pods/plant), small dark red seeds (11-11.5 g/100 seeds, 0.066 mm testa thickness, 9.14% testa/seed weight), 22.8% protein, cold tolerant, some resistance to virus infections, good lodging resistance, susceptible to *Pseudomonas adzukicola* and all 3 *Fusarium oxysporum* f.sp. *adzukicola* races, highly susceptible to *Phytophthora vignae* (Purss.) f. spp. *adzukicola*, M.A.F.F. registered. SEE: Murata, 1976; Murata *et al.*, 1977; Tsuchiya and Kodama, 1979; Chiba, 1982; Tanaka and Konno, 1982; Masaki *et al.*, 1986; Tsuchiya *et al.*, 1986; Ushirogi *et al.*, 1986; Hokkaido Ag. Affairs Bureau, 1987; Japan Bean Fund ..., 1987; Yoshida, 1987; Yoshida *et al.*, 1988; Kondo and Kodama, 1989; Kondo *et al.*, 1990.

Hebei No. 1 (Hebei, China)**:** Cultivar. SEE: Tian, 1990.

Heug sodu (Korea)**:** Cultivar. SEE: Han and Choe, 1977.

Heukdu (Korea)**:** Cultivar. SEE: Ahn *et al.*, 1986.

Higashinimen: 12% AzMV infection rate. SEE: Matsumoto, 1922.

Hikari (or **Toiku 7**)(Hokkaido)**:** Chagarawase x Wase daiyu 1 at Tokachi Agriculture Experiment Station, released in 1964-65, resistant to race 1 of *Fusarium oxysporum* f.sp. *adzukicola*, but not 2 and 3, early to intermediate maturity, medium height (52 cm), can become viney, light buff-colored pods, dark red seeds, small seed size, 29.1-32.7 pods/plant, 1990-2490 kg/ha yield, successfully introduced into India in 1970, grown predominately in Hokkaido, but also the Chuo region. SEE: Hokkaido Agricultural Experiment Station, 1967; Nomura, 1968; Narikawa, 1972; Satoh, 1974; Furusawa *et al.*, 1974; Thomas *et al.*, 1974; Ito *et al.*, 1975; Nomura and Satoh, 1977; Chiba, 1980; Hoshikawa, 1985; Yoshida, 1987; Kondo and Kodama, 1989; Kondo *et al.*, 1990.

Hoahm (Black) (Chungbug, Korea)**:** Black-seeded landrace, 26.9% *Phytophthora vignae* f.spp. *adzukicola* infection rate. SEE: Han *et al.*, 1982.

Hoahm (Red) (Chungbug, Korea)**:** Red-seeded landrace, 33.6% *Phytophthora vignae* f.spp. *adzukicola* infection rate. SEE: Han *et al.*, 1982.

Hokkai shiro shouzu: Shiro azuki (Kawanishi) x Chagarawase, medium maturity, medium height, small white seeds, M.A.F.F. registration in 1980. SEE: Murata, 1980; Japan Bean Fund ..., 1987; Yoshida, 1987.

Hokko No. 1: Intermediate maturity. SEE: Nomura and Satoh, 1977.

Hokui: Male parent for Daikan 3-go. SEE: Hoshikawa, 1985.

Hokuiku No. 3-3: Intermediate maturity. SEE: Matsukawa, 1973; Nomura and Satoh, 1977.

Hongcheon (Korea)**:** Cultivar, 94-105 day maturity time, 70.6 cm tall, 21 pods/plant, 4.3 seeds/pod, 12.5-14.3 g/100 seeds, 40.5% harvest index, grown throughout the country. SEE: Han and Choe, 1977; Han *et al.*, 1979, 1984; Hwang, 1982; Hong *et al.*, 1984, 1986.

Hongcheon jeoktu (Korea)**:** Cultivar, widely used in the late 1970s and early 1980s, may also be spelled Hongcheon jeukdu. SEE: Office ..., 1981; Hwang, 1982; Han *et al.*, 1984; Hong *et al.*, 1984, 1986; Ahn *et al.*, 1986.

Hong xiao dou No. 153 (Jilin, China)**:** Widely used cultivar in Baicheng region, developed in 1960s. SEE: Zhang, 1987.

Hua (N.E. China)**:** Early maturity, little daylength sensitivity, semi-prostrate, viney, highly branched. SEE: Tasaki, 1963, 1965.

Hung (N.E. China)**:** Early maturity, little daylength sensitivity, semi-prostrate, viney, highly branched, very temperature sensitive. SEE: Tasaki, 1963, 1965; Tasaki and Honma, 1965.

Hung Nos. 1 and 2 (N.E. China)**:** Cultivars, semi-prostrate, viney, highly branched. SEE: Tasaki, 1963.

Iryoo (Chungbug, Korea)**:** Landrace, red-seeded, 66.3% *Phytophthora vignae* f.spp. *adzukicola* infection rate. SEE: Han *et al.*, 1982.

Ishikawa (Hokkaido)**:** Grown in Kitami District, susceptible to bean yellow mosaic virus (BYMV) SEE: Ito *et al.*, 1975.

Iwate dainagon (Iwate)**:** Pure line selection from indigenous cultivars, released in 1965, late maturity, tall (53-89 cm), medium susceptibility to lodging, large scarlet seeds, 20.1 pods/plant, 1420-2470 kg/ha yield, highly susceptible to virus infections. SEE: Furusawa *et al.*, 1974; Narikawa, 1978; Tanaka and Konno, 1982; Masaki *et al.*, 1986; Japan Bean Fund ..., 1987; Yoshida, 1987; Japan ..., 1988.

Jangsu Jaerae (Korea)**:** SEE: Choi and Lee, 1989.

Table 7.4. Continued.

Jansen: SEE: Green *et al.*, 1986.
Jeogjung (Korea)**:** SEE: Han and Choe, 1977.
Jeokdu (Korea)**:** 94 days to maturity, 66 cm tall, 8.4 pods/plant, 5.3 seeds/pods, 14.8 g/100 seeds, 29% harvest index. SEE: Han and Choe, 1977; Han *et al.*, 1979, 1984; Hwang, 1982; Hong *et al.*, 1984, 1986; Ahn *et al.*, 1986.
Jeongpyeng (Chungbug, Korea)**:** Landrace, red-seeded, 50.6% *Phytophthora vignae* f.spp. *adzukicola* infection rate. SEE: Han *et al.*, 1982.
Jeougdu: Same as **Jeokdu.**
Ji County (China)**:** Developed and sold by producer. SEE: Duan, 1989.
Jihong No. 1 (Hebei, China)**:** Hybrid, 64-89 cm tall, matures in 86-90 days, 22-32 pods/plant, 2-3 seeds/pod, 15.4-16.3 g/100 seeds, higher overall seed protein and lysine content than other landraces in the province, 1500-2250 kg/ha yield. Two-year, three-site growout in Shanxi Province averaged 121 days to maturity, 49.8 cm tall, 4.1 branches/plant, 21.3 pods/plant, 5.4 seeds/pod, 13.6 g/100 seeds, 2122 kg/ha yield. SEE: Wang, 1986; Duan, 1989; Song *et al.*, 1989.
Jinan (China)**:** Developed and sold by producer. SEE: Duan, 1989.
Jincheon jeogdu (Korea)**:** SEE: Han and Choe, 1977.
Jingnong No. 1 (Beijing)**:** Early maturing, erect, drought resistant, high yield and good quality. SEE: Jin *et al.*, 1993.
Jingnong No. 2 (Beijing)**:** Early maturing, erect, drought resistant, high yield and good quality. SEE: Jin *et al.*, 1993.
Jingnongji No. 126 (China)**:** Cultivar collected by Beijing Agricultural College. SEE: Guo and Jin, 1988.
Jinheuhgdu (Korea)**:** SEE: Han and Choe, 1977.
Jinxain (Taiwan)**:** Paternal line for Kaohsiung 2. SEE: Liu *et al.*, 1981.
Jock doo (Korea)**:** Semi-prostrate, viney, highly branched. SEE: Tasaki, 1963.
Jokei dainagon (Tochigi)**:** Early maturity, bushy, stiff stem with thick base, few branches: SEE: Tasaki, 1963.
Juduk (Chungbug, Korea)**:** Landrace, black-seeded, 89.2% *Phytophthora vignae* f.spp. *adzukicola* infection rate. SEE: Han *et al.*, 1982.
Jukdoo (Korea)**:** Landrace. SEE: Chang *et al.*, 1968b; Chang *et al.*, 1969.
Jukoo (Korea)**:** SEE: Chang *et al.*, 1968b.
Jungweon-heukdu (Chungpuk, Korea)**:** Some resistance to disease. SEE: Office ..., 1981.
Jungwon gru (Korea)**:** Develops mosaic patterns on upper leaves from seed-borne AzMV. SEE: Ahn *et al.*, 1986; Choi and Lee, 1989.
Jyokei dainagon (Niigata)**:** Late maturity, autumn planted, very daylength sensitive. SEE: Tasaki, 1965.
K.S.S.1 (Taiwan)**:** Autumn planted, daylength sensitivity. SEE: Tsai, 1982.
KA 67-01-06 (Kaohsiung, Taiwan)**:** Breeding line, female parent for KA 77-05-122, autumn planting over two years in two regions averaged 92 days to maturity, 38.2 cm height, 15.4 pods/plant, 14.9 g/100 seeds, 2225 kg/ha yield. SEE: Hsu *et al.*, 1989ab.
KA 67-01-08 (Kaohsiung, Taiwan)**:** Breeding line, autumn planting over two years in two regions averaged 92 days to maturity, 42.3 cm height, 14.3 pods/plant, 15.6 g/100 seeds, 2008 kg/ha yield. SEE: Hsu *et al.*, 1989b.
KA 67-01-15 (Kaohsiung, Taiwan)**:** Breeding line, autumn planting over two years in two regions averaged 92 days to maturity, 43.1 cm height, 17.1 pods/plant, 16.2 g/100 seeds, 2473 kg/ha yield. SEE: Hsu *et al.*, 1989b.
KA 67-01-77 (Kaohsiung, Taiwan)**:** Breeding line, autumn planting over two years in two regions averaged 93 days to maturity, 38.7 cm height, 14.7 pods/plant, 14.1 g/100 seeds, 2027 kg/ha yield. SEE: Hsu *et al.*, 1989b.
KA 67-02-73 (Kaohsiung, Taiwan)**:** Breeding line, autumn planting over two years in two regions averaged 91 days to maturity, 41 cm height, 14.6 pods/plant, 15.7 g/100 seeds, 1886 kg/ha yield. SEE: Hsu *et al.*, 1989b.
KA 69-01-111 (Kaohsiung, Taiwan)**:** Breeding line, one year on three sites averaged 99 days to maturity, 31.8 cm height, 16.8 pods/plant, 6.4 seeds/pod, 12.6 g/100 seeds, 2144 kg/ha yield. SEE: Lee, 1989b.
KA 69-04-67 (Kaohsiung, Taiwan)**:** Breeding line, one year on three sites averaged 100 days to maturity, 35.6 cm height, 15.3 pods/plant, 6.8 seeds/pod, 13.2 g/100 seeds, 2341 kg/ha yield. SEE: Lee, 1989b.

Table 7.4. Continued.

KA 69-05-234 (Kaohsiung, Taiwan): Breeding line, one year on three sites averaged 97 days to maturity, 34.4 cm height, 14.7 pods/plant, 6 seeds/pod, 14.9 g/100 seeds, 2162 kg/ha yield. SEE: Lee, 1989b.
KA 69-05-36 (Kaohsiung, Taiwan): Breeding line, male parent for KA 77 06-123, autumn planting over two years in two regions averaged 91 days to maturity, 41.8 cm height, 17.2 pods/plant, 14.1 g/100 seeds, 2242 kg/ha yield. SEE: Hsu *et al.*, 1989ab.
KA 69-05-39 (Kaohsiung, Taiwan): Breeding line, autumn planting over two years in two regions averaged 92 days to maturity, 38.1 cm height, 12.2 pods/plant, 17 g/100 seeds, 1518 kg/ha yield. SEE: Hsu *et al.*, 1989b.
KA 70-05-98 (Kaohsiung, Taiwan): Breeding line, one year on three sites averaged 102 days to maturity, 39.2 cm height, 16.4 pods/plant, 6.1 seeds/pod, 13.1 g/100 seeds, 2251 kg/ha yield. SEE: Lee, 1989b.
KA 73-01-92 (Kaohsiung, Taiwan): Breeding line, Kaohsiung No. 2 x 68-I-L. SEE: Wu *et al.*, 1985b.
KA 73-02-93 (Kaohsiung, Taiwan): Breeding line, KS542 x 68-I-J. SEE: Wu *et al.*, 1985b.
KA 73-03-94 (Kaohsiung, Taiwan): Breeding line, KS540 x 68-I-A. SEE: Wu *et al.*, 1985b.
KA 73-04-95 (Kaohsiung, Taiwan): Breeding line, KS526 x Guang Xiao Dou. SEE: Wu *et al.*, 1985b.
KA 73-05-96 (Kaohsiung, Taiwan): Breeding line, Wu Feng Cai Lai x Da Guan. SEE: Wu *et al.*, 1985b.
KA 73-06-97 (Kaohsiung, Taiwan): Breeding line, 68-I-J x KS303. SEE: Wu *et al.*, 1985b.
KA 74-01-98 (Kaohsiung, Taiwan): Breeding line, KS658 x 69-I-L. SEE: Wu *et al.*, 1986.
KA 74-02-99 (Kaohsiung, Taiwan): Breeding line, 68-I-J x Kaohsiung No. 3. SEE: Wu *et al.*, 1986.
KA 74-03-100 (Kaohsiung, Taiwan): Breeding line, KS540 x 68-I-J. SEE: Wu *et al.*, 1986.
KA 74-04-101 (Kaohsiung, Taiwan): Breeding line, 68-I-A x Pingtung native. SEE: Wu *et al.*, 1986.
KA 74-05-102 (Kaohsiung, Taiwan): Breeding line, KS542 x Pingtung native. SEE: Wu *et al.*, 1986.
KA 74-06-103 (Kaohsiung, Taiwan): Breeding line, KS627 x Kaohsiung No. 2. SEE: Wu *et al.*, 1986.
KA 74-07-104 (Kaohsiung, Taiwan): Breeding line, 68-I-J x Kaohsiung No. 2. SEE: Wu *et al.*, 1986.
KA 77-01-118 (Kaohsiung, Taiwan): Breeding line, Kyoto dainagon x Kaohsiung No. 5. SEE: Hsu *et al.*, 1989a.
KA 77-02-119 (Kaohsiung, Taiwan): Breeding line, 68-I-J x Mikamo dainagon. SEE: Hsu *et al.*, 1989a.
KA 77-03-120 (Kaohsiung, Taiwan): Breeding line, Mikamo dainagon x KA 540. SEE: Hsu *et al.*, 1989a.
KA 77-04-121 (Kaohsiung, Taiwan): Breeding line, Kyoto dainagon x Kaohsiung No. 2. SEE: Hsu *et al.*, 1989a.
KA 77-05-122 (Kaohsiung, Taiwan): Breeding line, KA-01-06 x Kaohsiung No. 5. SEE: Hsu *et al.*, 1989a.
KA 77-06-123 (Kaohsiung, Taiwan): Breeding line, Erimo shouzu x KA 69-05-36. SEE: Hsu *et al.*, 1989a.
KA 77-07-124 (Kaohsiung, Taiwan): Breeding line, Kaohsiung No. 3 x Kaohsiung No. 5. SEE: Hsu *et al.*, 1989a.
Kage shouzu (Tottori): General term used to describe cultivars grown in the shade. SEE: Noda, 1952.
Kamisenda (Kochi): Late maturity, autumn planted, viney, prostrate, highly branched, very daylength sensitive. SEE: Tasaki, 1963, 1965.
Kamui dainagon (Hokkaido): Toiku 106 x Tokei 207, largest seeds of any Hokkaido cultivar, virus resistant. SEE: Shirai *et al.*, 1990.
Kanayama (Fukui): Late maturity, autumn planted, very daylength sensitive, little temperature sensitivity. SEE: Tasaki, 1965.
Kangwon jaerae No. 75 (Korea): Cultivar. SEE: Ahn *et al.*, 1986.
Kanto Nos. 1,2,3,4 (Ibaraki): Developed at the Ishioka Experiment Station between 1935 and 1950 for production in the Kanto Plain region. Intermediate maturity, bushy, moderate branching, little daylength sensitivity. SEE: Tasaki, 1963, 1965; Narikawa, 1978.
Kaochiao tsaosheng (Taiwan): Male parent in development of Kaohsiung No. 3. SEE: Wu *et al.*, 1985ab.

Table 7.4. Continued.

Kaohsiung No. 1 (Taiwan)**:** Popular cultivar developed by pure-line selection methods in 1974 from a cultivar introduced from Japan in 1969, grown in the Kaohsiung and Pingtung regions, one-year, three-site autumn growout averaged 30-35 days to flowering, 96 days to maturity, 29.5 cm height, 15.7 pods/plant, 6.5 seeds/pod, 10.7 g/100 seeds, bright red testa, 1738 kg/ha yield, highly resistant to *Rhizoctonia* infection and lodging, but very susceptible to powdery mildew infection, bushy, some pod-set high on the stem SEE: Ag. Develop. Committee, 1981; Liu *et al.*, 1981; Tsai, 1988; Lee, 1989b.

Kaohsiung No. 2 (Taiwan)**:** Cultivar, Pingtung local x Jinxian, released in 1980, former breeding line KS 280, one-year, three-site autumn growout averaged 96 days to maturity, 28.7 cm height, 16.9 pods/plant, 6.9 seeds/pod, 11.2 g/100 seeds, 2197 kg/ha yield, another five-year autumn growout averaged 2793 kg/ha yield, some resistance to powdery mildew infection. SEE: Liu *et al.*, 1981; Green *et al.*, 1986; Tsai, 1988; Lee, 1989b.

Kaohsiung No. 3 (Kaoping District, Taiwan)**:** Cultivar, Kaochiaotsaosheng x Dainagon, released in 1983, former breeding line KS 352, autumn growout in regional yield trials averaged 92 days to maturity, 37-38 cm height, 15.5 pods/plant, 14.2-15.9 g/100 seeds, red testa, zero hard seeds, 1948-2619 kg/ha yield, medium susceptibility to powdery mildew infection from *Erysiphe* spp., fast early season growth rate, high LAI, lodging resistant, indehiscent pods, first reproductive node 16-20 cm above ground suitable for direct combine harvest, a superior spring-planted cultivar, can also be autumn planted. SEE: Wu *et al.*, 1985ab; Green *et al.*, 1986; Tsai, 1988; Hsu, 1989ab; Hsu *et al.*, 1989ab; Lee, 1989b.

Kaohsiung No. 5 (Taiwan)**:** Cultivar, Pingtung native x Kaohsiung No. 1, released in 1989, early maturity, bushy, 86-95 days to maturity, 35-43 cm height, 10-13 pods/plant, 6-8 seeds/pod, 13.7-14.4 g/100 seeds, bright red testa, zero hard seeds, 2500-3000 kg/ha yield, first reproductive nodes 16-18 cm above ground, lodging resistant, sheds most leaves at maturity which is helpful for direct combine harvest, suited to loam type soils, moderately resistant to powdery mildew infection, male parent for breeding lines KA 77-01-118, KA 77-05-122 and KA 77-07-124. SEE: Hsu, 1989ab; Hsu *et al.*, 1989a.

Kaohsiung S-1 (Pingtung, Taiwan)**:** May be same as Kaohsiung No. 1. SEE: Chao, 1982; Green *et al.*, 1986.

Kari No. 47: Susceptible to races 1, 2 and 3 of *Fusarium oxysporum* f.sp. *adzukicola*. SEE: Kondo and Kodama, 1989.

Kari No. 63: Resistant to races 1, 2 and 3 of *Fusarium oxysporum* f.sp. *adzukicola*. SEE: Kondo and Kodama, 1989.

Kari No. 71: Used in *Fusarium oxysporum* f.sp. *adzukicola* race research. SEE: Kondo and Kodama, 1989.

Ken No. 3: Susceptible to races 1, 2 and 3 of *Fusarium oxysporum* f.sp. *adzukicola*. SEE: Kondo and Kodama, 1989.

Kenzaki: Early maturity, bushy, stiff stem with thick base, few branches. SEE: Tasaki, 1963.

Kenzaki (Hokkaido)**:** Pure line selection from Kamikawa District by Hokkaido Agricultural Experiment Station in 1906, medium-late maturity, medium height, blackish-brown pods, small seeds of good quality, smaller leaf area and lower yields than Maruba resulted in little dissemination, highly susceptible to AzMV, susceptible to minor *Heterodera schachtii* infestations, some resistance to *Phytophthora vignae* f.sp. *adzukicola*, recommended for the central and southern portions of Hokkaido. SEE: Matsumoto, 1922; Fujita and Miura, 1934; Hokkaido Agricultural Experiment Station, 1967; Narikawa, 1978; Yoshida, 1987; Tsuchiya, 1988.

Kenzaki (Shimane)**:** Late maturity, autumn planted, bushy, few branches, very daylength sensitive. SEE: Tasaki, 1963, 1965.

Kenzaki No. 7 (Hokkaido)**:** Early maturity, bushy, few branches, daylength sensitivity. SEE: Tasaki, 1963, 1965.

Ki (Iwate)**:** Intermediate maturity, bushy, few branches, little daylength sensitivity. SEE: Tasaki, 1963, 1965.

Kingi (Kanto region, Japan)**:** Early maturity in Kanto Plain and southern parts of Japan, regular seed size. SEE: Tsuchizaki *et al.*, 1970a; Hoshikawa, 1985.

Kintoki (Chiba)**:** Bushy, moderate branching. SEE: Tasaki, 1963.

Kintoki (Gunma)**:** Intermediate maturity, bushy, moderate branching, little daylength sensitivity, sensitivity to acidic soils, susceptible to common virus diseases, also grown in Tottori Prefecture. SEE: Noda, 1952; Tasaki, 1963, 1965; Mukoo, 1975.

Kintoki (Ishikawa)**:** Bushy, moderate branching. SEE: Tasaki, 1963.

Kiyohara haru: Male parent for Beni dainagon, strong resistance to virus infections. SEE: Masaki *et al.*, 1986.

Table 7.4. Continued.

Kogamae dainagon: Medium maturity, tall, viney, large seed size. SEE: Yoshida, 1987.
Kogane dainagon: SEE: Chiba, 1982.
Konishi: SEE: Nomura, 1968; Satoh, 1974; Nomura and Satoh, 1977; Yoshida, 1987.
Korea-Black (Korea)**:** Late maturity, autumn planted, very daylength sensitive. SEE: Tasaki, 1965.
Korea-Dark Red (Korea)**:** Late maturity, autumn planted, semi-prostrate, viney, highly branched, daylength sensitive. SEE: Tasaki, 1963, 1965.
Korea-Green (Korea)**:** Late maturity, autumn planted, semi-prostrate, viney and highly branched, very daylength sensitive. SEE: Tasaki, 1963, 1965.
Korea-Large Red (Korea)**:** Intermediate maturity, daylength sensitivity. SEE: Tasaki, 1965.
Korea-Light Red (Korea)**:** Intermediate maturity, some daylength sensitivity, semi-prostrate, viney, highly branched. SEE: Tasaki, 1963, 1965.
Korea-Red Large (Korea)**:** Semi-prostrate, viney, highly branched. SEE: Tasaki, 1963.
Korea-Yellow (Korea)**:** Late maturity, autumn planted, semi-prostrate, viney, highly branched, very daylength sensitive. SEE: Tasaki, 1963, 1965.
Koshi (Niigata)**:** Early maturity, bushy, stiff stem with thick base, few branches. SEE: Tasaki, 1963.
Koshi (Okayama)**:** Susceptible to common virus diseases, may be same as Shiro azuki. SEE: Mukoo, 1975.
Koshi ro: 16% AzMV infection rate. SEE: Matsumoto, 1922.
Koshi zairai (Niigata)**:** Intermediate maturity, with some daylength sensitivity. SEE: Tasaki, 1965.
Kotobuki shouzu: Noto shouzu x Wase dairyu 1, released in 1971, lodging resistant, early to medium maturity of 121-124 days on Hokkaido, 45-60.6 cm height, only 2.6 branches/plant, dark brown pods, 30.4-34.6 pods/plant, medium red to dark red seeds, 14-15.5 g/100 seeds, 21.6% protein, 2700-2780 kg/ha yield, 51.2-55.6% harvest index, highly responsive to high planting densities (550,000 plants/ha), some *Phytophthora vignae* (Purss.) f. sp. *adzukicola* and damping-off resistance, susceptible to races 1, 2 and 3 of *Fusarium oxysporum* f.sp. *adzukicola*. SEE: Koyama *et al.*, 1972; Tsuchiya and Kodama, 1979, 1981b; Chiba, 1982; Tanaka and Konno, 1982; Hoshikawa, 1985; Tsuchiya *et al.*, 1986; Ushirogi *et al.*, 1986; Hokkaido Ag. Affairs Bureau, 1987; Japan Bean Fund ..., 1987; Yoshida, 1987; Tsuchiya, 1988; Yoshida *et al.*, 1988; Kondo and Kodama, 1989; Kondo *et al.*, 1990; Tabiki and Tsuchiya, 1990.
KS 68-I-A (Taiwan)**:** Breeding line, female parent for KA 74-04-101, 101 days to maturity, 24.9 cm height, 14.1 pods/plant, 5.5 seeds/pod, 17.9 g/100 seeds, 1714 kg/ha yield. SEE: Lee, 1989b.
KS 68-I-J (Taiwan)**:** Breeding line, male parent for KA 73-02-93 and KA 74-03-100, female parent for KA 73-06-97, KA 74-02-99, KA 74-07-104 and KA 77-02-119, 93-104 days to maturity, 28.6-39 cm height, 12.4 pods/plant, 4.9 seeds/pod, 20.4-21.6 g/100 seeds, 1615-2474 kg/ha yield. SEE: Hsu and Lin, 1985; Lee, 1989b.
KS 68-I-L (Taiwan)**:** Breeding line, male parent for KA 73-01-92, two years at five locations averaged 89 days to maturity, 34.4 cm height, 14.1 g/100 seeds, 2203 kg/ha yield. SEE: Hsu and Lin, 1985.
KS 69-I-L (Taiwan)**:** Breeding line, male parent for KA 74-01098. SEE: Wu *et al.*, 1986.
KS 210 (Taiwan)**:** AVRDC accession number 5122, used in interspecific crosses. SEE: Chen, 1980; Chen *et al.*, 1983.
KS 526 (Taiwan)**:** Breeding line, female parent for KA 73-04-95, autumn planting averaged 93 days to maturity, 13.9 g/100 seeds, 2608 kg/ha yield. SEE: Wu *et al.*, 1985b.
KS 540 (Taiwan)**:** Breeding line, female parent for KA 73-03-94 and KA 74-03-100, autumn planting averaged 92 days to maturity, 37.1 cm height, 15.1 g/100 seeds, 2589 kg/ha yield. SEE: Wu *et al.*, 1985b.
KS 542 (Taiwan)**:** Breeding line, female parent for KA 73-02-93 and KA 05-102, 93-102 days to maturity, 32.3-36.5 cm height, 17.9 pods/plant, 6.9 seeds/pod, 11-13.8 g/100 seeds, 2074-2951 kg/ha yield. SEE: Wu *et al.*, 1985b; Lee, 1989b.
KS 627 (Taiwan)**:** Breeding line, female parent for KA 74-06-103, two-year growout averaged 91 days to maturity, 31.5 cm height, 11.6 g/100 seeds, 2250 kg/ha yield. SEE: Wu *et al.*, 1986.
KS 658 (Taiwan)**:** Breeding line, female parent for KA 74-01-98, one-year growout averaged 91 days to maturity, 38.1 cm height, 13.6 g/100 seeds, 2507 kg/ha yield. SEE: Wu *et al.*, 1985b, 1986.
Kum doo (Korea)**:** Semi-prostrate, viney, highly branched. SEE: Tasaki, 1963.

Table 7.4. Continued.

Kuma (Ehime): Late maturity, autumn planted, viney, prostrate, highly branched, very daylength sensitive. SEE: Tasaki, 1963, 1965.
Kuro (Okayama): *Phialophora gregata* and AzMV susceptible. SEE: Murayama, 1941; Chiba, 1982.
Kurodane (Niigata and/or Tochigi): Late maturity, autumn planted, bushy, few branches, very daylength sensitive. SEE: Tasaki, 1963, 1965.
Kuwagin (or **Kwa gin**): Susceptible to common virus diseases, i.e. 16% AzMV infection rate. SEE: Matsumoto, 1922; Mukoo, 1975.
Kyoto dainagon (Kyoto): Optimal yields at 66,667 plants/ha. SEE: Matsui, 1984; Japan Bean Fund ..., 1987.
Li (N.E. China): Early maturity, little daylength sensitivity, semi-prostrate, viney, highly branched. SEE: Tasaki, 1963, 1965.
Liao xiao 14 (China): Cultivar developed and sold by producer. SEE: Duan, 1989.
Long xiao dou (China): Pure line selection promoted for production. SEE: Duan, 1989.
Long xiao dou No. 1 (Heilongjiang, China): Cultivar selected from among 61 landraces near Jiamusi, 51 cm height, 16-20 pods/plant, 5-6.5 seeds/pod, 8.8-9.2 g/100 seeds, lodging resistant, 2000-2100°C total day degrees or 97 days to physiological maturity, 25°C optimum germination. SEE: Zhang *et al.*, 1981; Du, 1986; Editorial Com. ..., 1993.
Long xiao dou No. 2 (Heilongjiang, China): Slected from landrace grown near Qiqiarer, 50 cm height, 13/pods/plant, 10-12 g/100 seeds and 100-110 days to maturity. SEE: Du, 1986; Editorial Com. ..., 1993.
Lujiang (Anhui, China): mid-maturity, shatter resistant, 7.68 g/100 seeds. SEE: Zhou *et al.*, 1964.
Maji dainagon (Kyoto): Late maturity, viney, prostrate, highly branched. SEE: Tasaki, 1963.
Man azuki: Intermediate maturity. SEE: Matsukawa, 1973.
Mao ping (China): Cultivar developed and sold by producer. SEE: Duan, 1989.
Maro (Chungbug, Korea): Red-seeded landrace, 68.6% *Phytophthora vignae* f.spp. *adzukicola* infection rate. SEE: Han *et al.*, 1982.
Maruba (Hokkaido): Name is sometimes used to describe a specific ecotype from Hokkaido and Honshu islands, the cultivar was a pure line selection from Kamikawa District by the Hokkaido Agricultural Experiment Station in 1906, popular until release of Maruba 1 go, medium maturity, blackish-brown pods, 48.5% harvest index over 11-year average, some resistance to common virus diseases, 4% AzMV infection rate, no AzMV seed-borne transmission, susceptible to soybean cyst nematode (*Heterodera glycines*), susceptible to minor *Heterodera schachtii* infestations. SEE: Matsumoto, 1922; Murayama, 1941; Ichinohe, 1953; Hokkaido Agricultural Experiment Station, 1967; Mukoo, 1975; Narikawa, 1978.
Maruba (Iwate): Early maturity, little daylength sensitivity. SEE: Tasaki, 1965.
Maruba M 27 and M 38: Late maturity. SEE: Nomura and Satoh, 1977.
Maruba M 34: Intermediate maturity. SEE: Matsukawa, 1973.
Maruba M 66: Intermediate maturity. SEE: Nomura and Satoh, 1977.
Maruba No. 1 (Hokkaido): Pure line selection from Maruba made at the Tokachi Agricultural Experiment Station in 1938, intermediate to late maturity in Hokkaido, good seed quality, high yields, viney at maturity, aphid resistant, recommended for production in the central and southern parts of Hokkaido. Characteristics from growout in S. Korea averaged 88 days to maturity, 68.5 cm tall, 22 pods/plant, 4.8 seeds/pod, 9.9 g/100 seeds, 34.7% harvest index. SEE: Hokkaido Agricultural Experiment Station, 1967; Ito *et al.*, 1975; Narikawa, 1978; Han *et al.*, 1979, 1984; Hoshikawa, 1985.
Maruba No. 1 (Hokkaido and Iwate): Early maturity, bushy, few branches, little daylength sensitivity. SEE: Tasaki, 1963, 1965.
Maruba No. 1 (Tochigi): Late maturity, bushy, few branches, little daylength sensitivity, tall, small seeds, some *Phialophora gregata* tolerance, grown mainly in Tohoku District. SEE: Tasaki, 1963, 1965; Nomura, 1968; Satoh, 1974; Nomura and Satoh, 1977; Chiba, 1982; Yoshida, 1987.
Mejiro (Niigata and/or Tochigi): Intermediate maturity, bushy, few branches, little daylength sensitivity. SEE: Tasaki, 1963, 1965.
Midoriyo gorekei: Late maturity. SEE: Nomura and Satoh, 1977.
Mikamo dainagon: Japanese cultivar used as a parental line for Taiwanese KA 77-02 119 and KA 77-03 120. SEE: Hsu *et al.*, 1989a.
Minoka (USA): Released in 1980, 115-day maturity, not currently available. SEE: Robinson, 1980; Hardman, 1985, 1987.

Table 7.4. Continued.

Minzan (N.E. China)**:** Highly resistant to common virus diseases, no symptoms of AzMV, used in Japan. SEE: Matsumoto, 1922; Mukoo, 1975.
Miomote shiro (Niigata and/or Tochigi)**:** Late maturity, autumn planted, bushy, few branches, very daylength sensitive. SEE: Tasaki, 1963, 1965.
Miomoto Nos. 1, 2, 3 and 4 (Niigata and/or Tochigi)**:** Late maturity, autumn planted, bushy, few branches, very daylength sensitive. SEE: Tasaki, 1963, 1965.
Miyako (Kanto District, Japan)**:** Red color testa, medium brown pod color. SEE: Kakizaki, 1923.
Miyozawa (Yamanashi)**:** Intermediate maturity, bushy, moderate branching, some daylength sensitivity. SEE: Tasaki, 1963, 1965.
Miyun (**E 1037-1**) (China)**:** Early maturity (90 days), 1500-2250 kg/ha yield, disease resistance. SEE: Hu, 1986; Duan, 1989.
Monbetsu: Some resistance to common virus diseases. SEE: Mukoo, 1975.
Monbetsu 26 (Miyagi)**:** Intermediate/late maturity, bushy, few branches, daylength sensitivity, tall, viney, scarlet-color large seeds, 21.4 mature pods/plant, 1980 kg/ha yield, *Phialophora gregata* susceptible. SEE: Tasaki, 1963, 1965; Tsuchizaki *et al.*, 1970b; Furusawa *et al.*, 1974; Chiba, 1982.
Mumeimadara: 4% AzMV infection rate. SEE: Matsumoto, 1922.
Muroran: Some resistance to common virus diseases, 4% AzMV infection rate. SEE: Matsumoto, 1922; Mukoo, 1975.
Nabeyogoshi (Oita)**:** Late maturity, autumn planted, viney, prostrate, highly branched, very daylength sensitive. SEE: Tasaki, 1963, 1965.
Naebuki (Chungbug, Korea)**:** Landrace, black seed, 58.5% *Phytophthora vignae* f.spp. *adzukicola* infection rate. SEE: Han *et al.*, 1982.
Naganuma shouzu: Late maturity. SEE: Nomura and Satoh, 1977.
Nakate (Niigata)**:** Late maturity, autumn planted, daylength sensitive. SEE: Tasaki, 1965.
Nakate dainagon (Tochigi)**:** Early maturity, bushy, stiff stem with thick base, few branches. SEE: Tasaki, 1963.
Natsu (Fukushima)**:** Intermediate maturity, bushy, few branches, little daylength sensitivity. SEE: Tasaki, 1963, 1965.
Natsu (Gunma)**:** Intermediate maturity, bushy, moderate branching, little daylength sensitivity. SEE: Tasaki, 1963, 1965.
Natsu (Oita)**:** Intermediate maturity, bushy, moderate branching, some daylength sensitivity, some temperature sensitivity. SEE: Tasaki, 1963, 1965; Tasaki and Honma, 1965.
Natsu (Okayama)**:** Intermediate maturity, some daylength sensitivity. SEE: Tasaki, 1965.
Natsu (Okinawa)**:** Intermediate maturity, bushy, moderate branching, some daylength sensitivity, some temperature sensitivity. SEE: Tasaki, 1963, 1965; Tasaki and Honma, 1965.
Natsu (Tokushima)**:** Intermediate maturity, bushy, moderate branching, some daylength sensitivity. SEE: Tasaki, 1963, 1965.
Natsu azuki(Mie)**:** Late maturity. SEE: Matsukawa, 1973; Han *et al.*, 1984.
Nesumimochi: Growout in S. Korea averaged 92 days to maturity, 82.3 cm tall, 19.3 pods/plant, 4.7 seeds/pod, 11.6 g/100 seeds, 29.7% harvest index. SEE: Chang *et al.*, 1968b, 1969; Han *et al.*, 1979, 1984.
Niigata-zai Tochiohanzokin: *Phytophthora vignae* resistance. SEE: Tabiki and Tsuchiya, 1990.
Nishikawa shouzu: SEE: Nomura, 1968; Satoh, 1974; Nomura and Satoh, 1977; Yoshida, 1987.
Nishimoiwa shouzu: Intermediate maturity. SEE: Nomura, 1968; Satoh, 1974; Nomura and Satoh, 1977; Yoshida, 1987.
Nishine zairai: SEE: Furusawa *et al.*, 1974.
Nonomura shouzu: SEE: Nomura, 1968; Satoh, 1974; Nomura and Satoh, 1977; Yoshida, 1987.
Noto shouzu: Cultivar native to the Noto peninsula, some resistance to *Phytophthora vignae* f.sp. *adzukicola*. SEE: Satoh *et al.*, 1975b; Tsuchiya *et al.*, 1986; Tsuchiya, 1988; Tsuchiya and Kodama, 1981b.
Number 1 (Tochigi)**:** Intermediate maturity, little daylength sensitivity. SEE: Tasaki, 1965.
Odate No. 1 (Iwate)**:** Hybrid, released in 1948, late maturity, tall, bushy, viney, few branches, medium red seeds, 12.7 g/100 seeds, susceptible to AzMV infection. SEE: Tasaki, 1963; Furusawa *et al.*, 1974; Kobayashi *et al.*, 1976; Narikawa, 1978; Green *et al.*, 1986; Japan ..., 1988.

Table 7.4. Continued.

Odate No. 2 (Tohoku and Kanto regions, Japan)**:** Intermediate maturity, tall, bushy, few branches, 1.2 branches/plant, small but good quality red seeds, 10.4 g/100 seeds, 27.7 mature pods/plant, 2000 kg/ha yield, some sensitivity to temperature, highly susceptible to virus infections such as AzMV, some resistance to *Phytophthora vignae* f.sp. *adzukicola*, one of the main cultivars grown in Iwate in the 1970s. SEE: Tasaki, 1963, 1965; Tasaki and Honma, 1965; Tsuchizaki *et al.*, 1970ab, 1971; Furusawa *et al.*, 1974; Kobayashi *et al.*, 1976; Narikawa, 1978; Chiba, 1980; Tanaka and Konno, 1982; Masaki *et al.*, 1986; Tsuchiya, 1988.

Odate No. 3 (Yamagata)**:** Cross between Kingi x Hokui, intermediate maturity, 108 cm tall, bushy, lodging resistant, not viney, few branches, dark red intermediate-sized seed, resistance to azuki pod bug. SEE: Tasaki, 1963; Chiba, 1980; Hoshikawa, 1985.

Odate Nos. 1, 2 and 3 (Akita)**:** Intermediate maturity, some daylength sensitivity, developed at the Odate Experiment Station between 1936 and 1950. SEE: Tasaki, 1965; Tasaki and Honma, 1965; Narikawa, 1978.

Ogcheon gaeguri (Korea)**:** Cultivar. SEE: Han and Choe, 1977.

Ohunato shi: SEE: Furusawa *et al.*, 1974.

Okayama wase: Bushy, moderate branching. SEE: Tasaki, 1963.

Okute azuki: 8% AzMV infection rate. SEE: Matsumoto, 1922.

Otari azuki: SEE: Yoshida *et al.*, 1989abc.

Otoku (Mie)**:** Intermediate maturity, bushy, moderate branching, some sensitivity to daylength. SEE: Tasaki, 1963, 1965.

Otosara azuki: SEE: Yoshida *et al.*, 1989abc.

Ouda dainagon (Nara)**:** Late maturity, autumn planted, viney, prostrate, highly branched, very daylength sensitive. SEE: Tasaki, 1963, 1965.

Pai (N.E. China)**:** Early maturity, little daylength sensitivity, semi-prostrate, viney, highly branched, very temperature sensitive. SEE: Tasaki, 1963, 1965; Tasaki and Honma, 1965.

Pingtung local (Pingtung, Taiwan)**:** Introduced into Taiwan in the 1940s from Japan, 30-35 days to start of indeterminate flowering, 85-95 days to maturity, 28.7-59.7 cm height, semi-viney, 9-10 seeds/white colored pod, 10.3-14 g/100 seeds, dark red testa, 0.8% hard seeds, 1770-2342 kg/ha yield, susceptible to powdery mildew and *Rhizoctonia* infections, especially during pod development, low position of pods (15-17 cm) on stem, susceptibility to lodging. SEE: Ag. Develop. Committee, 1981; Liu *et al.*, 1981; Chao, 1982; Wu *et al.*, 1985ab; Green *et al.*, 1986.

Rei shouzu (Hokkaido)**:** Cultivar maintained at the Tokachi Agricultural Experiment Station, low harvest index. SEE: Chiba, 1980.

Ridonyo: No symptoms of AzMV infection. SEE: Matsumoto, 1922.

Rikuwandon (N.E. China)**:** Highly resistant to common virus diseases, used in Japan, no symptoms of AzMV infection. SEE: Matsumoto, 1922; Mukoo, 1975.

Rizaon-in (Rin-san-in) (N.E. China)**:** Highly resistant to common virus diseases, used in Japan, no symptoms of AzMV infection. SEE: Matsumoto, 1922; Mukoo, 1975.

Ryokuzu: No symptoms of AzMV infection. SEE: Matsumoto, 1922.

Sado No. 3 (Niigata)**:** Late maturity, autumn planted, daylength sensitive. SEE: Tasaki, 1965.

Sado Nos. 1, 2 (Niigata)**:** Late maturity, autumn planted, very daylength sensitive. SEE: Tasaki, 1965.

Sakae shouzu: (Chagarawase x Wase dairyu 1) F_6 x Wase dairyu 1, also called Toiku 70 in literature before release in 1973, medium maturity, tall, highly branched, large number pods/plant, small seeds, good seed quality, responsive to higher fertilizer levels, some cool weather tolerance, susceptible to races 1, 2 and 3 of *Fusarium oxysporum* f.sp. *adzukicola*, latent infection with bean yellow mosaic virus (BYMV) strain. SEE: Satoh *et al.*, 1975b; Chiba, 1980; Tanaka and Konno, 1982; Uyeda *et al.*, 1982; Ushirogi *et al.*, 1986; Kondo and Kodama, 1989.

Sapporo wase dainagon (Iwate)**:** Intermediate maturity, bushy, few branches, daylength sensitivity. SEE: Tasaki, 1963, 1965.

Sato azuki: Intermediate maturity. SEE: Matsukawa, 1973.

Seobyeong: Same as **Jeokdu.**

Shiboriwake: 8% AzMV infection rate. SEE: Matsumoto, 1922.

Shiiku 1 (Hokkaido)**:** Cultivar maintained at the Tokachi Agricultural Experiment Station; short (28 cm), few stem nodes (9.7). SEE: Chiba, 1980.

Shimo fuchi shouzu (Hokkaido)**:** Cultivar maintained at the Tokachi Agricultural Experiment Station; long pre-anthesis vegetative stage. SEE: Chiba, 1980.

Table 7.4. Continued.

Shin dainagon (Hokkaido)**:** Pure line selection from Wase dairyu in 1938 at the Tokachi Agricultural Experiment Station, late maturity, good seed quality, high yielding. SEE: Hokkaido Agricultural Experiment Station, 1967.
Shina 7: Intermediate maturity. SEE: Matsukawa, 1973.
Shira saya (Kumamoto)**:** Grown in the 1970s, very light brown pods, some resistance to common viruses, might also be called *Wase shira saya*. SEE: Mukoo, 1975; Narikawa, 1978; Tanaka and Konno, 1982.
Shiro (Kumamoto)**:** Late maturity, autumn planted, viney, prostrate, highly branched, very daylength sensitive. SEE: Tasaki, 1963, 1965.
Shiro azuki: 8% AzMV infection rate. SEE: Matsumoto, 1922.
Shiro kasuri: Susceptible to AzMV infection. SEE: Murayama, 1941.
Shiro saya: 4% AzMV infection rate. SEE: Matsumoto, 1922.
Shiro saya aka (Hokkaido)**:** Very light brown-white pods, released in 1923. SEE: Narikawa, 1978.
Shiro shouzu (Kawanihi) (N.E. China)**:** Several cultivars with natural mutations, some are late maturity, white seed, poor quality, resistance to common virus diseases. SEE: Mukoo, 1975; Murata, 1980.
Shiro zaya (Yamaguchi)**:** Intermediate maturity, bushy, moderate branching, some daylength sensitivity. SEE: Tasaki, 1963, 1965.
Shonagon: Susceptible to common virus diseases, high levels of seed-borne transmission. SEE: Murayama, 1941; Kawahara, 1962; Mukoo, 1975; Chiba, 1982.
Shouchouhin-10 (Nagano)**:** Source of dominant gene (Koshoshin 10) in Toiku 123 that provides resistance to all three races of *Fusarium oxysporum* f.sp. *adzukicola*, resistant to *Phialophora gregata*. SEE: Chiba, 1982; Kondo *et al.*, 1990.
Shouzu No. 1 (Shimane)**:** Late maturity, autumn planted, very daylength sensitive. SEE: Tasaki, 1965.
Shouzu 64-1 (Korea)**:** Cultivar with Japanese origins. SEE: Chiba, 1982.
Shouzu W 21: Intermediate maturity. SEE: Nomura and Satoh, 1977.
Shouzu W 59 (Hokkaido)**:** Cultivar maintained at the Tokachi Agricultural Experiment Station, high number of seeds/pod (8.93), high average bean weight 0.195 g. SEE: Chiba, 1980.
Shouzu W 69: Early maturity. SEE: Matsukawa, 1973.
Shouzu wase 1: Highly susceptible to *Phytophthora vignae* (Purss) f. spp. *adzukicola*. SEE: Tsuchiya and Kodama, 1979.
Shouzu wase kei 3 (Hokkaido)**:** Cultivar maintained at the Tokachi Agricultural Experiment Station, very small. SEE: Chiba, 1980.
Siyang (China)**:** Cultivar developed and sold by producer. SEE: Duan, 1989.
Soohahn (Black) (Chungbug, Korea)**:** Black-seeded landrace, 29.2% *Phytophthora vignae* f.spp. *adzukicola* infection rate. SEE: Han *et al.*, 1982.
Soohahn (Red) (Chungbug, Korea)**:** Red-seeded landrace, 67.4% *Phytophthora vignae* f.spp. *adzukicola* infection rate. SEE: Han *et al.*, 1982.
Soshokushu: Resistant to *Phytophthora vignae* f.sp. *adzukicola*. SEE: Tsuchiya, 1988.
Taiyuan 317 (China)**:** Cultivar developed and sold by producer. SEE: Duan, 1989.
Takachihoaki (Miyazaki)**:** Late maturity, autumn planted, viney, prostrate, highly branched, very daylength sensitive. SEE: Tasaki, 1963, 1965.
Takachihonatsu (Miyazaki)**:** Intermediate maturity, bushy, moderate branching, some daylight sensitivity. SEE: Tasaki, 1963, 1965.
Takahashi shouzu: Late maturity. SEE: Narikawa, 1972; Matsukawa, 1973.
Takahashi wase (Hokkaido)**:** Maruba x Buchi or Maruba x Beni Shouzu by Yoshinao Takahashi at the Hokkaido Agricultural Experiment Station, released in 1924, early maturity, bushy, medium height (57.3 cm), few branches, red seeds, 10.6 g/100 seeds, good quality, 2030 kg/ha yield, still cultivated in the 1960s, some resistance to common virus diseases and anthracnose, (*Colletotrichum phaseolorum* Takimoto), also grown in Miyagi Prefecture. SEE: Tasaki, 1963; Hokkaido Agricultural Experiment Station, 1967; Nomura, 1968; Matsukawa, 1973; Satoh, 1974; Furusawa *et al.*, 1974; Mukoo, 1975; Nomura and Satoh, 1977; Narikawa, 1978; Yoshida, 1987.
Takara shouzu (Hokkaido)**:** Native cultivar or pure line Shouzu W-45 selection made at the Tokachi Agricultural Experiment Station and released in 1959, early/intermediate maturity, little daylength sensitivity, medium height (48-60 cm), sparsely branched (3.6/plant), light brown pods, 10.4-13 g/100 seeds, 22.9% protein, 31-43 pods/plant, 6.66 seeds/pod, 0.127 g/seed, 10.14% testa to seed weight ratio, 62% harvest index, 1680-2740 kg/ha yield, red seeds which set the current Japanese standard for quality, some lodging and cool weather

Table 7.4. Continued.

resistance, susceptible to bean yellow mosaic virus (BYMV), cucumber mosaic virus (CMV), *Pseudomonas adzukicola* and *Phialophora gregata* infections, races 1, 2 and 3 of *Fusarium oxysporum* f.sp. *adzukicola*, once accounted for more than 50% of total prefectural production. SEE: Tasaki, 1963; Hokkaido Agricultural Experiment Station, 1967; Nomura, 1968; Narikawa, 1972; Satoh, 1974, 1982; Ito *et al.*, 1975; Satoh *et al.*, 1975ab; Uyeda *et al.*, 1975; Nomura and Satoh, 1977; Narikawa, 1978; Chiba, 1980; Ushirogi *et al.*, 1986; Hokkaido Ag. Affairs Bureau, 1987; Japan Bean Fund ..., 1987; Yoshida, 1987; Yoshida *et al.*, 1988; Kondo and Kodama, 1989.

Take: SEE: Chiba, 1982.

Takei No. 5 (Fukushima)**:** Intermediate maturity, little temperature sensitivity. SEE: Tasaki, 1965.

Takei Nos. 6, 7 (Gunma)**:** Intermediate maturity, daylength sensitivity. SEE: Tasaki, 1965.

Takeno shouzu: SEE: Nomura, 1968; Satoh, 1974; Nomura and Satoh, 1977; Yoshida, 1987.

Tamayama zairai: SEE: Furusawa *et al.*, 1974.

Tanaka shouzu: SEE: Nomura, 1968; Satoh, 1974; Nomura and Satoh, 1977; Yoshida, 1987.

Tanayen (Taiwan)**:** Female parent in cross leading to development of Kaohsiung No. 3. SEE: Wu *et al.*, 1985ab.

Tan'ba dainagon (Kyoto)**:** Late maturity, autumn planted, viney, prostrate, highly branched, very daylength sensitive, very large seed size, also grown in Tohoku region, some *Phialophora gregata* tolerance. SEE: Tasaki, 1963, 1965; Chiba, 1982.

Tane ichi zairai: SEE: Furusawa *et al.*, 1974.

Tianjin (E 1022-1) (China)**:** Medium maturity (105-115 days), semi-determinant, 1500-2250 kg/ha yield. SEE: Hu, 1986.

Tianjin (Hebei, China)**:** Famous cultivar group grown near the city of Tianjin, 99 days to maturity, 30-40 cm tall, 5-10 branches/plant, susceptible to virus infections. Two-year, three-site growout in Shanxi Province averaged 126 days to maturity, 51.4 cm tall, 4.6 branches, 16.7 pods/plant, 5.7 seeds/pod, 12.7 g/100 seeds, 1413 kg/ha yield. SEE: Lu and Ma, 1985; Song *et al.*, 1989; Zhu, 1989.

Tianjin red (Tianjin, China)**:** Very susceptible to *Uromyces azukicola*. SEE: Tian, 1990.

Tientsin pai Nos. 1 and 2 (China)**:** Cultivars, semi-prostrate, viney, highly branched. SEE: Tasaki, 1963.

Tochiki No. 1: Bushy, moderate branching, also grown in Korea. SEE: Tasaki, 1963; Han and Choe, 1977.

Toiku 7: See **Hikari**.

Toiku 24 (Hokkaido)**:** Cultivar maintained at the Tokachi Agricultural Experiment Station, high seed weight/plant. SEE: Chiba, 1980.

Toiku 68: Takara shouzu x Yamada shouzu. SEE: Nomura, 1968; Satoh, 1974; Nomura and Satoh, 1977; Yoshida, 1987.

Toiku 70: Later called **Sakae shouzu**.

Toiku 85: Early maturity. SEE: Chiba, 1982.

Toiku 91: SEE: Adachi *et al.*, 1979.

Toiku 106: Medium maturity, tall height, very large seeds. SEE: Yoshida, 1987.

Toiku 120: Susceptible to races 1, 2 and 3 of *Fusarium oxysporum* f.sp. *adzukicola*. SEE: Kondo and Kodama, 1989.

Toiku 122: Susceptible to races 2 and 3 of *Fusarium oxysporum* f.sp. *adzukicola* but not race 1. SEE: Kondo and Kodama, 1989.

Toiku 123: Erimo shouzu x Shochohin 10, resistant to all 3 races of *Fusarium oxysporum* f.sp. *adzukicola*. SEE: Kondo and Kodama, 1989; Kondo *et al.*, 1990.

Toiku 325: Resistant to races 1, 2 and 3 of *Fusarium oxysporum* f.sp. *adzukicola*. SEE: Kondo and Kodama, 1989.

Tone wase (Gunma)**:** Intermediate maturity, bushy, moderate branching, little daylength sensitivity. SEE: Tasaki, 1963, 1965.

Tsuru shouzu: General name for primitive rice bean cultivars, medium maturity, tall (70-163 cm), viney, small seeds of mixed colors, low yields (680 kg/ha), tolerant of acidic soils, low phosphorus availability and late plantings, grown in Hyogo, Tottori and Hokkaido. SEE: Noda, 1952; Ito *et al.*, 1975; Yoshida, 1987.

Tsurugi-3: Late maturity. SEE: Matsukawa, 1973.

TW Adzuki: Highly susceptible to *Ophiomyia phaseoli*, susceptible to *Liriomyia* spp., *Empoasca solana*, *Callosobruchus chinensis* and *Tetranychus telarius*, some resistance to *Adoretus sinicus* and *Trialeurodes vaporariorum*. SEE: Lin, 1981.

Umseong jeogdu (Korea)**:** Cultivar. SEE: Han and Choe, 1977.

Table 7.4. Continued.

Urasa (Niigata): Late maturity, autumn planted, daylength sensitivity, resistant to *Phytophthora vignae* f.sp. *adzukicola*. SEE: Tasaki, 1965; Tsuchiya, 1988.
Urasa (Shimane): Early maturity, bushy, stiff stem with thick base, few branches. SEE: Tasaki, 1963.
Urasa (Tochigi): Early maturity, bushy, stiff stem with thick base, few branches. SEE: Tasaki, 1963.
Urumi shouzu M 48: Late maturity. SEE: Nomura and Satoh, 1977.
Usu ginu: Susceptible to AzMV infection. SEE: Murayama, 1941.
Uzura (Niigata): Late maturity, autumn planted, bushy, few branches, very daylength sensitive, little temperature sensitivity. SEE: Tasaki, 1963, 1965; Tasaki and Honma, 1965.
Wase (Aichi): Intermediate maturity, bushy, moderate branching, little daylength sensitivity. SEE: Tasaki, 1963, 1965.
Wase (Miyagi): Intermediate maturity, bushy, moderate branching, little daylength sensitivity. SEE: Tasaki, 1963, 1965.
Wase (Shizuoka): Intermediate maturity, bushy, moderate branching, little daylength sensitivity. SEE: Tasaki, 1963, 1965.
Wase dainagon (Aomori and Hokkaido): Ecotype with many regional cultivars, intermediate to late maturity, dark red seeds, high yielding, susceptible to alfalfa mosaic virus (AMV), high levels of seed-borne transmission, some unidentified cultivar had no symptoms of AzMV, susceptible to races 2 and 3 of *Fusarium oxysporum* f.sp. *adzukicola* but not race 1. SEE: Matsumoto, 1922; Murayama, 1941; Matsumoto and Murayama, 1971; Hoshikawa, 1985; Kondo and Kodama, 1989.
Wase dainagon (Aomori): Late to intermediate maturity, tall, bushy, few branches, little daylength sensitivity in Niigata Prefecture, large seeds. SEE: Tasaki, 1963, 1965; Yoshida, 1987.
Wase dainagon (Hokkaido): Pure line selection from Muroran District in 1910, intermediate maturity, bushy, few branches, little daylength sensitivity, very temperature sensitive, large seeds of good quality, grown in southern Hokkaido. SEE: Tasaki, 1963, 1965; Tasaki and Honma, 1965; Hokkaido Agricultural Experiment Station, 1967.
Wase dairyu (Hokkaido): Pure line selection from a N.E. China cultivar by the Hokkaido Agricultural Experiment Station in 1915, early maturity, bushy, few branches, high yields, highly resistant to common virus diseases as well as anthracnose, *Colletotrichum phaseolorum* Takimoto, grown in northern Hokkaido. SEE: Tasaki, 1963; Hokkaido Agricultural Experiment Station, 1967; Mukoo, 1975.
Wase dairyu 1 (Hokkaido): Pure line selection from Wase tairyu made at the Tokachi Agricultural Experiment Station, registered in 1930, 110-day maturity, little daylength sensitivity, medium height (58.6-68.3 cm), viney, 3.6 branches/plant, large crimson-colored seeds (14-18.9 g/100 seeds), 36.3 pods/plant, 2430 kg/ha yield, 45.3% harvest index, good yield compensation to late planting, susceptible to race 3 of *Fusarium oxysporum* f.sp. *adzukicola* but not races 1 or 2. SEE: Horiguchi, 1950; Tasaki, 1965; Hokkaido Agricultural Experiment Station, 1967; Nomura, 1968; Narikawa, 1972; Ushirogi *et al.*, 1972; Satoh, 1974; Furusawa *et al.*, 1974; Nomura and Satoh, 1977; Chiba, 1980; Han *et al.*, 1984; Hoshikawa, 1985; Tsuchiya *et al.*, 1986; Yoshida, 1987; Kondo and Kodama, 1989.
Wase dairyu W 14 and M4: Intermediate maturity. SEE: Nomura and Satoh, 1977.
Wase kei 1: Early maturity. SEE: Matsukawa, 1973.
Wase kei shouzu 1,2,3 and 4: All four selections displayed early maturity, cold weather resistance and poor seed quality. SEE: Narikawa, 1972; Matsukawa, 1973; Nomura and Satoh, 1977.
Wase kotosubu: SEE: Kawahara, 1962.
Wase kurozaya (Niigata): Early maturity, bushy, few branches. SEE: Tasaki, 1963.
Wase maruba (Hokkaido): Pure line selection from Kamikawa District in 1915, early maturity, little daylength sensitivity, very temperature sensitive, grown in cold northern areas, bushy, few branches, susceptible to minor *Heterodera schachtii* infestations. SEE: Fujita and Miura, 1934; Takahashi, 1957; Tasaki, 1963, 1965; Tasaki and Honma, 1965; Hokkaido Agricultural Experiment Station, 1967.
Wase otsubu (Tohoku region, Japan): Early maturity, little daylength sensitivity. SEE: Kakuta and Satoh, 1953b; Tsuchizaki *et al.*, 1970a.
Wase shouzu: An unknown number of early maturity cultivars differentiated on the basis of area of selection and production, a geographically unidentified Wase cultivar is susceptible to common virus diseases with high levels of seed-borne transmission. SEE: Murayama, 1941; Tsuchizaki *et al.*, 1970a; Mukoo, 1975.

Table 7.4. Continued.

Wase shouzu (**M4**) (Hokkaido)**:** Cultivar maintained at the Tokachi Agricultural Experiment Station, highly branched (5.9 branches/plant). SEE: Chiba, 1980.
Wuqing (China)**:** Developed and sold by producer. SEE: Duan, 1989.
Xihong No. 1 (Hebei, China)**:** 102 days to maturity, 65.4 cm tall, 3.6 branches/plant, 20 pods/plant, 7.7 seeds/pod, 10 g/100 seeds, 551 kg/ha yield, bright red testa. SEE: Rong and Ma, 1986.
Xin 16-1 (Shanxi, China)**:** Growout in Shandong Province averaged 102 days to maturity, 41.2 cm tall, 3.8 branches, 6.4 pods, 6.8 seeds/pod, 10.6 g/100 seeds, 380 kg/ha yield, bright red testa. SEE: Rong and Ma, 1986.
Xingan (China)**:** Developed and sold by producer. SEE: Duan, 1989.
Yabutsuru: SEE: Chiba, 1982.
Yamada: SEE: Chiba, 1982.
Yamakoshi shouzu: SEE: Nomura, 1968; Satoh, 1974; Nomura and Satoh, 1977; Yoshida, 1987.
Yasuuchi kuromame (Korea)**:** Korean cultivar with Japanese origins. SEE: Chiba, 1982.
Yellow (Zhejiang and Hebei, China)**:** Susceptible to an unidentified seed-borne mosaic virus. SEE: Cheo and Tsai, 1959.
Yeongdong Jeogdu (Korea)**:** 91 days to maturity, 71 cm tall, 18.8 pods/plant, 5.4 seeds/pods, 11.8 g/100 seeds, 42.9% harvest index. SEE: Han *et al.*, 1979, 1984; Hwang, 1982; Hong *et al.*, 1984, 1986.
Yuseong jeogdu (Korea)**:** SEE: Han and Choe, 1977.
Zailaizhong (Taiwan)**:** 101 days to maturity (autumn planted), 30.3 cm stem, 18.2 pods/plant, 6.5 seeds/pod, 11.9 g/100 seeds, dark red seed testa, 2112 kg/ha yield. SEE: Lee, 1989b.
Zairai (Shimane)**:** Early maturity, bushy, stiff stem with thick base, few branches. SEE: Tasaki, 1963.
Zairai dainagon (Kumamoto)**:** Japanese, viney, prostrate, late maturity, highly branched. SEE: Tasaki, 1963.
Zairai dainagon (Niigata)**:** Late maturity, autumn planted, very daylength sensitive. SEE: Tasaki, 1965.
Zairai dainagon (Shimane)**:** Late maturity, autumn planted, very daylength sensitive. SEE: Tasaki, 1965.
Zairai dainagon (Tochigi)**:** Early maturity, bushy, stiff stem with thick base, few branches. SEE: Tasaki, 1963.
Zairai dainagon Nos. 1, 2 and 3 (Okayama)**:** Late maturity, autumn planted, viney, prostrate, highly branched, very daylength sensitive. SEE: Tasaki, 1963, 1965.
Zairai No. 1 (Miyazaki)**:** Late maturity, viney, prostrate, highly branched. SEE: Tasaki, 1963.
Zairai Nos. 1 and 2 (Yamanashi)**:** Intermediate maturity, bushy, moderate branching, little daylength sensitivity. SEE: Tasaki, 1963, 1965.
Zairai Nos. 5 and 6 (Gunma)**:** No. 6 has intermediate maturity and daylength sensitivity, both are bushy with moderate branching . SEE: Tasaki, 1963, 1965.
Zairai Nos. 5, 6 and 13 (Kumamoto)**:** Late maturity, autumn planted, viney, prostrate, highly branched, very daylength sensitive. No. 6 has little temperature sensitivity. SEE: Tasaki, 1963, 1965; Tasaki and Honma, 1965.
Zairai shu (Kumamoto and Oita)**:** Main cultivar on Kyushu island, grown as both a winter and summer crop. SEE: Japan Bean Fund ..., 1987.
Zairai wase (Niigata and Tochigi)**:** Intermediate maturity, bushy, few branches, little daylength sensitivity. SEE: Tasaki, 1963, 1965.

8

FOOD CHEMISTRY AND PROCESSING

Seeds are traditionally the most common part of the azuki plant used for human consumption. Thus, the biochemical characteristics of azuki seeds are important in understanding their nutritional value (Tables 8.1, 8.2 and 8.4), different uses and processing techniques. Japanese scientists and, to a lesser extent, American scientists have invested considerable energy into understanding the nutritional value of azuki and factors affecting processing quality. The most comprehensive Japanese reviews and reports on food quality characteristics of the azuki seed are included in two volumes published by the Hokkaido Central Agricultural Experiment Station in cooperation with the Japan Bean Fund Association (1989) and the Hokkaido Association for the Advancement of Beans (Soma, 1988). These include extensive information on azuki proteins, lipids, starches, sugars, seed color, water absorption, *an* yield, *an* particle size, etc.

SEED CHARACTERISTICS

Seed structure

Members of the Washington State University azuki research team used scanning electron microscopy (SEM) to examine the microstructure of azuki seeds (Chilukuri and Swanson, 1991). Azuki was found to have a seed coat surface pattern of convolutions and cracks, an epidermis with a thick layer of palisade cells underlain by multiple layers of tightly packed parenchyma cells and a hilum region with a double palisade layer above a sub-hilar plug (Fig. 8.1). Azuki cotyledons have spherical starch granules with a pitted surface and are loosely embedded in a protein matrix (Fig. 8.2). SEM was also used to compare the microstructure of five azuki cultivars: *Erimo*, *Express*, *Hatsune*, *Takara* and *VBSC* (Engquist and Swanson, 1992). These

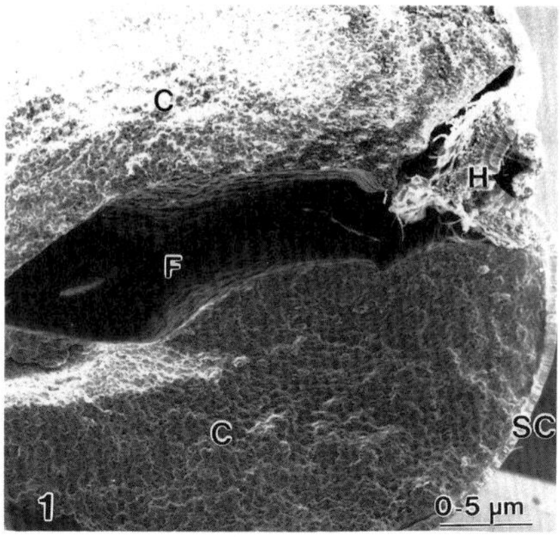

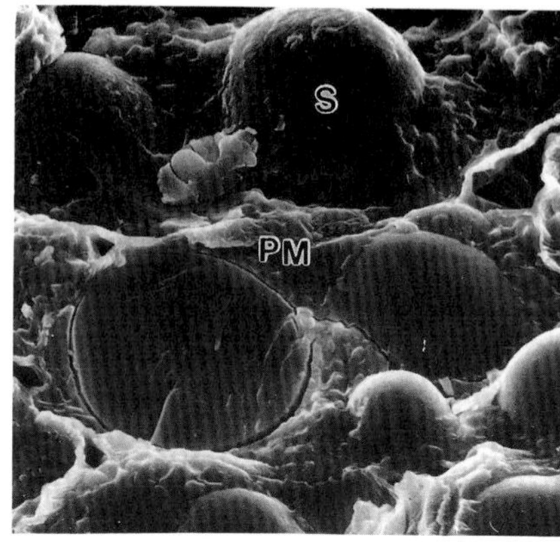

Fig. 8.1. (Left) A cross section of an azuki seed from Figure 1 in Chilukuri and Swanson (1991). The SEM photo shows the seed coat (SC), cotyledons (C), hilum (H) and fissure (F). The embryonic true leaves can be seen within the left side of the fissure.

Fig. 8.2. (Right) Starch granules (S) embedded in the protein matrix (PM) within a cotyledon of the cultivar *Erimo*. From Figure 14 in Engquist and Swanson (1992).

cultivars had anatomical differences in the protein matrices that hold the starch granules, shape of the starch granules and patterns of cracks, pits and deposits on the seed coat. Seed coat proteins can also be considered for differentiation (Lee *et al*., 1984).

Water imbibition

Azuki seeds are characterized by poor water imbibition which poses problems for crop establishment, processing and home consumption. Yoshida *et al.* (1989c; also Shiota *et al*., 1983) found that the imbibition rate was faster for regular-sized azuki than for the larger *dainagon* type. They also determined that the highest percentage of hard seeds consistently occurred with the smallest seeds; hard-seeded azuki ranged from 0 to 13% of the samples tested. This observation may serve as a partial explanation of why smaller azuki seeds have poor germination. Seven advanced azuki breeding lines and the check cultivar, *Kaohsiung* No. 3, averaged 2.1% hard seeds from both autumn and spring yield trials in the Kaoping region of Taiwan (Hsu *et al.*, 1989b). The use of the defoliant paraquat before harvest was shown to increase the percentage of hard seeds from 3% in the non-treated controls to an average of over 11% with the chemical treatments; the earlier the application time and higher the paraquat concentration, the higher the hard seed percentage (Huang, 1989).

Azuki demonstrated a unique water imbibition process and the slowest hydration curve of six legumes tested by Sefa-dedeh and Stanley (1979). The hydration curve was very gradual during the first six hours of soaking, with seed testa thickness and seed size being cited as possible reasons. They suggested that azuki seed softening was due to middle lamella deterioration and starch granule gelatinization and observed that azuki seeds exhibited a three phase softening process during soaking: a gradual softening during the first six hours; a rapid increase in softening from the 6th to the 12th hour of soaking; then almost no increase in seed softening after that. This S-curve phenomenon was also noted by Yoshida *et al.* (1989c) and Kubota (1979) who quantified it for soaking and cooking. Drager (1983) contends that azuki seeds will soften sufficiently only after they have absorbed enough water to double their dry weight.

Table 8.1. Nutritional components of dry and processed azuki seeds. All results are from Harukawa (1990) except the second data column (Kay, 1979) and are presented as g, mg or μg per 100 g of seed, except energy and vitamin A.

Component per 100 g	Dry seeds	Dry seeds (Kay)	Boiled	Boiled and canned	Koshi *an*	Sarashi *an*[*]
Energy, kcal	339		143	218	155	385
Energy, kJ	1418		598	912	649	1611
Water, g	15.5	13.0	64.8	45.3	62.0	5.0
Protein, g	20.3	25.3	8.9	4.4	9.8	26.2
Fat, g	2.2	0.6	1.0	0.4	0.6	1.0
Carbohydrate, g	54.4	57.1	22.3	48.3	26.2	65.2
Fiber, g	4.3	5.7	1.9	0.9	0.9	1.6
Ash, g	3.3	3.9	1.1	0.7	0.5	1.0
Ca, mg	75	253	30	13	25	60
P, mg	350		100	80	85	220
Fe, mg	5.4	7.6	1.7	1.3	2.8	7.4
Na, mg	1		1	90	3	11
K, mg	1500		460	160	60	180
Vitamin A, IU	0	15	0	0	0	0
Vitamin B_1, mg	0.45	0.57	0.15	0.02	0.02	0.01
Vitamin B_2, mg	0.16	0.18	0.06	0.04	0.05	0.03
Niacin, μg	2.2	3.2	0	0.03	0.1	0.8
Vitamin C, μg	0		0	0	0	0

[*]Also see Ogawa *et al.*, 1983.

Table 8.2. Nutritional and elemental analysis of azuki beans from four sources. Also see Hsieh *et al.*, 1992 and Aykroyd and Doughty, 1982.

Components and elements	Robinson, 1983 (Irrigated)	Robinson, 1983 (Dryland)	Kim and Choi, 1987	Duke, 1981	Dong and Liang, 1980
Water %			11.6	10.8	
Protein %	19.8-25.5	24.0-31.2	23.6	19.9	20.7-29.2
Fat %			0.8	0.6	0.4-3.6
Carbohydrate %			38.8	64.4	55.9-61.0
Fiber %			15.3	7.8	4.9
Ash %			10.0	4.3	3.3
Phytic acid %			0.8		
N %	3.6	4.2			
P %	0.5	1.6		0.3	
K %	1.6	0.5			
S %	0.2	0.2			
Ca %	0.1	0.1		0.1	
Mg %	0.2	0.2			
Fe μg/g	7	75		98	
Zn μg/g	35	42			
Mn μg/g	20	21			
Al μg/g	14	8			
B μg/g	17	12			
Cu μg/g	9	9			
Na μg/g	11	12			
Ni μg/g	3	4			
Pb μg/g	3	1			
Cr μg/g	<1	<1			
Cd μg/g	<1	<1			

Storage conditions also affect soaking rates, cooking rates and the percentage of hard azuki seeds. Azuki seed moisture can affect quality characteristics over time, e.g. seed stored at less than 13% moisture maintained quality for up to four years and a high germination rate for five years (Duan, 1989). Hayakawa and Breene (1982) observed a highly positive correlation between the relative humidity of azuki seed storage and cooking properties. As relative humidity of storage increased from 12.4% to 98.2%, so did the percentage of completely cooked seeds. Ozawa (1978) and Drager (1983) attributed the degree and development rate of azuki seed testa impermeability to seed age and the humidity and temperature of storage. Drager (1983) suggested that protein, lipid or phytate changes in azuki cotyledons during cooking might impede necessary cell separation and that imbibition of water could be increased with a scarification treatment, especially in old seed (Hayakawa and Breene, 1992). Hu *et al.* (1985) observed a positive and significant correlation between increasing levels of gamma irradiation applied to azuki seeds for azuki bean weevil control and seed splitting during cooking.

The stage of seed maturity has also been shown to affect imbibition and cooking time. Hsieh *et al.* (1992) analyzed and compared the composition (Table 8.5) and cooking time of seeds of four azuki cultivars and two kidney bean cultivars (*P. vulgaris*) harvested at three stages of maturity. In general, overmature azuki seeds required longer cooking time and were less affected by presoaking than immature and mature seeds.

Six dimeric procyanidin types from azuki seed were isolated, characterized and tested for organoleptic astringency by Ariga and Asao (1981). The total dimeric procyanidin content in azuki was found to be greater than 0.011% which could be a source of azuki astringency.

Color

Characteristics of the azuki seed testa are very important quality considerations. Sasanuma *et al.* (1966) identified one outer soluble pigment of azuki seeds as a 3-monoglucoside of delphinidin. Azuki seeds contain approximately 21.7 mg catechin equivalent (CE/g) of condensed tannin (Laurena *et al.*, 1984). Goto *et al.* (1984) found differences in azuki seed color as a result of different drying methods, e.g. traditional field stacks or drying on paddy field racks resulted in superior seed testa color as compared to forced air drying at 30°C. The effects of three storage temperatures over seven to nine months were tested for their effects on color, seed hardness and *an* yield (Hatai, 1982). A high storage temperature (35°C) caused less seed brightness (L*), less *an* yield and harder seeds.

Colorimetry has been used to measure the degree of yellowness (b*), brightness (L*) and redness (a*) of azuki seed. These were found to vary 34%, 14% and 8% respectively, primarily as a result of the region of production, grower and year, and only slightly as a result of harvesting and drying procedures (Yoshida *et al.*, 1989b, 1991). These researchers felt that seed testa color could be improved through both breeding and management. The color, protein and size of 380 azuki samples produced in Hokkaido were evaluated by Kato *et al.* (1992). The redness of azuki seeds was found to be negatively correlated to seed size. The seed coat of the cultivar *Erimo* had the highest values for brightness and yellowness among five cultivars; cultivars from the Tokachi district of Hokkaido had the highest values for these components of color compared to other districts on the island. Compared to the seed coat, brightness for unsweetened *an* was much higher and redness and yellowness values were lower; sweetened *an* had much lower brightness and slightly higher redness values.

Pesticide residues

Japan has set pesticide residue standards for raw azuki seeds. A listing of restricted materials and their maximum residue levels was compiled from several sources and includes: benomyl (0.5 ppm); BHC (0.2 ppm); captan (5.0 ppm); chloropropham (0.05 ppm); DDT (0.2 ppm); disulfoton (0.1 ppm); diuron (0.05 ppm); endosulfan (0.5 ppm); fentin acetate (0.02 ppm); fentin chloride (0.02 ppm); fentin hydroxide (0.02 ppm); parathion (0.3 ppm); phenthoate (0.05 ppm); prometryne (0.05 ppm); malathion (0.5 ppm); thiophanate-methyl (1.0 ppm); trifluralin (0.01 ppm); and, simazine (not to be detectable) (Soil and Pesticide Division of the Japanese Environmental Agency, 1975, 1977; Tomizawa, 1977; Japan Plant Protection Association, 1978, 1980).

FOOD CHEMISTRY

An production

Currently azuki is principally used to make a sweetened paste called *an* in Japanese, pronounced as 'ahn', composed of nearly equal parts azuki paste and sugar (Piper and Morse, 1914; Insull *et al.*, 1968; Sacks, 1977; Cox, 1978; Hernandez *et al.*, 1978; Scarbnick, 1979; Duke, 1981; Watanabe *et al.*, 1983; Hoshikawa, 1985; Hammer *et al.*, 1986; Sakurai, 1986; Breene and Hardman, 1987; McClary *et al.*, 1989; van der Maesen and Somaatmadja, 1989). Nissan Sugar Manufacturing Company (1989) in Japan described the following generalized steps for the production of smooth azuki *an* (*koshi an*) (Fig 8.3) from small azuki beans: dust separation, stone separation, metal separation, vibrating screen for bean size separation, weighing, washing, soaking, boiling twice in a pressure cooker, rinsing with water to remove antidigestive compounds, crushing, removing seed testas, removing excess water in leaching tank, dehydrating, weighing, combining with sugar and various stabilizing ingredients such as agar agar or even bacterial cellulose (Okiyama *et al.*, 1993), kneading, weighing and straining. The production of *an* containing whole and crushed beans (*tsubu-an*, *ogura*) is the same until after boiling, when the large cooked beans are soaked in a sweetening tank. Various systems have been developed to dispose of waste generated by this process (Kawase, 1993).

Factors affecting the processing of *an*, the sweetened azuki paste, have been the subject of considerable food science research, especially regarding starch, starch granules, proteins, lipids, fiber, color, aromatic compounds and taste.

Fig. 8.3. The cleaning and processing of azuki: (top left) hand picking of Hokkaido azuki; (top right) draining of cooked azuki in cloth bags followed by pressing of the filled bags, center of photo; (center row left) an American color sensor used in Japan to remove defective seeds; (center row center) azuki cooking pots; (center row right) mixers for combining cooked azuki and sugar; (bottom left) trays of *koshi an*; (bottom right) tubes of *an*.

Starch and sugar

The amylose, amylopectin, oligosaccharides and α-amylase inhibitor composition of several *Phaseolus* and *Vigna* species was analyzed by Sathe *et al.* (1982). They reported that azuki starch is 34.9% amylose, the seed is 0.3% raffinose and 2.8 to 4.1% stachyose (flatulence factors), and α-amylase inhibitor activity is about 64 to 92 units/g of bean. Others (Bae and Lee, 1987) also studied the content of the flatulence factors in azuki and other legume seeds and report that azuki contains 2.9% stachyose and 0.5% raffinose. They concluded that azuki and mungbean seeds had less of these factors than kidney bean and soybean. A flurometric assay for trypsin inhibitor in azuki has been developed (Jung and Lee, 1982). Stachyose is considered to be the major oligosaccharide in azuki seed (Tanusi *et al.*, 1972; Kay, 1979; Saini, 1988). Japan Bean Fund Association (1989) lists a range of 2.0 to 2.9% for stachyose in different azuki cultivars grown at different locations. The Association also reports that the cultivar *Erimo* grown on Hokkaido contained free sugars ranging from 5.8 to 6.3% (Japan Bean Fund ..., 1989). Sucrose was 0.9 to 1.2% and galactosyl was 0.2 to 0.3%. Galactosyl cyclitol has also been found (Yasui, 1980).

The moisture content of azuki has been found to be related to starch. Ishibashi *et al.* (1992) studied the equilibrium moisture contents of azuki, soybean and kidney bean under a range of temperatures and relative humidities. They found that azuki had a higher moisture content and took the longest, about 100 days, to reach a static moisture, which was attributed to its relatively high starch content.

Storage temperatures (five treatments over two years) have been found to affect the properties of pectic polysaccharides and hemicelluloses in azuki seeds (Shiota *et al.*, 1982; Shiota *et al.*, 1991). Using DEAE-cellulose chromatography, seeds stored at 30°C were found to have low uronate contents in all acidic fractions and increased amounts and smaller molecular sizes in the III and IV acidic fractions. Neutral sugar compositions in the II, III and IV acidic fractions and hemicelluloses showed no difference as a result of the two storage temperatures. The major neutral sugar constituent was xylose in the II through IV fractions. An α-glucan was purified from azuki seed extract through a column of DEAE-Sephadex A-50 and was found to have an average molecular weight of about 20,000 with a 2.1 ratio of A chain to B chain (Wei *et al.*, 1993).

Starch granules and processing

Some studies have concentrated on tests for determining amylase activity, starch gelatinization, composition values and new processing procedures. Starch isolated from azuki cv *Takara* produced in Minnesota (Tjahjadi and Breene, 1984) has granules that are oval to kidney shaped and 15 to 45 μm in diameter with 28.8% amylose and a gelatinization temperature range of 64-66-68°C. Gel strength was similar to maize and wheat starch. A new method to isolate starch granules from azuki seeds by using cellulases isolated by fractionation from *Fusarium moniliforme* was developed by Cho and Park (1991).

A method for determining the proportion of ruptured starch granules in standard *an* through the use of α-amylase was described by Kugimiya and Hirata (1987a) and Kugimiya (1987ab). α-Amylase was found to solubilize 2% of the starch in intact granules and 98% in ruptured granules. This method for determination was developed further (Kugimiya and Hirata, 1987b) by including an approach to estimate the proportion of damaged starch granules in *an*, in addition to the intact and ruptured granules. The average composition of 20 *an* samples from commercial processors was 61.7% intact granules, 31.3% damaged granules and 6.9% ruptured granules. Measureable differences were found in the proportion of ruptured and damaged *an* starch granules among 12 cultivars of azuki (Kugimiya, 1992a). Also, prolonged soaking before cooking was found to increase damaged granules and a pH higher or lower than 6 or a cooking temperature of 120°C compared to 100°C, increased both ruptured and damaged granules. The amount of ruptured and damaged granules resulting from each stage of processing of *neri-an* in commercial *an* producing factories was highest during the kneading process and least during cooking, grinding, sieving, leaching and dehydrating (Kugimiya, 1992b).

Others have also studied starch in intact and ruptured granules. Fujimura and Kugimiya (1993a) separated and compared isolated azuki seed starch and starch inside cells. The solubility and swelling power of starches inside cells was extremely low. When starch was extracted from cells that had been previously heated in water, it still had less solubility, swelling power and birefringence than isolated starches. Later, Fujimura and Kugimiya (1993b) compared the gelatinization of starches inside cotyledon cells of azuki and isolated starches using differential scanning calorimetry (DSC). Cotyledon cells subjected to a DSC thermogram exhibited an endothermic transition with a peak at 70°C trailing a shoulder to 98°C, while isolated starches had a single endothermic transition with a peak at 71°C. Their results suggest that gelatinization of starches inside cotyledon cells was suppressed, possibly due to cell wall strength and limited supply of intra-cellular water.

Rate changes in the hydrolysis of azuki *an* starch granules by β-amylase pullulanase were found to be a better indicator of changes in starch gelatinization or retrogradation than the gluco-amylase method (Watanabe *et al.*, 1983).

An improved method for cooking azuki increased *an* yield from the usual 67.5% to 68.1% (Abe, 1987a). This improved yield was accomplished by designing a cooking container that limited seed expansion to 2.22 times and increasing the cooking time from 60 minutes to 90 minutes. Azuki cooked by the new method had more free starch which was excessively sticky (Abe,

1987b). Stickiness of the sweetened *an* was reduced by slowing the stirring speed of the kneader by about half. Kneading time did not influence this quality.

The qualities of *an* prepared from *Erimo* azuki and several cultivars of *Phaseolus vulgaris*, *P. lunatus* and *P. coccineus* were compared (Kobayashi *et al.*, 1992a). Azuki and a cultivar of *P. vulgaris* (cv *Otebo*) each yielded *an* that was softer and more dispersible in the mouth. Kobayashi *et al.* (1992b) also prepared *an* from azuki and three species of *Phaseolus* for light and scanning electron microscopy of its particles (see Table 8.3). The cell walls of azuki were thinner than the *Phaseolus* species and could be removed by an alkaline solution followed by cellulase.

An experimental device for freezing fresh azuki *an* using cold air was tested for freezing times and rates of heat transfer (Tojo *et al.*, 1975). *An* shaped into spheres with a diameter of 30 and 45 mm froze to the center in 10 to 60 minutes when exposed to temperatures of -1 to -5°C with cold air velocities ranging from 1 to 13 m/s.

Table 8.3. The moisture content and particle size distribution of *Nama an* from four cultivars of azuki forced through five different mesh screens (Soma, 1988).

Moisture	100 Mesh	150 Mesh	200 Mesh	250 Mesh	<250 Mesh	Median Mesh
62.6-64.1%	5.3-7.8%	31.1-51.1%	33.0-48.7%	4.0-9.5%	5.1-6.8%	142-165

Protein

Azuki seed protein content (Tables 8.1 and 8.2) is influenced by seed size, cultivar differences and environmental factors during growth. Seed protein was found to be positively correlated with seed size (Kato *et al.*, 1992). Yoshida *et al.* (1988) earlier reported a positive relationship ($r = 0.508$) between seed size and protein content from seed grown in seven Hokkaido production areas. Protein ranged from 18 to 26% for the azuki cultivar *Erimo* grown in these Hokkaido production areas. The average protein content of 1,461 samples tested in one study was 22.6% (Lin and Chen, 1993). When 207 Chinese accessions were tested, 14.5% had protein levels greater than 24% and six had a group protein average of 32.9% (He and Hao, 1987). Others (Yoshida *et al.*, 1988) have also identified cultivar differences, e.g. *Hatsune* averaged 26.6% protein in contrast to 21.6% for *Kotobuki adzuki*. Twenty-three azuki cultivars grown in the Philippines averaged 19.0% protein under dry season conditions (Hernandez *et al.*, 1978). Minnesota azuki produced under irrigated and dryland conditions contained 16.5% and 20.4% seed protein, respectively (Robinson, 1983). Tjahjadi (1983)

Table 8.4. Amino acid composition of azuki seed from three sources. Results are presented on the basis of mg/g N (Harukawa and Duke), percent of seed weight (Robinson) and percent of seed protein (Fu and Li).

Amino acids	Harukawa 1990 (mg/g N)	Duke 1981 (mg/g N)	Robinson 1983 (Irrigated) % of seed	Robinson 1983 (Dryland) % of seed	Fu and Li 1993 % of protein
Total protein			16.49	20.38	22.56
Isoleucine	270	244	0.81	0.93	3.95
Leucine	480	450	1.38	1.71	7.53
Lysine	450	456	1.28	1.65	6.64
Methionine	98	81	0.18	0.21	1.06
Cystine	99	56	0.20	0.20	1.06
Phenylalanine	350	338	0.89	1.16	5.63
Tyrosine	160	213	0.35	0.52	
Threonine	210	213	0.63	0.76	3.36
Tryptophan	64	106	0.18	0.18	0.79
Valine	320	275	1.04	1.10	4.88
Histidine	200	206	0.56	0.68	2.86
Arginine	410	394	1.09	1.75	6.56
Alanine	260	250	0.74	0.89	
Aspartic acid	660	613	2.00	2.42	
Glutamic acid	950	1075	2.85	3.42	
Glycine	240	213	0.66	0.85	
Proline	260	294	0.80	0.93	
Serine	290	269	0.85	1.04	

determined that complete extractable protein was 21.6% for the cultivar *Takara* grown in Minnesota.

Azuki protein is composed of 47% albumins and 37% globulins (Danimihardja and Lester, 1974; Tjahjadi, 1983; Liang *et al.*, 1988). Approximately 80% of all globulins were classified by Saianova and Reva (1977) as vicilin-like (V) and the rest as legumin-like (L). Vicilins from azuki are not degraded by protease C1 (Qi *et al.*, 1994). Azuki seeds lack the amino acid, pipecolic acid, and the acidic peptide, γ-glutamyl-S-methylcysteine, according to Casimir and Le Marchand (1966) and Bell (1971), but have the γ-glutamyl peptide of L-β-phenyl-β-alanine.

The main azuki storage protein is substantially less heterogeneous than the seed proteins of other legumes. The 7S protein-I has a molecular weight of 150,000 $\pm$ 15,000 daltons including three subunits of 55,000 $\pm$ 3000, 28,000 $\pm$ 1400 and 25,000 $\pm$ 1300 daltons (Sakakibara *et al.*, 1979). Chen *et al.* (1984) only identified two carbohydrate-containing 7S protein-I subunits, designated as alpha and beta$_1$, with relative masses of 35,000M_r and

55,000M_r, respectively. When azuki bean, soybean, cowpea and mungbean are immersed in water at 50 to 60°C, they were found to release proteins like 7S globulin (Hirano *et al*., 1992).

A haemagglutinin taxonomic marker was used to determine that azuki seed extracts are haemagglutinin negative in albumin and saline tests; extracts did produce weak anti-A lectin type reactions with ficin- and trypsin-treated erythrocyte suspensions (Toms, 1981).

Protein extracted from the azuki cv *Takara* grown in Minnesota was compared with soybean protein (Tjahjadi *et al*., 1988). Azuki protein extract, isolate, albumin and globulin were 78.8, 69.4, 89.2 and 91.3%, respectively. Hunter color values, water and fat adsorption capabilities, emulsifying capacity and emulsion stability, foaming capacity and foam stability, and influence of pH on solubility and emulsifying capacity were measured. All azuki fractions had superior or comparable essential amino acid contents compared to the soy protein; most showed higher solubility at acidic and alkaline pH, greater fat and water adsorption and foaming, and lower emulsifying capacities and stabilities. Germination increases the content of essential amino acids by about 25% (Lee and Karunanithy, 1990).

An taste and processing adaptability are partially influenced by protein content (Yoshida *et al*., 1988). Hirata and Kugimiya (1985) observed that small amounts of azuki protein inhibited the swelling of *an* starch granules at 60 to 95°C in a way characteristically different from other proteins such as ovalbumin, polypeptone, casein and gelatin. Ogawa *et al.* (1983) suggested that *an* starch granule cell walls, as well as protein, suppress starch gelatinization.

Lipids

Azuki lipids are composed of 75% unsaturated lipids (linoleic, oleic and linolenic) and 25% saturated lipids (stearic, palmitic and carnaubic) (Kay, 1979). Lipids constitute 1.0% or less of azuki seed weight, thus azuki has one of the lowest percentages of 22 grain legumes (Duke, 1981; van der Maesen and Somaatmadja, 1989). Kojima *et al.* (1991e) analyzed the total lipids from 16 legume species. These species were divided into two groups based on the contents of oleic acid. Group A consisted of azuki, rice bean, cowpea, green gram, black gram, lima bean, kidney bean, scarlet runner bean, soybean and hyacinth bean. Group B consisted of pea, broad bean, peanut, white lupine and winged bean.

Kojima *et al.* (1991c) studied the lipids in six species of legumes. They reported that the major lipid classes in azuki were triacylglycerol, phosphatidylcholine and phosphatidylethanolamine. In five of the six legume species, including azuki, free sterols, sterylglycosides and acylsterylglycosides showed the same pattern, but acylsterols were different. Triacylglycerol showed molecular differences useful to differentiate species

(Kojima *et al.*, 1991e). They (Kojima *et al.*, 1992) later found that the acylsterol content of azuki seeds did not change during maturation, but the proportion of stigmasterol increased and sitosterol decreased in free sterol, sterylglycosides and acylsterylglycoside. Two digalactosyldiacylglycerols, A and B isomers, have been found in the seeds of azuki and other common

Table 8.5. Fatty acids, cholesterol and vitamin E (tocopherols) composition of dry azuki beans.

Lipids and fractions	Fu and Li, 1993	Hsieh *et al.*, 1992*	Harukawa 1990	
Lipids			2.2	g/100 g beans
TFA fatty acids			0.89	g/100 g beans
SFA fatty acids			0.27	g/100 g beans
MUFA fatty acids			0.07	g/100 g beans
PUFA fatty acids			0.55	g/100 g beans
TFA			404	mg/g lipid
SFA			122	mg/g lipid
MUFA			31	mg/g lipid
PUFA			251	mg/g lipid
SFA			30.1	% of fatty acids
MUFA			7.6	% of fatty acids
PUFA			62.2	% of fatty acids
Cholesterol			0	mg/100 g beans
α Tocopherol			0.2	mg/100 g beans
β Tocopherol			0.2	mg/100 g beans
δ Tocopherol			3.0	mg/100 g beans
γ Tocopherol			8.1	mg/100 g beans
Vitamin E (E-Po.)			0.5	mg/100 g beans
6:0		2.2-5.5%		
12:0		0.2-0.4%		
Myristic, 14:0		0.3-0.4%	0.2	g/100 g fatty acid
Palmitic,16:0	27.9%	8.6-11.5%	24.9	g/100 g fatty acid
Palmitoleic, 16:1			1.2	g/100 g fatty acid
Heptadecanoic, 17:0			0.3	g/100 g fatty acid
Stearic, 18:0	3.2%	1.3-1.8%	2.7	g/100 g fatty acid
Oleic, 18:1	6.0%	2.2-4.0%	6.2	g/100 g fatty acid
Linoleic, 18:2, n-6	45%	13.0-20.2%	43.1	g/100 g fatty acid
Linolenic, 18-3, n-3	17.3%	14.1-17.8%	19.1	g/100 g fatty acid
Arachidic, 20:0	0.6%		0.4	g/100 g fatty acid
Eicosenoic, 20:1			0.2	g/100 g fatty acid
Behenic, 22:0			1.6	g/100 g fatty acid

*Percent of free lipid

beans and include several moieties each (Kojima *et al.*, 1991c). Four classes of non-acylated and acylated triterpenoids have also been identified and described (Ito *et al.*, 1985) as have glycosylsterols, saponin (Kojima *et al.*, 1989; Kitagawa *et al.*, 1983abc), glyceroglycolipids (Kojima *et al.*, 1989) and other glycerolipids (Kojima *et al.*, 1990a).

Cerebroside and ceramide were isolated and characterized from the lipid fraction of azuki by Ohnishi and Fujino (1981). Azuki, soybean and two *Phaseolus* spp. have similar major ceramide residues in cerebrosides of their seeds (Kojima *et al.*, 1991b). Kojima *et al.* (1991a) analyzed the sterol lipids and sphingolipids in the plant parts and seeds of azuki. Stigmasterol and sitosterol sterollipids are the major constituents in all parts plus acylsterol in the stems, roots and seeds. The major associated sphingoids are dihydroxy bases for cerebroside and trihydroxy bases for ceramide. The major associated fatty acids in cerebroside are 1-O-glucosyl-N-2'-hydroxy palmitoyl-sphingadienine and in ceramide are N-2'-hydroxylignoceroyl-4-hydroxy-trans-8-sphingenine.

Lee *et al.* (1992b) studied the tocopherols, lipids and compositions of fatty acids in 13 species of food legumes used in Korea. They found that the tocopherol concentrations of azuki differed from the other species. Concentrations in azuki were ranked as delta > gamma > alpha tocopherol, compared to a ranking of gamma > delta > alpha. Azuki contains 2.8 mg of total tocopherols per 100 g of seeds (Ito *et al.*, 1981). Approximately 76% of the total tocopherols were in the form of the tocopherol homologue sigma tocopherol; neither beta tocopherol nor beta tocotrienol were isolated.

Lipoxygenase from azuki acts on the free form of polyunsaturated fatty acids. The enzyme is activated by calcium, has optimum activity at pH 6.5 and yields 13 to 9 hydroperoxides from linoleic acid at a ratio of 48:52 (Yamamoto *et al.*, 1991a). Activity of lipoxygenase is greater in overmature seeds than immature or mature azuki seeds (Hsieh *et al.*, 1994).

Fiber

Azuki was found to contain less soluble and insoluble dietary fiber, ash and free lipid than kidney bean, and these did not change consistently during seed maturation. Nearly all insoluble fiber in both azuki and kidney beans is derived from the seed coat (Hsieh *et al.*, 1992). The fiber content of azuki seed ranges from 4.3% (Hayakawa and Breene, 1982) to 7.8%, one of the highest values for 22 grain legumes tested (van der Maesen and Somaatmadja, 1989), and one report claims 15.8% (Table 8.2).

Honda *et al.* (1992) noticed a discrepancy between the dietary fiber in raw and processed beans of five species when the AOAC method was used. They tried solubilizing the samples of cooked beans with dimethylsulfoxide and then analyzing fiber content with a modified Prosky method. This approach yielded results comparable to the analysis of raw beans. The original discrepancy of higher fiber values in cooked beans is thought to be caused by a retrograded starch.

Aromatic compounds

Alkanols and aromatic alcohols are two of the most important classes of 70 compounds (Table 8.6) identified by Tokitomo and Kobayashi (1988) to comprise the aroma of cooked azuki, with maltol contributing the most. Tokitomo (1988) has also reported on the odor components of raw azuki bean and odor at 10 and 60 minutes after cooking.

Table 8.6. Odor concentrates of cooked azuki and *Koshi an* (Tokitomo, 1992).

Peak no.	tR (min)	Compound	Cooked beans	*Koshi an*	Dried *Koshi an*, 0 days	Dried *Koshi an*, 40 days	Dried *Koshi an*, 90 days
1	4.9	Acetone	5.56		4.53	3.54	4.03
2	5.4	Methanol	6.67	1.17	5.64	2.88	0.12
3	5.74	Dichloromethane (solvent)					
4	6.39	2, 3-Butanedione	0.93	0.62			
5	6.5	*n*-Pentanal	0.41	0.53	1.17	0.53	0.53
6	7.66	2, 3-Pentanedione	0.20	0.40	4.04	0.16	
7	8.61	*n*-Hexanal	0.79	1.85	2.87	4.00	2.30
8	9.8	*(E)*-3-Penten-2-one		0.42			
9	10.69	3-Heptanone	0.35	0.50	0.97	0.72	0.19
10	11.79	2-Heptanone	2.45	0.42	0.54	0.30	0.28
11	11.9	Pyridine	13.95	1.84	0.24	0.21	0.19
12	12.56	3-Methyl-1-butanol	2.07	0.75			
13	13.26	*(E)*-2-hexenal	0.32	0.44	0.68	0.61	0.40
14	14	2-Pentylfuran	0.37	0.60	1.90	1.51	0.83
15	14.47	*n*-Pentanol	0.33	0.65	0.66	0.92	0.70
16	15.08	Styrene		1.59			
17	16.21	3-Hydroxy-2-butanone	2.68	2.89	0.55	0.26	
18	19.88	*n*-Hexanol	0.22	0.58	0.27	0.22	0.19
19	22.3	3-Octanol		0.17			
20	25.3	Acetic acid	1.05	0.68	1.04	1.30	0.80
21	25.7	1-Octen-3-ol	+	0.65	0.44	0.64	0.82
22	25.9	*n*-Heptanol		0.18			
23	26.47	2-Furfural	1.66	1.06	2.33	2.26	
24	28.18	2-Ethyl-1-hexanol	0.59	2.73	3.30	6.11	7.68
25	30	Benzaldehyde	2.17	12.51	13.40	8.10	5.49
26	32.37	*n*-Octanol		0.36			
27	33.25	Ethyl furyl ketone	0.89	0.37	1.81	5.81	12.98
28	34.2	Dimethyl sulfoxide	0.26				
29	37.64	2-Acetylthiazole	1.20	2.01			
30	38.72	Acetophenone		0.51			
31	39.61	Furfuryl alcohol	15.11	1.77	0.54	0.90	0.14
32	40.7	4-Hexanolide	+	+			
33	45.22	Sesquiterpene hydrocarbon		0.32			

Table 8.6. Continued.

34	47.4	***n*-Octadecane**		3.02			
35	48.07	**1-Phenylethanol**		0.17			
36	50.01	**α-Ionone**		0.15			
37	50.25	**Geranyl acetone**		0.66			
38	50.75	**Guaiacol**					
39	50.96	**Benzyl alcohol**	1.67	0.93	0.64	1.08	0.82
40	52.92	**2-Phenylethanol**	1.94	0.77	+	0.25	0.13
41	55.15	**Benzothiazole**	0.24	0.55			
42	55.55	**Maltol**	7.69	2.89	1.78	+	0.72
43	58	**Phenol**	0.27	0.59	0.51	0.47	0.49
44	60.53	**Nerolidol**	+	1.55			
45	68.88	**2, 3-Dimethyl-2-nonen-4-olide**		0.87			
46	69.8	**2-Methoxy-4-vinylphenol**	3.18	1.19	3.15	2.11	1.66
47	89.74	**4-Vinylphenol**	0.68	0.45	2.04	0.52	+

\+ = trace

SPROUTS

Azuki is one of the preferred seeds in Asia for bean sprouts. Azuki sprouts have good consumer acceptance according to Chang (1986). Research has been conducted to address production and packaging concerns. The application of artificial light (380 to 700 nm; 500 Lux) for one hour per day over two days resulted in azuki hypocotyls that were longer and harder, and that had greater diameter, weight and vitamin C content than azuki sprouts produced in dark growth chambers (Tajiri, 1981). In the same study, light-induced azuki seedling characteristics were found to be more desirable because of recent consumer demand for thicker sprouts. Azuki sprouts had a shelf life of 10 to 11 days at 0 to 5°C when packaged in either polyvinyliden chloride laminated ‘Cellophane’ or polyvinyliden chloride ‘Saran Wrap’ (Tajiri, 1979).

9

USES AND MARKETING

HISTORY

Azuki is consumed in China, Korea and Japan in a myriad of ways that take advantage of the seed's red color, smooth texture and delicate flavor. It is traditionally served in a sweetened form on festive days such as weddings, birthdays or New Year parties (Sasanuma *et al.*, 1966; White, 1972; Tanaka and Konno, 1982; Vietmeyer, 1986; McClary *et al.*, 1989). Uses of azuki have been documented in Chinese books for nearly 2000 years, beginning with the 1st century book *Fan Sheng Zhi Shu* (氾勝之書) which has been lost but was excerpted in the 6th century book *Qi Min Yao Shu* (齊民要術). In this book, azuki leaves were reportedly sold as greens. Yang and Lu (1985) reported that *Qi Min Yao Shu* states that diseases could be prevented by eating azuki seed during the first and seventh lunar months that had been stored with castor bean seed and some other ingredients in a well. Hu (1984) reported that the book *Jing Chu Sui Shi Ji* (荊楚歲時記), AD 502 to 557, records that people ate azuki to ward off ghosts.

USES

Food products

Azuki *an*, either in a smooth, whole-bean or intermediate form, is used in numerous East Asian foods, especially East Asian and Western-style desserts, e.g. steamed, *an*-filled buns (*manju* [Japanese], *baozi* [Chinese]) and Asian versions of sweet rolls, donuts and ice cream (Sperbeck, 1981; McClary *et al.*, 1989). Molded desserts are produced from *yokan*, *an* gelatinized with agar agar. A smooth *an* without the bean seed coat (*neri an*) is used as a filling for products such as *taiyaki* (a filled, fish-shaped waffle). Saito (1988) has published one of the most informative resources about Japanese confectionary products, including some recipes.

Sugar-infused whole azuki beans with some bean sauce (*tsubu an*) are used as a topping for ice cream and snow cones (Figs. 9.1 and 9.2). *An* is

Fig. 9.1. Japanese azuki confectioneries: (top left) a traditional sweet with *koshi an* in the center covered with *mochi* rice paste and wrapped in a leaf; (top right) gift packages of *koshi an* and *tsubu an* sweets; (center left) a *taiyaki* waffle stand at a festival (*matsuri*) in Kyoto; (center right) *neri an* filled *taiyaki* waffles served with tea in a famous Tokyo *taiyaki* shop; (bottom left) shaved ice and ice cream desserts with *ogura, amanatto* and *mochi* covered *koshi an* toppings; (bottom right) a Chinese Moon Festival cake filled with a dry type of *tsubu an*.

also used as a base for a soft drink served hot from vending machines (*shiruko*) (Motomiya and Ito, 1972; Narikawa, 1972; Scarbnick, 1979; Nakayama and Shinya, 1981; Matsuda *et al.*, 1982; Chen *et al.*, 1984; Vietmeyer, 1986; Duan, 1989; McClary *et al.*, 1989). *An* can be flavored with sweet syrups or with soy sauce (Insull *et al.*, 1968). A white-seeded azuki is used to make high quality white *an* for specialty Japanese bakery products (Tanaka and Ohno, 1895; Narikawa, 1972). Kamimura and Maeda (1991) recently received a patent for the development of an azuki *an*-based candy with a sugar alcohol-treated soft center. Many American health and

Fig. 9.2. Azuki bean products: (top left) dry beans for home use produced by the Hokuren Cooperative on Hokkaido; (top center) *an*-filled Danish pastry produced in Torrance, California for the Asian-American market; (top right) a bread stick with an azuki bean layer through the center, baked in Los Angeles; (bottom left) precooked and dried *sekihan*, rice with azuki beans, for home preparation; (bottom center) an azuki bean ice cream bar sold in Malaysia; (bottom right) an *ogura an* sweet being produced on Hokkaido.

oriental food stores sell whole azuki seeds and various products containing *an* (Hardman and Meronuck, 1982). The rice bean, various kidney beans, lima beans, Austrian winter peas, etc. are occasionally used for Japanese *an* production, but azuki is the preferred seed for high quality *an* (McClary *et al.*, 1989).

Azuki beans are also candied whole (*amanatto*), used as a component in sweet soups (*zenzai, sarashi ame*), mixed with rice (*azuki-mochi* and *sekihan*), sprouted (*moyashi*), made into a flour, popped like maize or used as a substitute for coffee (Tanaka and Ohno, 1895; Sacks, 1977; Kubota, 1979; Hoshikawa, 1985; Shi, 1988; Duan, 1989; McClary *et al.*, 1989). Noodles are made from azuki flour mixed with wheat flour in Gansu Province, China (Shi, 1988). Some of the earliest azuki cultivars characterized in the United States produced immature pods that were judged inedible (Strand, 1943). The harvest, quick-freezing and marketing of immature azuki pods as a new frozen vegetable was unsuccessful due to the fibrousness of the pods (Piper and Morse, 1914; White, 1972; Kay, 1979). However, their use as a green vegetable is implied by Smartt (1985) and Fernandez and Talekar (1990).

Other uses

Azuki is mainly produced for food, but other uses include: cosmetics (facial creams and shampoos), medicines, a nitrogen-fixing green manure, livestock feed and a soil stabilizer (Georgeson *et al.*, 1891; Tanaka and Ohno, 1895; Piper and Morse, 1914; Zhukovsky, 1950; Plant Production ..., 1959; Thomas *et al.*, 1974; Sacks, 1977; Hernandez *et al.*, 1978; Kay, 1979; Duke, 1981; Ci Hai ..., 1983; Hoshikawa, 1985; Duan, 1989; van der Maesen and Somaatmadja, 1989). A traditional Japanese facial scrub called *azuki no kona* is made from briefly roasted, crushed and sifted azuki which is applied directly to the face with water or in a moistened silk or cotton scrubbing bag (Anderson-Leigh, 1989).

Azuki (usually *chi dou* cultivars) is used in traditional Chinese medicine for the treatment of kidney ailments, fever, *Staphylococcus aureus* infections, typhoid, dysentary, various difficulties of pregnancy and delivery, and for improvement in blood circulation and urination (Sacks, 1977; Ag. Lexicon ..., 1979; Kay, 1979; Duke, 1981; Tong, 1981; Ci Hai ..., 1983; Chinese Ag. Resources ..., 1987; Jin and Jiang, 1988; Duan, 1989; van der Maesen and Somaatmadja, 1989). Soma *et al.* (1992b) recently received a patent for the identification of lipopolisaccharides from azuki and other plants that inhibit inflammation, serving as anti-rheumatics. In the *yin* and *yang* view of life in Asian cultures, azuki seeds are used as a strongly *yang* medicinal food to bring *yin* forces into balance. Leaves have *yin* powers

(Sacks, 1977). Azuki flour has also been used as an ingredient in culture media. Kamyama *et al.* (1993) were the inventors of a patented process to prepare L-carnitine from filamentous fungi growing on a medium containing azuki.

MARKETING

Consumption patterns

Statistics on azuki consumption trends and patterns in most Asian countries are generally not available (see production figures in Chapter 4) although they are known to differ widely by country and region. Japanese consumption can be classified into the following categories: *an* paste (68.9%), candied seeds (12.8%), boiled seeds (2.4%) and other (15.9%) (Japan Bean Fund ..., 1987). Approximately 30% of all *an* is used by the Japanese and Korean ice cream industries (Hayakawa and Breene, 1982). In the mid-1970s, 85% of Japanese domestic production and imports of azuki were used in the production of *an* (Konno and Narikawa, 1978). The Taiwanese use azuki in processed food products and in a whole, cooked form; azuki constituted about 4.3% of all grain legumes consumed directly between 1964 and 1978 (Chao, 1982). Chinese and Korean azuki use characteristics are more traditional, with less adaptation to Western confectionery as in Japan (Mun-Chan *et al.*, 1986).

Japanese consumption has been stable or increasing in large part due to three societal trends: an aging population, increased concerns about health and a higher standard of living. Current and projected increases in Japanese azuki consumption are generally related to demographics and season. Young, working women are the major consumers of azuki confections. The gelatinous *yokan* is mostly consumed during the summertime by Japanese between the ages of 55 and 64; the same segment of society eats *manju*, usually during late autumn and early winter complemented by hot tea (McClary *et al.*, 1989).

Azuki consumed in the United States comes from the sale of imported or domestically-produced azuki in Asian-American communities and organic food stores (Sperbeck, 1981; Tevis, 1987; McClary *et al.*, 1989; National Research Council, 1989). However, for the majority of the American population azuki does not seem to have any unique appeal.

Grading

Japanese classify raw azuki seeds into four grades based on percentages of moisture, foreign material and damaged/immature seeds. Grade one, with the highest price, allows 10% damaged/immature seeds, no foreign matter or seeds of a different type and requires a 15% seed moisture (Japan Bean Fund ..., 1987). Azuki that meets the grading criteria of large seed size (*dainagon*) and scarlet seed testa color, is used for high quality confection products and customarily commands a price 40 to 90% higher than standard grade azuki (Furusawa *et al.*, 1974).

China uses a three-grade system based on criteria for percentage of moisture, foreign material and damaged/immature seeds, as well as seed testa color and smell. The Chinese grade one requires a minimum of 96% azuki seeds larger than a diameter of 2.0 mm, not more than 2% broken or immature azuki seeds smaller than a diameter of 2.0 mm, 14% seed moisture and normal testa color and smell (Agriculture Publishing House, 1982). Grade one must be 98% pure, grade two 95%, grade three 92% and all three grades must have not more than 1% foreign material and 13.5% water (Zhao and Tong, 1993). Chinese grade one azuki commands a higher producer-level price and is almost exclusively exported while grades two and three are sold predominately on the domestic market (Mou and He, 1985). In Hebei province, grade one has >55% starch, >73% starch+protein+fat and >20% protein (Fan *et al.*, 1992). According to Guo (1993) the price of azuki paid to Chinese farmers reached 4.6 to 5.8 yuan/jin (>$1/kg) during October of 1993, which was more than the price of pork.

Little economic data is available on farmer-level azuki production decisions in East and Southeast Asia, with the exception of Japan (Statistics and Information ..., 1992). Azuki usually provides a higher return to the Oriental producer than other crop options; in Japan, this is principally attributable to official restrictions on imported azuki. The profitability of azuki production has encouraged production in the marginally adapted cool region of Hokkaido (Sunada *et al.*, 1975). An azuki production survey conducted in Hokkaido by the Asian Productivity Organization in the late 1970s found that average total fixed and variable costs were $1548 per ha, gross revenue was $2655 per ha for a net revenue of $1107 per ha. This net return to the producer from azuki was greater than that for soybean, sugarbeet and potato production but less than for paddy rice (Suzuki and Konno, 1982).

Azuki prices in Japan are influenced mostly by domestic production and less so by production in exporting countries such as China (Motomiya and Ito, 1972). Producers in Japan received about 20,000 to 24,000 ¥/60 kg during the late 1980s (Economic Info. ..., 1991), but wholesale prices in Japan reached a record high of 90,000 ¥/60 kg for Hokkaido azuki during November and December of 1993 as a result of poor yields from a cold summer. Prices for 60 kg units of Japanese and Chinese azuki are listed in the financial section of major Japanese newspapers.

Azuki was determined by Park and Hong (1970) to be the second most profitable of four potential catch crops evaluated in paddy rice fields throughout Korea. However, a survey conducted later by the Asian Productivity Organization in Korea showed that total expenses were greater than gross income for a negative net revenue (Suzuki and Konno, 1982). Hwang (1982) attributed the outcome to the power of middle men to lower farm prices despite the existence of a Korean government price support program for azuki. The system of azuki price supports was discontinued in 1981 with a final release of 3169 mt of azuki into the marketplace (Choi and Lee, 1987).

International trade

International trade in azuki is split into four product categories. Raw azuki seed is the most well documented (see Chapter 4). Other azuki products that are traded include various forms of *an*, flour, desserts, snack foods, cosmetics and medicines.

Japan is the dominant importer of azuki in its various forms although several other countries such as Korea import limited amounts (Kay, 1979; van der Maesen and Somaatmadja, 1989). Japanese institutions play a major role in the foreign trade and marketing channels for azuki (Fig. 9.3) (McClary *et al.*, 1989). For example, the Japanese azuki bean import market is regulated by means of a quota which is determined on a biannual basis by the Ministry of Agriculture, Forestry and Fisheries (MAFF) (Cox, 1978). MAFF determines the imported quantity based on projected domestic azuki production levels (McClary *et al.*, 1989). Imports of azuki *an* entering Japan are not restricted by quota. During the 1980s, domestic azuki production annually supplied an average of 73% of the total quantity of azuki used in Japan with the remaining amount provided by imports (27%) (Japan Peas and Beans Importer Association, 1987).

An example of another organization is the Japan Bean Fund Association, which is a semi-governmental organization established in 1965 to protect and promote the production and consumption of miscellaneous legumes, such as azuki, through the dissemination of information and support of approved research (Tanaka and Konno, 1982). This organization receives half of the import tariff of 10% administered to imported azuki (McClary *et al.*, 1989) and played a role in the initial seed increase and promotion of the Japanese azuki cultivar, *Iwate dainagon* (Furusawa *et al.*, 1974).

World prices for azuki are largely determined by supply and demand in the Japanese market, which consumes about 110,000 mt annually (JETRO ..., 1991). During the early 1990s, export prices for Chinese azuki tended to range between about $400 and $800 per mt ($890 in 1989), $1000 and $1400 for U.S. production. Forty-three trading corporations are allowed to import azuki based on a quota issued by the government. Importing is quite lucrative for those importers who receive government patronage in the form of these quotas. Imported azuki is often resold to Japanese processors for three or more times the purchase price.

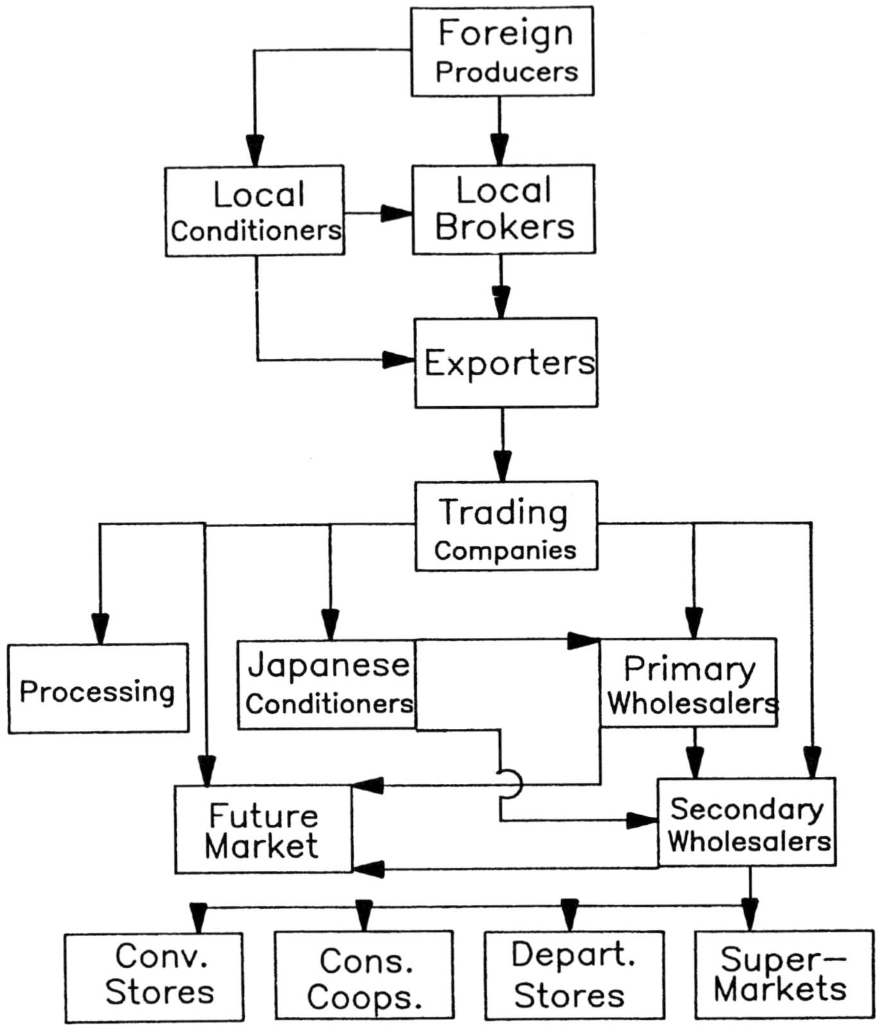

Fig. 9.3. Marketing channels for foreign azuki entering the Japanese market. From the top, the azuki moves from foreign producers, conditioners, brokers and exporters to Japanese trading companies, processors, wholesalers and retailers (adapted from Muto, 1987).

As this book goes to press the government of Japan is preparing a new policy for azuki imports. The azuki import quota based on the size of the deficit in domestic azuki production is expected to be eliminated. Azuki may be grouped with all other pulses in a multi-tiered tariff system, with possibly 120,000 mt of all pulses allowed in under a low tariff for the first tier.

REFERENCES

Abe, S. (1987a) A modification of the boiled beans for the raw ann manufacture of azuki (studies on improvement of sweetened adzuki ann Part II). *Nippon Shokuhin Kogyo Gakkaishi* 34(1), 14-19.

Abe, S. (1987b) The kneading conditions to produce sweetened adzuki ann (red bean jam) from the cooked beans prepared by the modified cooking method (studies on the quality improvement of sweetened adzuki ann Part III). *Nippon Shokuhin Kogyo Gakkaishi* 34(6), 380-385.

Abe, N., Sato, H. and Sakamura, S. (1987) Antifungal stress compounds from adzuki bean, *Vigna angularis*, treated with *Cephalosporium gregatum* type B. *Agriculture and Biological Chemistry* 51(2), 349-353.

Abivardi, C. and Rahimian, M. K. (1977) Further studies on camphor as a promising candidate insecticide for control of the azuki bean weevil. *Agrochimica* 21(4), 219-225.

Abo, M. and Mori, K. (1993) Synthesis of R-callosobruchusic acid from methyl R-3-carboxybutanoate. *Bioscience, Biotechnology and Biochemistry* 57(2), 265-267.

Adachi, T., Chiba, I., Murata, K., Satoh, H. and Narikawa, T. (1980) A new adzuki bean variety "Hokkai-shiroshozu" (in Japanese, English summary). *Bulletin Hokkaido Agricultural Prefecture Experiment Station* 43, 80-87.

Adachi, T., Murata, K. and Narugawa, T. (1979) Azuki growth research. III. Varietal responses to planting density (in Japanese). *Hokkaido Assembly Breeding Crop Institute* 19, 47 (Abstr.).

Agricultural Development Committee, Agriculture and Forestry Bureau. (1981) Small red bean. In: *An Introduction to the Cultivars of Minor Crops in Taiwan (Continued)* (in Chinese). Agriculture and Forestry Bureau of Taiwan, Taipei, pp. 139-149.

Agricultural Publishing House Editorial Committee. (1980) *Chinese Agricultural Proverbs*. Agricultural Publishing House, Beijing, 731 pp.

Agriculture, Forestry, and Fisheries Technique Convention Executive Office. (1967) Part III. Azuki (in Japanese). In: *Compilation of Miscellaneous Grain Crops Research Experiment Results, Peanuts, Kidney Bean, Azuki, Broad Bean, Peas and Other Miscellaneous Beans* (Showa 27-38th year). pp. 95-108.

Agriculture Lexicon Editing Committee. (1979) Red beans (in Chinese). In: *Agriculture Lexicon*, 1st ed. Science and Technology Publishing House of Jiangsu Province, Nanjing, China, p. 710.

Agriculture Publishing House. (1982) Azuki (in Chinese). In: *Chinese Agriculture Yearbook 1982* (in Chinese). Agriculture Publishing House, Beijing, p. 530.

Aguilar, H. G., Tanigoshi, L. K. and Lumpkin, T. A. (1994a) Antibotic responses of azuki bean accessions to the two spotted spider mite. In: *Research Reports: 53rd*

Annual Northwest Insect Management Conference. January 10 & 11, 1994, Portland, Oregon. Oregon State University Extension Service, p. 3 (Abstr.).

Aguilar, H. G., Tanigoshi, L. K. and Lumpkin, T. A. (1994b) Control of two spotted spider mite on azuki bean. In: *Research Reports: 53rd Annual Northwest Insect Management Conference*. January 10 & 11, 1994, Portland, Oregon. Oregon State University Extension Service, p. 4 (Abstr.).

Ahn, C. S. (1988) Improvement of adzuki bean (*Vigna angularis* (Willd.) Ohwi and Ohashi) through interspecific hybridization between rice bean (*Vigna umbellata* (Thunb.) Ohwi and Ohashi) and adzuki bean (in Korean, English summary). *Research Report of the Korean Rural Development Association* 31, 257-262.

Ahn, C. S. and Hartmann, R. W. (1978a) Interspecific hybridization among four species of the genus *Vigna* Savi. In: Cowell, R. (ed.), *Proceedings of the First International Mungbean Symposium*. Asian Vegetable Research and Development Center, Shanhua, Taiwan, pp. 240-246.

Ahn, C. S. and Hartmann, R. W. (1978b) Interspecific hybridization between mungbean bean (*Vigna radiata* (L.) Wilczek) and azuki bean (*V. angularis* (Willd.) Ohwi and Ohashi). *Journal of the American Society of Horticultural Science* 103(1), 3-6.

Ahn, C. S. and Hartmann, R. W. (1978c) Interspecific hybridization between rice bean (*Vigna umbellata* (Thunb.) Ohwi and Ohashi) and azuki bean (*V. angularis* (Willd.) Ohwi and Ohashi). *Journal of the American Society of Horticultural Science* 103(4), 435-438.

Ahn, C. S., Min, K. S., Chung, S. J., Koh, G. C. and Park, S. H. (1986) Interspecific hybridization between *Vigna angularis* (Willd.) Ohwi and Ohashi and *Vigna umbellata* (Thunb.) Ohwi and Ohashi through *in vitro* culture of embryos (in English, Korean summary). *Korean Journal of Breeding* 18(4), 283-291.

Ahn, C. S., Park, H. G. and Tickoo, J. L. (1981) Genetic resources of mungbean, *Vigna radiata* (L.) Wilczek, at AVRDC. Presentation at the FAO/IBPGR Working Group on Genetic Resources of *Vigna*. September 17-19, New Delhi, India.

Ahn, W. S. (1987) Genetic Resources in the Republic of Korea. *FAO/IBPGR Plant Genetic Resources Newsletter* 69(12), 39-43.

Akai, J. (1988) Azuki diseases: *Pseudomonas syringae* pv. *syringae* van Hall, *P. adzukicola* Tanii et Baba and *Epicoccum nipponica* (Hiura) Hara (in Japanese). In: Kishi, K. (ed.), *Plant Diseases in Japan*. Zenkoku Noson Kyoiku Kyokai Co. Ltd., Tokyo, pp. 119-122.

Akai, J., Iizuka, N. and Obata, K. (1988) Azuki diseases (in Japanese). In: Kishi, K. (ed.), *Plant Diseases in Japan*. Zenkoku Noson Kyoiku Kyokai Co., Ltd., Tokyo, pp. 117-123.

Akai, J., Kajino, Y., Kodama, F., Sawasaki, A., Tomioka, Y., Tsuboki, K., Ui, T. and Yamada, E. (1986) *Guide for the Control of Diseases and Pests of Hokkaido Prefecture* (in Japanese). Hokkaido Prefecture Plant Protection Association, Sapporo, Japan, 495 pp.

Akai, J., Tsuboki, K. and Shiritsuki, T. (1971) Concerning the outbreak and damage of brown stem rot that multiplies in the Tokachi area (in Japanese). *Annals of the Phytopathological Society of Japan* 37, 168 (Abstr.).

Akiyama, Y. and Torikura, H. (1984) Notes on the injury of upland crops by a broad-nosed weevil, *Meotiorhynchus querendus* (in Japanese). *Annals of the Report of Plant Protection for North Japan* 35, 122-125.

Alberto, S. and Taylor, B. (1966) Estudios sobre polen de *Phaseolus* (in Spanish, English summary). *Turrialba* 16(1), 7-14.

Alexopoulos, C. J. and Mim, C. W. (1979) *Introductory Mycology,* 3rd ed. John Wiley and Sons, New York, 632 pp.

Alford, D. K. (1970) *Endogenous short period and circadian rhythms in movements of the leaves of Phaseolus angularis Wight grown under controlled humidity and carbon dioxide conditions.* Unpublished PhD thesis, University of Wisconsin, 86 pp.

Alford, D. K. and Tibbitts, T. W. (1970) Circadian rhythms of leaves of *Phaseolus angularis* plants grown in a controlled carbon dioxide and humidity environment. *Plant Physiology* 46, 99-102.

Alford, D. K. and Tibbitts, T. W. (1971) Endogenous short period rhythms in the movements of unifoliate leaves of *Phaseolus angularis* Wight. *Plant Physiology* 47, 68-70.

Allard, H. A. and Zaumeyer, W. J. (1944) Responses of beans (*Phaseolus*) and other legumes to length of day. *US Department of Agriculture Technical Bulletin* 867, 1-24.

Allen, D. J. (1983) *The Pathology of Tropical Food Legumes: Disease Resistance in Crop Improvement.* John Wiley and Sons Ltd., New York, pp. 33, 37, 192.

Allen, O. N. and Allen, E. K. (1981) *Phaseolus* L. *Vigna* Savi. In: Allen, O. N. and Allen, E. K. (eds.), *The Leguminosae: A Source Book of Characteristics, Uses, and Nodulation.* The University of Wisconsin Press, Madison, Wisconsin, pp. 512-515, 682-687.

Al-Yasiri, S. A. and Coyne, D. P. (1966) Interspecific hybridization in the genus *Phaseolus. Crop Science* 6, 59-60.

American Embassy Office of Agricultural Affairs. (1987) *Japan: Importer of Pulses.* Office of Agricultural Affairs, American Embassy, Tokyo.

Anderson-Leigh, M. (1989) Bean powder an effective face-wash. *The Japan Times,* June 15, p. 15.

Anishetty, N. M. and Moss, H. (1988) *Vigna* genetic resources: current status and future plans. In: *Proceedings of the Second International Mungbean Symposium.* Asian Vegetable Research and Development Center, Shanhua, Tainan, Taiwan, pp. 13-18.

Anonymous. (1960) *Index of Plant Disease in the United States. US Department of Agriculture Handbook.* Washington D. C., No. 165, 531 pp.

Anonymous. (1986) Five bronze age grains. *Korea Today* 10, 44.

Aota, T., Tanii, A. and Akai, J. (1981) Concerning the relationship between azuki brown stem rot disease and nematodes (in Japanese). *Annals of the Phytopathological Society of Japan* 47, 99-100 (Abstr.).

Aota, T., Tsuboki, K. and Akai, J. (1983) The effects that corn and other crops, as preceding crops, have on azuki brown stem rot symptoms (in Japanese). *Annals of the Phytopathological Society of Japan* 49(1), 116 (Abstr.).

Applebaum, S. W., Marco, S. and Birk, Y. (1969) Saponins as possible factors of resistance of legume seeds to the attack of insects. *Journal of Agricultural and Food Chemistry* 17(3), 618-622.

Arias, E. A. (1975) Estudio de fechas de siembra del frijol rojo Japones. *Agricultura Technica en Mexico* 3(10), 393-396.

Ariga, T. and Asao, Y. (1981) Isolation, identification and organoleptic astringency of dimeric proanthocyanidins occurring in azuki beans. *Agricultural Biology and Chemistry* 45(12), 2709-2712.

Arora, R. K. (1985) Diversity and collection of wild *Vigna* species in India. *FAO/IBPGR Plant Genetic Resources Newsletter* 63(12), 26-33.

Austin, M. E. (1948) *Reconnaissance Soil Survey of Japan: Kanto Plain Area.* SCAP, Natural Resources Section Report 110-A, Washington D. C., 73 pp.

Ayad, W. G. and Anishetty N. M. (1980) *Directories of Germplasm Collections, 1. Food Legumes.* International Board for Plant Genetic Resources, Rome, 22 pp.

Aykroyd, W. R. and Doughty, J. (1982) *Legumes in Human Nutrition.* FAO, Rome, pp. 115, 118.

Ayonoadu, U. W. U. (1974) Nuclear DNA variation in *Phaseolus. Chromosoma* (Berl.) 48, 41-49.

Bae, J. H. and Lee, S. R. (1987) Quantitative determination of flatulence factors in legume seeds and soy products. *Korean Journal of Food Science Technology* 19(5), 387-390.

Bailey, C. J. and Boulter, D. (1971) Urease, a typical seed protein of the Leguminosae. In: Harborne, J. B., Boulter, D. and Turner, B. L. (eds.), *Chemotaxonomy of the Leguminosae.* Academic Press, London and New York, pp. 485-502.

Bassiri, A. and Adams, M. W. (1978) An electrophoretic survey of seedling isozymes in several *Phaseolus* species. *Euphytica* 27, 447-459.

Baudoin, J. P. and Maréchal, R. (1988) Taxonomy and evolution of the genus *Vigna.* In: *Proceedings of the Second International Mungbean Symposium.* Asian Vegetable Research and Development Center, Shanhua, Tainan, Taiwan, pp. 2-12.

Bell, E. A. (1971) Comparative biochemistry of non-protein amino acids. In: Harborne, J. B., Boulter, D. and Turner, B. L. (eds.), *Chemotaxonomy of the Leguminosae.* Academic Press, London and New York, pp. 179-206.

Bennett, M. D. and Smith, J. B. (1976) Nuclear DNA amounts in angiosperms. *Philosophical Transactions of the Royal Society of London, Series B* 274, 227-274.

Bettencourt, E., Konopka, J. and Damania, A. B. (1989) *Directory of Germplasm Collections: 1. I. Food Legumes: Arachis, Cajanus, Cicer, Lens, Lupinus, Phaseolus, Pisum, Psophocarpus, Vicia, and Vigna.* International Board for Plant Genetic Resources, Rome, 190 pp.

Bhadra, S. K., Hammatt, N. and Davey, M. R. (1989) Prospects for the use of *in vitro* techniques in the improvement of *Vigna* pulses. *Sabrao Journal* 21(2), 75-91.

Biliaderis, C. G. (1980) *Physicochemical Studies on Legume Starches.* Unpublished PhD thesis, University of Saskatchewan, Canada.

Birch, N., Southgate, B. J. and Fellows, L. E. (1985) Wild and semi-cultivated legumes as potential sources of resistance to bruchid beetles for crop breeder: a study of

Vigna/Phaseolus. In: Wickens, G. E., Goodin, J. R. and Field, D. V. (eds.), *Plants for Arid Lands*. George Allen and Unwin, London, pp. 303-320.

Blackhurst, H. T. and Miller, J. C. (1980) Chapter 21, Cowpea. In: Fehr, W. R. and Hadley, H. H. (eds.), *Hybridization of Crop Plants*. American Society of Agronomy and Crop Science Society of America, Madison, Wisconsin, pp. 327-337.

Blumler, M. A. (1991) Modelling the origins of legume domestication and cultivation. *Economic Botany* 45(2), 243-250.

Borkovec, A. B., Nagasawa, S., Nakayama, I. and Terry, P. H. (1968) Sterilization of *Callosobruchus chinensis* by hexamethylphosphorothioc triamide. *Journal of Economic Entomology* 61(5), 1295-1298.

Borrow, D. J., De Long, D. M. and Triplehorn, C. A. (1981) *An Introduction to the Study of Insects*, 5th ed. Saunders College Publishing, New York.

Botanical Institute, Chinese Academy of Sciences (ed.). (1972) *Iconographia Cormophytorum Sinicorum, Tomus II*. Scientific Publishing House, Beijing, 1312 pp.

Bradbury, J. F. (1986) *Guide to Plant Pathogenic Bacteria*. CAB International Mycological Institute, Kew, Surrey, England, 332 pp.

Branca, C., Ricci, D. and Bassi, M. (1988) Epidermis integrity and epicotyl growth in azuki bean. *Journal of Plant Growth Regulation* 7(2), 95-109.

Brandenburg, W. A. (1986) Objectives in classification of cultivated plants. In: Styles, B. T. (ed.), *Infraspecific Classification of Wild and Cultivated Plants*. Systematic Assoc., Clarendon Press, Oxford, pp. 87-98.

Bray, F. (1984) Part II, Agriculture. In: Needham, J. (ed.), *Science and Civilisation in China, Volume 6, Biology and Biotechnology*. Cambridge University Press, 724 pp.

Breene, W. M. and Hardman, L. L. (1987) Anatomy of a specialty crop - the adzuki bean experience. In: *Grain Legumes as Alternative Crops*. Center for Alternative Crops and Products, University of Minnesota, pp. 67-77.

Bretschneider, E. (1934) Red bean (in Chinese). In: Shi, S. (trans.), *A Review of Chinese Botanical Documents*. National Edition and Translation Bureau, Beijing, pp. 3-7, 11, 50, 72.

Brockwell, J., Andrews, J. A., Gault, R. R., Gemell, L. G., Griffith, G. W., Herridge, D. F., Holland, J. F., Karsono, S., Peoples, M. B., Roughley, R. J., Thompson, J. A., Thompson, J. A. and Troedson, R. J. (1991) Erratic nodulation and nitrogen fixation in field-grown pigeonpea (*Cajanus cajan*, L. Mill. sp.). *Australian Journal of Experimental Agriculture* 31(5), 653-661.

Brown, J. K., Chapman, M. A. and Nelson, M. R. (1990) Bean Calico Mosaic, a new disease of common bean caused by a whitefly-transmitted geminivirus. *Plant Disease* 74(1), 81 (Abstr.).

Browne, W. P. (1855) Report on the seeds and cuttings recently introduced into the United States. *United States Commercial Patents Report for Agriculture 1854*, pp. 10-35.

Cao L. (1981) The history of fertilizers. In: *Popular Series of Chinese Agronomy* (in Chinese). Agriculture Press, Beijing.

Cao, R. and Van, C. (1991) Identification of rust resistance in azuki and *Vigna* (in Chinese). *Crop Germplasm Resources* (*Ziwupinzhongziyuan*) 1, 34-35.

Casimir, J. and Le Marchand, G. (1966) Distribution and taxonomic importance of amino acids and free peptides in *Phaseolinae* (in French). *Bulletin du Jardin Botanque de l'Etat Bruxelles* 36, 53-56.

Cerana, R., Bonetti, A., Marre, M. T., Romani, G. and Lado, P. (1983) Effects of a brassinosteroid on growth and electrogenic proton extrusion in azuki bean epicotyls (*Vigna angularis*). *Physiologia Plantarum* 59(1), 23-27.

Chandel, K. P. S. (1982) Evolutionary studies in *Vigna radiata, V. mungo* and other related species. *International Board for Plant Genetic Resources Newsletter* 6(4), 7-8.

Chandel, K. P. S. (1984) Role of wild *Vigna* species in the evolution and improvement of *Mung* (*Vigna radiata* (L.) Wilczek) and *Urd* bean (*V. mungo* (L.) Hepper). *Annals of Agricultural Research* 5, 98-111.

Chandel, K. P. S., Arora, R. K. and Pant, K. C. (1988) *Rice Bean: A Potential Grain Legume*. National Bureau of Plant Genetic Resources Science Monograph No. 12, Pusa Campus, New Delhi, India.

Chandel, K. P. S. and Pant, K. C. (1982) Genetic resources of *Vigna* species in India, their distribution, diversity and utilization in crop improvement. *Annals of Agricultural Research* 3(1-2), 19-34.

Chang, C. A. (1980a) An unknown virus causing infections in azuki and peanut. I. Study of virus characteristics (in Chinese). *Plant Protection Bulletin* (*Taiwan*) 22, 426 (Abstr.).

Chang, C. A. (1980b) An unknown virus causing infections in azuki and peanut. II. Purification of the virus (in Chinese). *Plant Protection Bulletin* (*Taiwan*) 22, 426 (Abstr.).

Chang, C. A., Purcifull, D. E. and Zettler, F. W. (1990) Comparison of two strains of peanut stripe virus in Taiwan. *Plant Disease* 74(8), 593-596.

Chang, D. C. N. (1986) Cultivation of short-rooted new sprouts (in Chinese, English summary). *Journal of the Chinese Society of Horticultural Science* 32(1), 54-60.

Chang, K. Y. (1969a) Studies on the correlation and path-coefficient analysis of some yield components in an adzuki bean cross (in Korean, English summary). *Korean Journal of Breeding* 1(1), 42-48.

Chang, K. Y. (1969b) Selection indexes and genetic advances for the agronomic characters and their applications to adzuki bean breeding (in Korean, English summary). *Korean Journal of Breeding* 1(1), 55-61.

Chang, K. Y. and Han, K. S. (1965) Studies on the adzuki bean varieties in Korea: ecotype classification and relationships between various characteristics (in Korean, English summary). *Bulletin of Chinju Agricultural College* 4, 15-27.

Chang, K. Y., Han, K. S. and Park, J. C. (1968a) Studies on the selection in adzuki bean breeding. I. Heritabilities of some characters in the population of adzuki bean varieties (in English, Korean Summary). *Bulletin of Chinju Agricultural College* 7, 29-32.

Chang, K. Y., Han, K. S. and Park, J. C. (1968b) Studies on the selection in adzuki bean breeding. II. Heritabilities of some characters in an adzuki bean cross (in English, Korean summary). *Bulletin of Chinju Agricultural College* 7, 33-37.

Chang, K. Y., Han, K. S. and Park, J. C. (1968c) Studies on the selection in adzuki bean breeding. III. Phenotypic and genotypic correlations among some characters in the population of adzuki bean varieties (in English, Korean summary). *Bulletin of Chinju Agricultural College* 7, 39-44.

Chang, K. Y., Han, K. S. and Park, J. C. (1969) Studies on the selection in adzuki bean breeding. IV. Phenotypic and genotypic correlations in adzuki bean cross (in English, Korean summary). *Korean Journal of Crop Science* 5, 51-56.

Chang, N. (1989) Pests in azuki (in Chinese). In: *Flourishing Agriculture*. Flourishing Agriculture Publishing House, Taipei, pp. 65-71.

Chang, N. T. (1987a) Seasonal abundance and developmental biology of thrips *Megalurothrips usitatus* on soybean at southern area of Taiwan. *Taiwan Plant Protection Bulletin* (*Zhiwu baohuxue hui huikan*) 29, 165-173.

Chang, N. T. (1987b) The damage and control of thrips (Insecta: Thysanoptera) on root crops, pulses, and other grain crops (in Chinese, English summary). *Chinese Journal of Entomology, Special Publication* 1, 55-72.

Chang, N. T. (1988a) The preference of thrips, *Megalurothrips usitatus* (Bagnall), for three leguminous plants (in English, Chinese summary). *Taiwan Plant Protection Bulletin* (*Zhiwu baohuxue hui huikan*) 30(1), 68-77.

Chang, N. T. (1988b) Population trends of *Megalurothrips usitatus* (Bagnall) (Thysanoptera: Thripidae) on adzuki bean and soybean examined by four sampling methods (in English, Chinese summary). *Taiwan Plant Protection Bulletin* (*Zhiwu baohuxue hui huikan*) 30(3), 289-302.

Chang, N. T. (1990) Color preference of thrips (Thysanoptera: Thripidae) in the adzuki bean field. *Taiwan Plant Protection Bulletin* (*Zhiwu baohuxue hui huikan*) 32, 307-316.

Chang, N. T. (1992) Dispersion patterns of bean flower thrips *Megalurothrips usitatus* (Bagnall) (Thysanoptera: Thripidae) on flowers of adzuki bean. *Plant Protection Bulletin* (*Taiwan*) 34(1), 41-53.

Chao, C. K. (1982) Country reports on the status of grain legume production: Republic of China. In: *Grain Legumes Production in Asia.* Asian Productivity Organization, Tokyo, Japan, pp. 281-296.

Charles-Edwards, D. A. and Lawn, R. J. (1984) Light interception by grain legume row crops. *Plant Cell and Environment* 7(4), 247-251.

Chen, C. S. (1956) Crop combinations and crop regions in Taiwan (in Chinese, English summary). *Fu-Min Institute of Agricultural Geography Research Report* 62, 1-22.

Chen, G. F. (1979a) Adzuki bean (*Phaseolus angularis* Wight) (in Chinese). In: Liang, E. (ed.), *Beans and Peas, Harvest Farm Magazine #794.* Taipei, Taiwan, pp. 49-69.

Chen, L. B. and Wei, S. H. (1991) Identification for azuki resistance leaf spot diseases (in Chinese). *Heilongjiang Agricultural Science* 1, 49-51.

Chen, L. S. (1973) The occurrence of the bruchids in the imported beans and the ecological study on *Callosobruchus maculatus* Fab. (in Chinese, English summary). *Taiwan Plant Protection Bulletin* (*Chih wu pao hu hsueh hui hui kan*) 15(2), 53-60.

Chen, L. S. (1979b) Thrips of leguminous plants in Taiwan. *Proceedings of the National Science Council, R. O. C.* 3(4), 414-428.

Chen, N. C. (1980) *Interspecific hybridization in the genus Vigna*. Unpublished PhD thesis, Michigan State University, 122 pp.

Chen, N. C., Baker, L. R. and Honma, S. (1983) Interspecific crossability among four species of *Vigna* food legumes. *Euphytica* 32, 925-937.

Chen, N. C., Parrot, J. F., Jacobs, T., Baker, L. R. and Carlson, P. S. (1978) Interspecific hybridization of food legumes by unconventional methods of plant breeding. In: Cowell, R. (ed.), *Proceedings of the First International Mungbean Symposium*. Asian Vegetable Research and Development Center, Shanhua, Taiwan, pp. 247-252.

Chen, T. H. H., Gusta, L. V., Tjahjadi, C. and Breene, W. M. (1984) Electrophoretic characterization of adzuki bean (*Vigna angularis*) seed proteins. *Journal of Agricultural Food Chemistry* 32(2), 396-399.

Cheng, T. and Qou, Q. (1967) Azuki and mungbean (in Chinese). In: *Agronomy*. Fuxing Book Bureau, Taipei, Taiwan, pp. 355-360.

Cheo, C. C. and Tsai, S. L. (1959) Virus diseases of legumes (Annual Report, 1957-1958) (in Chinese, English summary). *Acta Phytopathologica Sinica* 5(1), 7-11.

Chiba, I. (1980) Cultivar differentiation and breeding of adzuki bean (in Japanese). *Recent Advances in Breeding* 21, 59-64.

Chiba, I. (1982) Breeding for resistance to brown stem rot (BSR) in adzuki beans. I. On the varietal differences in resistance to BSR. (in Japanese, English summary). *Bulletin of the Hokkaido Prefectural Agricultural Experiment Station* 48, 56-63.

Chiba, I. (1985) Breeding for resistance to brown stem rot (BSR) in adzuki beans. II. Formation of varietal differences (in Japanese, English summary). *Bulletin of the Hokkaido Prefectural Agricultural Experiment Station* 52, 79-84.

Chiba, I., Narikawa, T., Murata, K. and Adachi, T. (1987) Breeding for resistance to brown stem rot (BSR) in adzuki bean. III. The inheritance of resistance to BSR and effect of its introduction (in Japanese, English summary). *Bulletin of the Hokkaido Prefectural Agricultural Experiment Station* 56, 1-7.

Chilukuri, A. and Swanson, B. G. (1991) Microstructure of adzuki beans (*Vigna angularis* cv. Express). *Food Structure* 10(2), 131-135.

Chinese Agricultural Resources and Regionalization Editing Committee (eds.). (1987) Edible legumes. In: *Outlines of Agricultural Resources and Regionalization in China* (in Chinese). Publishing House of Industry and Commerce, Beijing, pp. 161-164.

Cho, Y. and Park, K. (1991) Purification of cellulase from *Fusarium moniliforme* and its application to the production of red bean starch precipitate. *Journal of Food Biochemistry* 15(4), 295-313.

Choi, Y. K. (ed.). (1988) *Crop Statistics, 1987, No. 126-11-04*. Ministry of Agriculture and Fisheries, Republic of Korea. pp. 136.

Choi, Y. K. and Lee, S. M. (eds.). (1987) *Statistical Yearbook of Agriculture, Forestry and Fisheries*. Agriculture and Fisheries Statistics Bureau, Ministry of Agriculture and Fisheries, Republic of Korea.

Choi, Y. M. and Lee, S. H. (1989) Identification of azuki bean mosaic virus on *Phaseolus angularis* (in Korean). *Korean Journal of Plant Pathology* 5(1), 49-53 (Abstr.).

Choopanya, D. and Kiratiya-angul, K. (1980) Peanut mottle virus in Thailand. In: *Southeast Asian Symposium on Plant Diseases in the Tropics.* October 20-26, Bangkok, Thailand, p. 80.

Chun, Y. S. and Ryoo, M. I. (1992) Oviposition response of azuki bean weevil (Coleoptera: Bruchidae) to the densities of azuki bean and influence of intraspecific competition on the response. *Korean Journal of Entomology* 22(2), 73-80.

Ci Hai Editing Committee. (1983) Red beans, *Phaseolus angularis* (in Chinese). In: *Ci Hai Agriculture Lexicon,* 3rd ed. Shanghai Lexicology and Book Publishing House, Shanghai, p. 176.

Commonwealth Mycological Institute. (1958) *Index of Fungi: Published Since the Beginning of 1940* 2(17), 428.

Cotter, P. F. (1991) Seed extracts as hemagglutination and wattle reaction reagents in chickens. *Poultry Science* 70(Suppl. 1), 30.

Cox, M. (1978) Minnesota farmers may have found a lucrative new crop: adjuki [sic] beans, sought by Japanese, are seen as potential for high return. *Wall Street Journal* October 5, p. 38, col. 4.

Coyne, D. P., Schuster, M. L. and Al-Yasiri, S. (1963) Reaction studies of bean species and varieties to common blight and bacterial wilt. *Plant Disease Reporter* 47(6), 534-537.

Cramer, C. (1986) Putting the future to the test. *The New Farm* 8(7), 14-17.

Cramer, C. (1991) Quality times three: clean crops and savvy marketing help this organic seedman triple his net. *The New Farm* 13(6), 18-22.

Crawford, G. W. and Yoshizaki, M. (1987) Ainu ancestors and prehistoric Asian agriculture. *Journal of Archaeological Science* 14(2), 201-214.

Cultural Relic Publishing House Committee. (1978) Azuki (*Phaseolus angularis* Wight). In: *Study on Animal and Plant Samples from 'Number 1 Mawangdui' Han dynasty tomb in Changsha, Hunan Province* (in Chinese). Cultural Relic Publishing House, Beijing, pp. 8-9.

Cummins, G. B. and Ling, L. (1950) An index of the plant rusts recorded for continental China and Manchuria. *Plant Disease Reporter Supplement 196*, p. 536.

Dai, Y. and Zhang, C. (1982) Classification standards and characters of edible legumes (in Chinese). In: *Preliminary Observations and Classifications of Edible Legume Germplasm Resources in Shaanxi Province, National Proceeding.* Shaanxi, China, pp. 3-7.

Dana, S. (1976) Origin, evolution and distribution of some grain legumes. *Indian Journal of Genetics and Plant Breeding* 36(1), 143-145.

Dana, S. (1980) Genomic relationship in the genus *Vigna* and its implications in breeding programme. In: Gill, K. S. (ed.), *Breeding Methods for the Improvement of Pulse Crops.* Ludhiana, Punjab, India.

Dana, S. and Karmakar, P. G. (1990) Species relation in *Vigna* subgenus *Ceratotropis* and its implications in breeding. In: Janick, J. (ed.), *Plant Breeding Reviews Volume 8.* Timber Press, Inc., Portland, pp. 19-42.

Danimihardja, S. and Lester, R. N. (1974) Comparative studies of some Asiatic species of *Phaseolus* using electrophoretic analyses of seed proteins. *Annales Bogorienses* 5(4), 159-171.

Darlington, C. D. and Wylie, A. P. (1956) *Chromosome Atlas of Flowering Plants.* George Allen and Unwin Ltd., London, pp. 170-172.

DeWet, J. M. J. and Harlan, J. R. (1975) Weeds and domesticates: evolution in the man-made habitat. *Economic Botany* 29, 99-107.

Ding, J. and Zhang, Y. (1984) Current azuki production and prospects in Tianjin (in Chinese). *Tianjin Agricultural Sciences* 1, 21-25.

Dong, J. and Liang, Y. (1980) Tianjin azuki (in Chinese). *Agricultural Science and Technology Communication* 6, 11.

Dong, Y. (trans.). (1982) The origins of Chinese crop cultivation. In: *Cultivation Origins of Major World Crops* (in Chinese). Agriculture Publishing House, Beijing, pp. 15-16.

Dorer, M. J. (1973) Adzuki beans - jewels from Japan. *Organic Gardening and Farming* 20, 48-49.

Dou, Z., Hu, S., Li, C. and Ye, Z. (1989) Application of B9 (plant hormone) to"Ji No 1" (in Chinese). *Bulletin of Agricultural Science and Technology* 3, 5.

Drager, H. (1983) *Influence of storage environment upon imbibition and thermal softening of adzuki beans* (*Vigna angularis*). Unpublished MSc thesis, University of Minnesota.

Du, B. (1987) Utilizing regional improvements and growing legume crops (in Chinese). *Agricultural Science and Technology* 4, 11-12.

Du, Z. (1986) Production and breeding selection of edible legumes in Heilongjiang Province (in Chinese). *Heilongjiang Agricultural Science* 6, 35-36.

Du, Z. and Zhang, M. (1987) Introduction on the selection of azuki cultivar "Baihong No. 1" (in Chinese). *Baicheng Agricultural Science and Technology* 2, 11-13.

Duan, H. (1989) Small bean (in Chinese). In: Long, J., Lin, L., Xushen, H. and Duan, H. (eds.), *Edible Bean Crops.* Science Publishing House, Beijing, pp. 160-171.

Duan, X. (1982) Research on edible legume resources in China (in Chinese). *Crop Germplasm Resources* 2, 6-12.

Duke, J. A. (1981) *Vigna angularis* (Willd.) Ohwi and Ohashi. In: Duke, J. A. (ed.), *Handbook of Legumes of World Economic Importance.* Plenum Press, New York, pp. 288-293.

Durgan, B. R., Gunsolus, J. L., Becker, R. L. and Dexter, A. G. (1989)*Cultural and Chemical Weed Control in Field Crops.* University of Minnesota Extension Service AG-BU-3157.

Economic Information Bureau. (1991) 4. Miscellaneous grain crops, beans-pulse crops (in Japanese). In: *Agriculture, Forestry and Fisheries Production Statistics.*, Ministry of Agriculture, Forestry and Fisheries. Agriculture and Forestry Statistics Association, Tokyo, pp. 201-205.

Economic Information Bureau. (1989) *Agriculture, Forestry and Fisheries Production Statistics. Heisei 3rd Year Edition.* Ministry of Agriculture, Forestry and Fisheries Statistics Information Bureau, Agriculture and Forestry Statistics Association, Tokyo, pp. 200-239.

Economic Daily News. (1991) *Economic Yearbook of the Republic of China.* Economic Daily News, Taipei, p.179.

Editorial Committee of Heilongjiang Agricultural Encyclopedia (ed.) (1993) *Heilongjiang Agricultural Encyclopedia* (Heilongjiang Nongye Baikequanshu) (in Chinese). Chinese Encyclopedia Publishing, Beijing, pp. 76, 105, 371, 405.

Edwards, D. I. and Malek, R. B. (1971) Cyst production by *Heterodera lespedezae* on selected legumes. *Plant Disease Reporter* 55(10), 927-929.

Edwards, D. I., Noel, G. R. and Sikora, E. J. (1990) The soybean cyst nematode problem (revised). *Report on Plant Diseases* 501, 1-13.

Egawa, Y. (1988) Phylogenetic differentiation between three Asian *Vigna* species, *V. radiata, V. mungo* and *V. umbellata* (in English, Japanese summary). *Bulletin of the National Institute of Agrobiological Resources* 4, 189-200.

Egawa, Y. (1991) Phylogenetic relationship in Asian *Vigna* species. In: Thavarasook, C. (ed.), *Proceedings of The Mungbean Meeting 90.* February 23-24, 1990, Chiang Mai, Thailand. Tropical Agriculture Research Center, Japan, pp. 87-94.

Egawa, Y., Horikoshi, H., Tomooka, N. and Nakano, H. (1991) Collection of wild relatives of adzuki bean, *Vigna riukiuensis* and *V. reflexo-pilosa* on Yonaguni Island, Okinawa Prefecture (in English, Japanese summary). *Annual Report on Exploration and Introduction of Plant Genetic Resources*, National Institute of Agrobiological Resources, Tsukuba, Japan, 7, 13-20.

Egawa, Y., Miyazaki, S. and Nakagahara, M. (1986) Cross compatibility between four *Vigna* leguminous species (in Japanese). *Japanese Journal of Plant Breeding* 36(Suppl. 1), 260-261.

Egawa, Y., Nakagahara, M. and Fernandez, G. C. J. (1988a) Cytogenetical analysis of tetraploid *Vigna glavrescens* by interspecific hybridization involving diploid Asian *Vigna* species. In: *Proceedings of the Second International Mungbean Symposium.* Asian Vegetable Research and Development Center, Shanhua, Taiwan, pp. 200-204.

Egawa, Y., Nakagahara, M. and Fernandez, G. C. J. (1990a) Cross compatibility and cytogenetical relationships among Asian *Vigna* species. In: Fujii, K., Gatehouse, A. M. R., Johnson, C. D., Mitchel, R. and Yoshida, T. (eds.), *Bruchids and Legumes: Economics, Ecology and Coevolution.* Kluwer Academic Publishers, Dordrecht, Netherlands, pp. 201-208.

Egawa, Y., Siriwardhane, D., Yagasaki, K., Hayashi, H., Takamatsu, M., Saito, M., Nomura, Y., Okabe, T., Idezawa, F. and Miyazaki, S. (1990b) Collection of millets and grain legumes in the Shimoina District of Nagano Prefecture, 1989 (in English, Japanese summary). *Journal of Plant Exploration* (*Shokubutsu Tansaku Houkoku*) 6, 1-22.

Egawa, Y., Yamashita, M., Tomooka, N., Kitamura, K. and Nakagahara, M. (1988b) Interspecific variation of seed storage protein in Asian *Vigna* species by SDS-polyacrylamide gel electrophoresis (in Japanese). *Japanese Journal of Plant Breeding* 38(Suppl. 2), 442-443.

Engquist, A. and Swanson, B. G. (1992) Microstructural differences among adzuki beans (*Vigna angularis*) cultivars. *Food Structure* 11, 171-179.

Evans, A. M. (1976a) Species hybridization in the genus *Vigna.* In: *Proceedings of the IITA Collaboration Meeting on Grain Legume Improvement.* International Institute of Tropical Agriculture, Ibadan, Nigeria, pp. 31-34.

Evans, A. M. (1976b) Beans. In: Simmonds, N. W. (ed.), *Evolution of Crop Plants.* Longman, London, pp. 168-172.

Fagbenle, H. H., Edwards, D. I. and Malek, R. B. (1986) Cyst production by two geographical isolates of *Heterodera lespedezae* on selected legumes. *Plant Disease* 70(7), 643-644.

Fan, Z., Xing, S. and Liu Y. (1992) The relationship among azuki quality, yields and climate. *Chinese Agricultural Meteorology* 13(4), 12-14.

Farr, D. F., Bills, G. F., Chamuris, G. P. and Rossman, A. Y. (1989) *Fungi on Plants and Plant Products in the United States.* The American Phytopathological Society, St. Paul, Minnesota, 1252 pp.

Fehr, W. R. and Hadley, H. H. (1980) *Hybridization of Crop Plants.* American Society of Agronomy and Crop Science Society of America, Madison, Wisconsin, 765 pp.

Fernandez, G. C. J. and Talekar, N. S. (1990) Genetics and breeding for bruchid resistance in Asiatic *Vigna* species. In: Fujii, K., Gatehouse, A. M. R., Johnson, D., Mitchel, R. and Yoshida, T. (eds.), *Bruchids and Legumes: Economics, Ecology and Coevolution.* Kluwer Academic Publishers, Dordrecht, Netherlands, pp. 209-217.

Fery, R. L. (1980) Genetics of *Vigna.* In: Janick, J. (ed.), *Horticulture Reviews, Volume 2.* AVI Publishing Company, Inc., Westport, Connecticut, pp. 314-394.

Foschine, A., da Silva, E. A. B., Cardin, M. C. F., Leonel, M. L. M. and Calafiori, M. H. (1983) Resistance of varieties of the common bean (*Phaseolus vulgaris* L.) and azuki bean (*Phaseolus angularis* Wight) to attacks of *Liriomyza* sp. (Diptera: Agromyzidae) (in Portuguese, English summary). *Ecossistema* 8, 94-96.

Fox, C. W. (1993) A quantitative genetic analysis of oviposition preference and larval performance on two hosts in the bruchid beetle, *Callosobruchus maculatus. Evolution* 47(1), 166-175.

Francki, R. I. B., Milne, R. G. and Hatta, T. (1985) *Atlas of Plant Viruses, Volume II.* CRC Press, Inc., Boca Raton, Florida, 222 pp.

Fu, C. and Li, A. (1993) Nutritional characters of mungbean (in Chinese, English summary). In: CAAS, MAPRC and AVRDC (eds.), *The Study and Application of AVRDC Mungbean in China.* Beijing, pp. 218-222.

Fujii, K. (1969) Studies on the interspecies competition between the azuki bean weevil and the southern cowpea weevil. IV. Competition between strains. *Researches on Population Ecology (Kyoto)* 11(1), 84-91.

Fujii, K. (1970) Studies on the interspecies competition between the azuki bean weevil, *Callosobruchus chinensis*, and the southern cowpea weevil, *C. maculatus. Researches on Population Ecology (Kyoto)* 12(2), 219-232.

Fujii, K. (1983) Resource dependent stability in an experimental laboratory resource-herbivore-carnivore system (in English, Japanese summary). *Research on Population Ecology* (Suppl. No. 3), 15-267.

Fujii, K. and Miyazaki, S. (1987) Infestation resistance of wild legumes (*Vigna sublobata*) to azuki bean weevil, *Callosobruchus chinensis* (L.) (Coleoptera: Bruchidae) and its relationship with cytogenetic classification. *Applied Entomology and Zoology* 22(2), 229-230.

Fujimura, T. (1976) Oviposition and host selection by the spotted cutworm, *Amathes c-nigrum* L. (in Japanese, English summary). *Research Bulletin of the Hokkaido National Agricultural Experiment Station* 113, 79-85.

Fujimura, T. and Kugimiya, M. (1993a) Gelatinization of starch inside cotyledon cells of adzuki bean (in Japanese). *Journal of the Japanese Society for Food Science and Technology* 40(7), 490-495.

Fujimura, T. and Kugimiya, M. (1993b) Gelatinization of starches inside cotyledon cells of adzuki bean studied by differential scanning calorimetry (in Japanese). *Journal of the Japanese Society for Food Science and Technology* 40(10), 702-707.

Fujita, K. and Miura, O. (1934) On the parasitism of *Heterodera schachtii* Schmidt on beans. *Transactions of the Sapporo National Historical Society* 13(3), 359-364.

Fujita, Y. (1974) *Panicum crusgalli* Linn. var. *frumentaceum* Trin., *P. miliaceum* Linn. and *Phaseolus angularis* W. F. Wight: an ethno-botanical treatise (in Japanese). *Biological Science* 26(2), 102-105.

Fukunaga, K. (ed.). (1987) *Japanese Pesticides Guide.* Japan Plant Protection Association, Tokyo.

Funk, A., Weder, J. K. P. and Belitz, H. D. (1993) Primary structures of proteinase inhibitors from *Phaseolus vulgaris* var. *nanus* (cv. Borlotto) (in English, German summary). *Zeitschrift fuer Lebensmittel-Untersuchung und-Forschung* 196(4), 343-350.

Furusawa, F., Maita, S., Sato, T., Ono, Y., Kamada, N., Kamiyama, Y. and Takahashi, Y. (1974) The new azuki bean variety "Iwate-dainagon" (in Japanese). *Bulletin of the Iwate Ken Agricultural Experiment Station* 18, 11-21.

Gai, J. and Jin, W. (1994) Production state and developmental strategies of edible legumes in China. *Theses in Association of Crop Science,* Beijing, 8 pp.

Galloway, B. T. (1909) Seeds and plants imported during the period from January 1 to March 31, 1908, Inventory No. 14, Nos. 21732 to 22510. Washington D. C. *USDA Bureau of Plant Industry Bulletin* 137, 7-9.

Gams, W. (1971) *Cephalosporium-artige Schimmelpilze* (*Hypomycetes*). Gustav Fischer Verlag, Stuttgart, Germany, 262 pp.

Georgeson, C. C., Burtis, F. C. and Shelton, W. M. (1891) Test of some Japanese beans. *Bulletin of the Kansas Agricultural Experiment Station* 32, 233-238.

Giga, D. P. and Smith, R. H. (1984/85) Oviposition markers in *Callosobruchus maculatus* F. and *Callosobruchus rhodesianus* Pic. (Coleoptera, Bruchidae): asymmetry of interspecific responses. *Agriculture, Ecosystems and Environment* 12(3), 229-233.

Giga, D. P. and Smith, R. H. (1987) Egg production and development of *Callosobruchus rhodensianus* (Pic.) and *Callosobruchus maculatus* (F.) (Coleoptera: Bruchidae) on several commodities at two different temperatures. *Journal of Stored Products Research* 23(1), 9-15.

Gill, H. S. and Sharma, S. R. (1980) Blight resistant pulses for hilly areas. *Indian Farming* 29(2), 23.

Gossrau, R. (1991) Catalytic histochemistry of acid and neutral hydrolases in plant seedlings. *Histochemical Journal* 23, 483-489.

Goto, K., Takahashi, M., Nishiiri, K. and Abe, K. (1983) Effect of overhead flooding on growth and yield of soybeans and azuki beans (in Japanese). *Research Bulletin of Hokkaido National Agriculture Experiment Station* 23,12.

Goto, K., Takahashi, M., Nishiiri, K. and Abe, K. (1984) Survey on cool weather injuries to soybeans and azuki beans in 1983 in Abashiri District (in Japanese). *Research Bulletin of Hokkaido National Agriculture Experiment Station* 26, 107-114.

Goto, K., Takahashi, M., Nishiiri, K. and Abe, K. (1985) Effect of overhead flooding on growth and yield of soybeans and azuki beans (in Japanese, English summary). *Research Bulletin of Hokkaido National Agriculture Experiment Station* 141, 127-147.

Gray, L. E. and Hepburn, A. G. (1989) Restriction fragment length polymorphisms between soybean and adzuki bean *Phialophora gregata* isolates. *Phytopathology* 79(10), 1172-1173 (Abstr.).

Gray, L. E. and Hepburn, A. G. (1992) Mitochondrial DNA restriction patterns of *Phialophora gregata* isolates from soybean and adzuki bean. *Phytopathology* 82(2), 211-215.

Green, S. K., Lee, D. R. and Horne, N. M. (1988) Survey for viruses of soybean, mungbean, and peanut in Java, Indonesia. *Plant Disease* 72(11), 994 (Abstr.).

Green, S. K., Lee, D. R., Vetten, H. J. and Lesemann, D. E. (1986) Occurrence of an unidentified potyvirus of soybean in Taiwan. In: *International Symposium of Virus Diseases of Rice and Leguminous Crops in the Tropics.* Proceedings of a Symposium on Tropical Agriculture Research, October 15, 1985, Yatabe, Tsukuba, Ibaraki, Japan, pp. 108-115.

Grierson, A. J. C. and Long, D. G. (1987)*Flora of Bhutan, Vol. 1, Part 3.* Royal Botanic Garden, Edinburgh.

Guo, A. (1993) Azuki price is more than pork price (in Chinese). *People's Daily* (Overseas Edition) November 26, 4.

Guo, S. and Jin, W. (1988) Analysis of azuki karyotype (in Chinese). *Beijing Agricultural Sciences* 2, 19-22.

Guo, W. (1980) Azuki (in Chinese). *New Agriculture* 9, 7.

Gupta, P. K. and Bahl, J. R. (1983) Cytogenetics and origin of some pulse crops. In: Swaminathan, M. S., Gupta, P. K. and Sinha, U. (eds.), *Cyogenetics of Crop Plants.* Macmillan India, New Delhi, pp. 405-440.

Hachiya, K. (1990) Effect of injury by the flower thrips, *Frankliniella intonsa* Trybom on pod formation and yield of adzuki bean, *Phaseolus angularis* L. (in Japanese, English summary). *Japanese Journal of Applied Entomology and Zoology* 34(4), 326-329.

Hamada, S., Ezaki, S., Hayashi, K., Toko, K. and Yamafuji, K. (1992) Electric current precedes emergence of a lateral root in higher plants. *Plant Physiology* (*Bethesda*) 100(2), 614-619.

Hammer, K. and Mun-Chan, B. (1986) Variable landraces collected in Democratic People's Republic of Korea. *FAO/IBPGR Plant Genetic Resources Newsletter* 67(12), 19-21.

Hammer, K., Dong-Chol, W. and Un-Xoan, H. (1989a) Report on a mission for the collection of plant genetic resources to the Democratic People's Republic of Korea, 1988 (in English, Russian summary). *Kulturpflanze* 37, 391-399.

Hammer, K., Hjong-Gun, O. and Un-Xoan, H. (1988) Report on a mission for the collection of plant genetic resources to the Democratic People's Republic of Korea, 1987 (in English, Russian summary). *Kulturpflanze* 36, 465-473.

Hammer, K., Laghetti, G. and Perrino, P. (1989b) Collection of plant genetic resources in South Italy, 1988 (in English, Russian summary). *Kulturpflanze* 37, 401-414.

Hammer, K., Mun-Chan, B. and Un-Xoan, H. (1986) Notes on a mission to the Democratic People's Republic of Korea in 1985 for the collection of land-races of cultivated plants (in English, Russian summary). *Kulturpflanze* 34, 293-304.

Han, E. D. and An, J. H. (1990) Studies on ecology and control of the azuki bean weevil, *Callosobruchus chinensis* (L.), in Korea (in Korean, English summary). *Research Reports of the Rural Development Administration (Suweon)-Crop Protection* 32(1), 24-59.

Han, E. H., Lee, C. H., Sin, C. S. and Lee, E. K. (1982) An investigation on the stem rot red-bean caused by *Phytophthora vignae* Purss (in Korean, English summary). *Research Report of the Office of Rural Development* 24, 69-71.

Han, K. S. (1983) Genetic analysis on quantitative characters by diallel crosses of adzuki bean (*Phaseolus angularis* W. F. Wight) (in Korean, English summary). *Journal of Gyeong Sang National University* 22, 1-10.

Han, K. S. and Chang, K. Y. (1976) Studies on the breeding of constantly high yielding adzuki bean varieties. II. Changes in heritability, genotypic and phenotypic correlations and path coefficients in the different cultivated seasons (in Korean, English summary). *Journal of Gyeongsang National University* 15, 47-53.

Han, K. S. and Choe, Z. R. (1975) Studies on the breeding of adzuki bean (*Phaseolus angularis* W. F. Wight) with a stable and high yielding capacity. I. Changes in heritabilities, genotypic and phenotypic correlations and path coefficients of each character of *Phaseolus angularis* cultivated under the different meteorological conditions (in Korean, English summary). *Journal of Gyeongsang National University* 14, 171-176.

Han, K. S. and Choe, Z. R. (1977) Studies on the breeding of constantly high yielding adzuki bean (*Phaseolus angularis* W. F. Wight) varieties. III. Utilization of selection index of several useful characters of adzuki bean cultivars (in Korean, English summary). *Journal of Gyeongsang National University* 16, 291-294.

Han, K. S., Chang, K. Y. and Bin, Y. H. (1979) Genetic studies on some adzuki bean characters by the diallel cross (in Korean, English summary). *Journal of Gyeongsang National University* 18, 61-70.

Han, K. S., Chang, K. Y. and Kim, J. H. (1984) Genetic studies on some quantitative characters of adzuki bean by the diallel crosses. I. Distribution of genes for various characters in F_1 and F_2 generations (in Korean, English summary). *Korean Journal of Breeding* 16(3), 316-322.

Han, K. S., Chang, K. Y. and Ko, M. S. (1970) Effect on some yield components influenced upon the yield of the adzuki beans (in Korean, English summary). *Bulletin of Chinju Agricultural College* 9, 19-26.

Han, S., Tan, S. and Hu, J. (1987) Cultivation principles for high yielding azuki (in Chinese). *Bulletin of Agriculture Science and Technology* 8, 13.

Hanada, S., Nakano, M., Saitoh, H. and Mochizuki, T. (1975) Studies on the pollution of apple orchard surface soil and its improvement in relation to inorganic spray residues: arsenic injury to azuki bean (in Japanese, English summary). *Bulletin of the Faculty of Agriculture, Hirosaki University* 25, 13-24.

Hang, A. H., McClary, D. C., Gilliland, G. C. and Lumpkin, T. A. (1993) Plant configuration and population effects on yield of azuki bean in Washington State. In: Janick, J. and Simon, J. (eds.), *New Crops.* John Wiley and Sons, New York, pp. 588-590.

Harada, K. (1953) The blooming and maturing of red bean as related to various sowing times (in Japanese). *Japanese Journal of Crop Science* 22(1/2), 101-102.

Harborne, J. B. (1971) Distribution of flavonoids in the Leguminosae. In: Harborne, J. B., Boulter, D. and Turner, B. L. (eds.), *Chemotaxonomy of the Leguminosae.* Academic Press, London and New York, pp. 31-71.

Hardman, L. L. (1985) Adzuki bean cultural information. *Crop News: Agronomy and Plant Genetics No. 58* (Revised). University of Minnesota Agricultural Extension Service.

Hardman, L. L. (1987) *Varietal Trials of Farm Crops. University of Minnesota Agricultural Experiment Station Report 24*, 44 pp.

Hardman, L. L. and Meronuck, R. A. (1982) A guide to dry edible bean production and pest management in Minnesota. *University of Minnesota Agricultural Extension Service Special Report 103-1982*, 72 pp.

Harlan, J. R. and DeWet, M. J. (1971) Toward a rational classification of cultivated plants. *Taxonomy* 20, 509-517.

Harrison, H. F. Jr. (1988) Metribuzin tolerance within the genus *Vigna. Weed Technology* 2, 59-63.

Hartmann, R. W. (1969) Photoperiod responses of *Phaseolus* plant introductions in Hawaii. *Journal of the American Society for Horticultural Science* 94, 437-440.

Harukawa, A. (1990) *Standard Tables of Food Consumption in Japan.* Joshi Eiyo Daigaku (Ladies Nutrition University), Shuppan Bu (Publishing Department), Tokyo, 512 pp.

Hasegawa, S., Kodama, F., Nakajima, M., Akai, J. and Morooka, H. (1987) Biological control of *Fusarium* wilt of adzuki bean: *Microellobosporia-flavea* Y-1 (in Japanese, English summary). *Bulletin of the College of Agriculture and Veterinary Medicine, Nihon University* 44, 80-86.

Hasegawa, S., Kondo, N. and Kodama, F. (1991) Suppression of Fusarium wilt of adzuki bean by rhizosphere microorganisms. In: Hedin, P. A. (ed.), *Naturally Occurring Pest Bioregulators.* American Chemical Society, Washington D. C., pp. 407-416.

Hatai, A. (1982) Influences of storage temperatures on cooking characteristics of adzuki beans, *Vigna angularis*, Ohwi et Ohashi. *Japanese Home Economics Association* 33, 621-627.

Hattori, T. and Ohta, Y. (1985) Induction of phenylalanine ammonia-lyase activation and isoflavone glucoside accumulation in suspension-cultured cells of red bean, *Vigna* *angularis*, by phytoalexin elicitors, vanadate, and elevation of medium pH. *Plant and Cell Physiology* 26(6), 1101-1110.

Hayakawa, I. and Breene, W. M. (1982) A study on the relationship between cooking properties of adzuki bean and storage conditions. *Journal of the Faculty of Agriculture Kyushu University* 27(1/2), 83-88.

Hayashi, J. and Noda, A. (1961) Research on drought resistance of rice bean (in Japanese). *Japanese Journal of Crop Science* 29, 447 (Abstr.).

He, W. X. and Hao, H. B. (1987) Identification of nutritional quality in grain legumes (in Chinese). *Crop Germplasm Resources* 1, 13-17.

Hedrick, U. P. (1931) Beans of New York. In: *The Vegetables of New York, Vol. I, Part II*. J. B. Lyon Co., New York, p. 13.

Hermawan, W., Tsukuda, R., Fujisaki, K., Kobayashi, A. and Nakasuji, F. (1993) Influence of crude extracts from tropical plant *Andrographis paniculata* Acanthaceae on suppression of feeding by the diamondback moth *Plutella xylostella* Lepidoptera Yponomeutidae and oviposition by the azuki bean weevil *Callosobruchus chinensis* Coleoptera Bruchidae. *Applied Entomology and Zoology* 28(2), 251-254.

Hernandez, J. E., Carpena, A. L., Lantican, R. M. and Navarro, R. S. (1978) Yield performance and other agronomic characters of twenty-three varieties of adzuki bean (*Vigna angularis* (Willd.) Ohwi and Ohashi) under Philippine conditions. *Philippine Journal of Crop Science* 3(1), 13-18.

Hino, T. (1960) Studies on the asparagus-bean mosaic virus (in English, Japanese summary). *Annals of the Phytopathology Society of Japan* 25(4), 178-186.

Hino, T. (ed.). (1990) *Phaseolus angularis* (Willd.) Wight (=*Vigna angularis*). In: *Catalogue of Plant Germplasm*, 3rd ed. National Institute of Agrobiological Resources (NIAR), Tsukuba, Japan, pp. 779-782.

Hirano, H., Kagawa, H. and Okubo, K. (1992) Characterization of proteins released from legume seeds in hot water. *Phytochemistry* 31(3), 731-735.

Hirata, S. (1952) On the rust fungi of cowpea, kidney bean and small red bean (in Japanese). *Annals of the Phytopathology Society of Japan* 16(1), 13-18.

Hirata, T. and Kugimiya, M. (1985) Effect of azuki bean protein on gelatinization of azuki bean starch (in Japanese, English summary). *Journal of the Japanese Society of Food Science and Technology* 32(1), 35-42.

Hogetsu, T., Shibaoka, H. and Shimokoriyama, M. (1974) Involvement of cellulose synthesis in actions of gibberellin and kinetin on cell expansion. Gibberellin-coumarin and kinetin-coumarin interaction on stem elongation. *Plant and Cell Physiology* 15, 265-272.

Hokkaido Agricultural Affairs Bureau, Upland Farming Promotion Section. (1987) Important azuki and kidney bean variety characteristics and dissemination rate (in Japanese). In: *Showa 62 Bean Producing Training School Text*. Japan Bean-Class Endowment Society, Tokyo, pp. 75-76.

Hokkaido Agricultural Experiment Station. (1967) Beans (in Japanese). In: *Hokkaido Agricultural Techniques Research History*. pp. 194-223.

Hokkaido Agricultural Experiment Station. (1977) Azuki (in Japanese). In: *The Cultivation Methods for Main Crops in Hokkaido, Agronomy Series No. 2*. pp. 67-98.

Hokkaido Central Experiment Station, Breeders Agriculture Station. (1976) Seed storage experiments. *North Agriculture* 43(11), 45-63.

Hokkaido Prefecture, Tokachi Agricultural Experiment Station. (1972a) (10) Azuki cultivation, the influence that missing plants exert on adjacent plants (in Japanese). *Hokkaido Prefectural Tokachi Agricultural Experiment Data* 4, 74.

Hokkaido Prefecture, Tokachi Agricultural Experiment Station. (1972b) (11) An experiment on planting density of azuki varieties versus fertilizer application dose (in Japanese, English summary). *Hokkaido Prefectural Tokachi Agricultural Experiment Data* 4, 75.

Hokkaido Prefecture, Tokachi Agricultural Experiment Station. (1978) Intervarietal differences in the transition to number of pods reaching maturity (flowering pattern) (in Japanese). *Showa 52 Bean Experiment Results Report*, pp. 46-48.

Holloway, G. J. (1986) The potency and effect of phytotoxins with yellow split-pea (*Pisum sativum*) and adzuki bean (*Vigna angularis*) on survival and reproductive potential of *Sitophilus oryzae* (L.) (Coleoptera: Curculionidae). *Bulletin of Entomological Research* 76(2), 287-295.

Holloway, G. J. and Mackness, M. I. (1988) The evolutionary adaptation of enzyme systems in *Sitophilus oryzae* fed on toxic legumes. *Entomologia Experimentalis et Applicata* 48(2), 165-172.

Homma, Y. (1991) Present situation of soilborne diseases and their control and related problems in Japan. *Japan Pesticide Information* 58, 12-19.

Honda, C., Nanba, T., Asaoka, O., Yumoto, K., Hayashi, T., Ikegami, S. and Takai, Y. (1992) Modification of methods for determination of dietary fiber in raw and processed pulses (in Japanese, English summary). *Journal of the Food Hygienic Society of Japan* 33(1), 46-51.

Honda, H. and Ohsawa, K. (1990) Chemical ecology for stored product insects (in Japanese). *Journal of Pesticide Science* 15(2), 263-270.

Honda, H., Yamamoto, I. and Sakai, A. (1989) Analysis of the oviposition preference from the egg-distribution in *Callosobruchus chinensis* L. (in Japanese, English summary). *Journal of Agricultural Science* (*Tokyo Agricultural University*) 33(3), 239-245.

Honda, Y. and Ikegami, M. (1986) Mungbean yellow mosaic virus. In: *Descriptions of Plant Viruses.* Association of Applied Biologists, Warwick, United Kingdom, Set 20, No. 323.

Honda, Y., Iwaki, M. and Saito, Y. (1983) Mechanical transmission, purification, and some properties of whitefly-borne mung bean yellow mosaic virus in Thailand. *Plant Disease* 67(7), 801-804.

Honda, Y., Iwaki, M., Thongmeearkom, P., Kiratiya-angul, K., Kiratiya-angul, S., Srithongchai, W., Prommin, M., Kittipakorn, K., Sarindu, N., Deema, N., Syamananda, R., Hibi, T. and Saito, Y. (1986) Mungbean yellow mosaic virus isolated from mungbean in Thailand. *Technical Bulletin of the Tropical Agriculture Research Center* 21, 189-202.

Hong, E. H. and Park, K. Y. (1982) Changes in variety and cultural practices of soybean, sweet potato and corn since 1962 in Korea (in Korea, English summary). *Korean Journal of Crop Science* 27(4), 462-469.

Hong, E. H., Kim, S. D. and Hwang, Y. H.(1987) Bean (azuki) variety improvement (in Korean). In: Committee of Memorable Paper (ed.), *Memorable Paper of Dr Bea Sungho, 60th Birthday: Varieties Improvement of Main Crop in Korea.* Sang Rok Sa Publishing, Soo Won City, 89-118.

Hong, E. H., Kim, S. D., Hwang, Y. H., Lee, Y. H., Moon, Y. H. and Park, R. K. (1986) A new good quality and high yielding adzuki bean variety "Chungjupat" (in Korean, English summary). *The Research Reports of the Office of Rural Development* 28(1), 148-150.

Hong, E. H., Kim, S. D., Lee, Y. H., Hwang, Y. H., Moon, Y. H. and Ham, Y. S. (1984) A new late planting tolerant and high yielding azuki bean variety "Chungwonpat" (in Korean, English summary). *The Research Reports of the Office of Rural Development* 26 2(C), 82-84.

Hong, E. H., Lee, Y. H., Kim, S. D., Hwang, Y. H. and Moon, Y. H. (1983) Selection of appropriate herbicides for establishment of weed control system in adzuki bean and mungbean (in Korean, English summary). *Korean Journal of Weed Science* 3(2), 199-207.

Horiguchi, I. (1950) The right seeding season for crops seeded in spring in the Tokachi district (in Japanese). *North Agriculture (Japan)* 17(7), 1-9.

Horvat, F. and Stainier, F. (1980) Study of the exine of the *Phaseolus-Vigna* complex and related genera. IV. (in French). *Pollen and Spores* 22(2), 139-172.

Horvath, J. and Beezner, L. (1982) Virus susceptibility and virus resistance to bean (*Phaseolus* L.) species: new compatible and incompatible host-virus relations (in Hungarian, English summary). *Kertgazdasag* 14(2), 71-76.

Hoshikawa, K. (1985) Azuki beans (in Japanese). In: *Edible Crops.* Yokendo, Tokyo, pp. 460-471.

Hoshikawa, K. (1987) Azuki beans (in Japanese). In: *Origin and Spread of Cultivated Crops.* Ninomiya Shoten Publishing House, Tokyo, Japan, pp. 60-61.

Hoson, T. and Masuda, Y. (1991a) Inhibition of auxin-induced elongation and xyloglucan breakdown in azuki bean epicotyl segments by fucose-binding lectins. *Physiologia Plantarum* 82(1), 41-47.

Hoson, T. and Masuda, Y. (1991b) Effect of xyloglucan nonasaccharide on cell elongation induced by 2, 4-D and IAA. *Plant and Cell Physiology* 32(6), 777-782.

Hoson, T., Kamisaka, S. and Masuda, Y. (1991a) Growth of higher plants in a simulated weightless environment. *Physiologia Plantarum* 82(3), B2.

Hoson, T., Kamisaka, S., Masuda, Y. and Yamashita, M. (1992) Changes in plant growth processes under microgravity conditions simulated by a three-dimensional clinostat. *Botanical Magazine Tokyo* 105(1077), 53-70.

Hoson, T., Masuda, Y., Sone, Y. and Misaki, A. (1991b) Xyloglucan antibodies inhibit auxin-induced elongation and cell wall loosening of azuki bean epicotyls but not of oat coleoptiles. *Plant Physiology* 96(2), 551-557.

Hoson, T., Sone, Y., Misaki, A. and Masuda, Y. (1993) Role of xyloglucan breakdown in epidermal cell walls for auxin-induced elongation of azuki bean epicotyl segments. *Physiologia Plantarum* 87(2), 142-147.

Hsieh, H. M., Pomeranz, Y. and Swanson, B. G. (1991) Composition and functional properties of azuki beans harvested at three moisture levels. *Cereal Foods World* 36(8), 696.

Hsieh, H. M., Pomeranz, Y. and Swanson, B. G. (1992) Composition, cooking time, and maturation of azuki (*Vigna angularis*) and common beans (*Phaseolus vulgaris*). *Cereal Chemistry* 69(3), 244-248.

Hsieh, H. M., Swanson, B. G. and Pomeranz, Y. (1994) Lipoxygenase activity and apparent K_m in maturing azuki. *Italian Journal of Food Science* 2, 243-247.

Hsieh, L. C., Chang, Y. C. and Chien, C. C. (1988) Studies on the ecology of powdery mildew of beans (in Chinese, English summary). *Journal of Agricultural Research of China* 37(1), 76-85.

Hsu, C. C. (1989a) The new cultivar of azuki: "Kaohsiung No. 5" (in Chinese). *Harvesting Year* 39(3), 22-23.

Hsu, C. C. (1989b) Breeding of the new cultivar "Kaohsiung No. 5" (in Chinese). *Bulletin of the Kaohsiung Agricultural Experiment Station* pp. 1-17.

Hsu, C. C. and Lin, C. L. (1985) District yield trial of newly-developed adzuki bean strains (in Chinese, English summary). *Annual Report of Dryland Food Crops Improvement* 27, 144-149.

Hsu, C. C., Chen, Y. R. and Lin, C. L. (1989a) Regional yield trial of newly-developed adzuki bean lines (in Chinese, English summary). *Annual Report of Dryland Food Crops Improvement* 31, 251-256.

Hsu, C. C., Chen, Y. R., Tseng, Y. T., Lin, C. L. and Wu, Y. L. (1989b) Breeding of adzuki bean (in Chinese, English summary). *Annual Report of Dryland Food Crops Improvement* 31, 243-249.

Hsu, C. C., Wang, M. M., Cheng, Y. R. and Lin, C. L. (1989c) Studies on the cultural techniques of adzuki bean (in Chinese, English summary). *Annual Report of Dryland Food Crops Improvement* 31, 257-261.

Hsu, S. H. (1983) Studies on dew: dew on winter crops of dryland in southwest area of Taiwan (in Chinese, English summary). *National Science Council Monthly* 11(6), 527-536.

Hu, D. (1982) Research on cultivation practices for azuki (in Chinese). *Bulletin of Agricultural Science and Technology* 11, 53-54.

Hu, J. P. (1981) Preliminary conclusions about azuki germplasm resources (in Chinese). *Journal of the Chinese Academy of Agricultural Sciences* 3, 1-29.

Hu, J. P. (1984) Preliminary report on azuki germplasm resources (in Chinese). *Crop Germplasm Resources* 1, 21-25.

Hu, J. P. (1986) A rough introduction to several excellent varieties and growth methods of red bean (in Chinese). *Bulletin of Agricultural Science and Technology* (5), 23.

Hu, T., Tsai, L. T., Fu, Y. K. and Shen, M. L. (1985) Utilization of gamma rays for insect pest control of bagged mungbean and adzuki bean (in Chinese, English summary). *Taiwan Plant Protection Bulletin* (*Chih wu pao hu hsueh hui hui kan*) 27(1), 63-68.

Hu, W., Wang, J. L. and Wang, J. P. (1987) Study on the edible legume resources and production in Inner Mongolia (in Chinese). *Inner Mongolia Agricultural Sciences and Technology* 3, 21-26.

Hu, X. (1989) Corn-Azuki (in Chinese). In: Hu, X. (ed.), *Multiple Planting Systems.* Academic Publishing House, Beijing, pp. 139-142.

Hu, Y. S. and Ma, Z. (1985) Primary observation on growth and development of Tianjin azuki. *Acta of Hebei Agricultural Science* 10(1), 76.

Huang, C. S., Mew, I. P. and Yang, C. Y. (1976) Screening for varietal resistance to *Rhizoctonia* damping-off in mungbean and other legumes (in Chinese). *Taiwan Plant Protection Bulletin* (*Chih wu pao hu hsueh hui hui kan*) 17(4), 8 (Abstr.).

Huang, M. T. (1989) Studies on hard seeds of adzuki bean. II. Influences of drying agent on the occurrence of hard seeds (in Chinese, English summary). *Annual Report of Dryland Food Crops Improvement* 31, 263-268.

Huang, S. S. and Lie, Y. C. (1987) Effects of rates of missing plants on the yield and production profit of leguminous crops (in Chinese, English summary). *Annual Report of Dryland Food Crops Improvement* 29, 185-200.

Hugenot, C., Furneaux, M. T., Thottappilly, G., Rossel, H. W. and Hamilton, R. I. (1993) Evidence that cowpea aphid-borne mosaic and blackeye cowpea mosaic viruses are two different potyviruses. *Journal of Genetic Virology* 74(3), 335-340.

Hwang, Y. H. (1982) Country reports on the status of grain legumes production: Republic of Korea. In: *Grain Legumes Production in Asia.* Asian Productivity Organization, Tokyo, pp. 431-448.

Hymowitz, T. (1987) The grain legumes: an overview of crops and species. In: *Grain Legumes as Alternative Crops.* Center for Alternative Crops and Products, University of Minnesota, pp. 32-44.

Ichinohe, M. (1953) On the parasitism of the soy bean nematode, *Heterodera glycines* (in Japanese, English summary). *Research Bulletin of the Hokkaido National Agricultural Experiment Station* 64, 113-124.

Ichinohe, M. (1988) Current research on the major nematode problems in Japan. *Journal of Nematology* 20(2), 184-190.

Igata, S. (1977) Azuki beans (in Japanese). In: *Research on the History of Cereal Crops.* Yoshikawa, Kobunkan, Tokyo, pp. 173-193.

Iida, W. (1972) Major diseases of leguminous crops in Japan. In: *Symposium on Food Legumes, Proceedings of a Symposium on Tropical Agriculture Research.* 12-14 September 1972. Tropical Agriculture Research Series No. 6, pp. 101-107.

Iizuka, T. (1988) Azuki diseases: bean common mosaic virus (BCMV) and peanut stunt virus (PnSV) (in Japanese). In: Kishi, K. (ed.), *Plant Diseases in Japan.* Zenkoku Noson Kyoiku Kyokai Co., Ltd., Tokyo, pp. 118-119.

Ikeda, M. and Hirokawa, F. (1977) Weeding season and weed damage in azuki fields. *North Agriculture* 501, 11-20.

Imagawa, K., Toko, K., Ezaki, S., Hayashi, K. and Yamafuji, K. (1991) Electrical potentials during gravitropism in bean epicotyls. *Plant Physiology* 97(1), 193-196.

Imazeki, H. (1993) Cloning of cDNA for ethylene-inducible acidic chitinase of *Vigna angularis* (in Japanese). *Patent: Japan Kokai Tokkyo Koho JP 9384087 A2; JP 0584087.* Date issued: 930406. 6 pp.

Imura, O. (1991) A comparative study of the feeding habits of *Tribolium freemani* Hinton and *Tribolium castaneum* (Herbst) (Coleoptera: Tenebrionidae). *Applied Entomology and Zoology* 26(2), 173-182.

Inouhe, M. and Yamamoto, R. (1991) Effects of 2-deoxygalactose on auxin-induced growth and levels of UDP-sugars in higher plants. *Plant and Cell Physiology* 32(3), 433-438.

Inouye, N., Maeda, T. and Mitsuhata, K. (1982) Cucumber mosaic virus isolated from *Calanthe discolor*. *Agricultural Research* (*Nogaku Kenkyu*) 60(1), 1-11.

Institute of Agricultural Science in Nanjing, Jiangsu. (1981) Study of azuki cultivation technology for high yields (in Chinese). *Nanjing Agricultural Science and Technology* 1, 9.

Institute of Developing Economies. (1969) *One Hundred Years of Agricultural Statistics in Japan*. Kabushiki Kaisha Sangyo Tokei Kenkyu-sha, Tokyo, 270 pp.

Insull, W. Jr., Oiso, T. and Tsuchiya, K. (1968) Diet and nutritional status of Japanese. *The American Journal of Clinical Nutrition* 21(7), 753-777.

International Board for Plant Genetic Resources. (1988) Report on an IBPGR mission in northern Thailand for the collection of *Vigna umbellata* var. *gracilis*. *FAO/IBPGR Plant Genetic Resources Newsletter* 73/74, 58.

Ishibashi, K. I., Hironaka, K. and Nakagawa, M. (1992) Equilibrium moisture content of beans (in Japanese, English summary). *Research Bulletin of Obihiro University Series I* 17(4), 385-389.

Ishige, F., Mori, H., Yamazaki, K. I. and Imaseki, H. (1993a) Identification of a basic glycoprotein induced by ethylene in primary leaves of azuki bean [*Vigna angularis*] as a cationic peroxidase. *Plant Physiology* 101(1), 193-199.

Ishige, F., Mori, H., Yamazaki, K. and Imaseki, H. (1993b) Cloning of a complementary DNA that encodes an acidic chitinase which is induced by ethylene and expression of the corresponding gene. *Plant and Cell Physiology* (*Kyoto, Japan*) 34(1), 103-111.

Ishige, F., Yamazaki, K., Mori, H. and Imaseki, H. (1991) The effects of ethylene on the coordinated synthesis of multiple proteins: accumulation of an acidic chitinase and a basic glycoprotein induced by ethylene in leaves of azuki bean, *Vigna angularis*. *Plant and Cell Physiology* 32(5), 681-690.

Ishii, Y. and Inoue, N. (eds.). (1968) Azuki *Phaseolus radiatus* L. var. *aurea* Prain (in Japanese). In: *Latest Horticulture Large Dictionary*. Tokyo, Seibundo Shinkosha, Japan, p. 196.

Ishikawa, C., Watanabe, K. and Sakata, N. (1985) Azuki bean (*Vigna angularis*) protease inhibitors: isolation and amino acid sequences. *Journal of Biochemistry* (*Tokyo, Japan*) 97(1), 55-70.

Ishimoto, M. and Kitamura, K. (1991) Effect of absence of seed α-amylase inhibitor on the growth inhibitory activity to azuki bean weevil (*Callosobruchus chinensis*) in common bean (*Phaseolus vulgaris* L.). *Japanese Journal of Breeding* 41(2), 231-240.

Ishimoto, M. and Kitamura, K. (1992) Tolerance to the seed α-amylase inhibitor by the two insect pests of the common bean (*Zabrotes subfasciatus*) and *Acanthoscelides obtectus* (Coleoptera: Bruchidae). *Applied Entomology and Zoology* 27(2), 243-251.

Ishimoto, M. and Kitamura, K. (1993a) Inhibitory effects of Adzuki bean weevil-resistant mungbean seeds on growth of the bean bug. *Japanese Journal of Plant Breeding* 43(1), 75-80.

Ishimoto, M. and Kitamura, K. (1993b) Specific inhibitory activity and inheritance of an α-amylase inhibitor in a wild common bean accession resistant to the Mexican bean weevil (in English, Japanese summary). *Japanese Journal of Plant Breeding* 43(1), 69-73.

Ishiyama, Y. and Tamura, G. (1985) Isolation and partial characterization of homogeneous nitrite reductase from etiolated bean shoots (*Phaseolus angularis* W. F. Wight). *Plant Science Letters* 37, 251-256.

Ishiyama, Y., Shinoda, I., Fukushima, K. and Tamura, G. (1985) Some properties of ferredoxin-nitrite reductase from green shoots of bean and an immunological comparison with nitrite reductase roots and etiolated shoots. *Plant Science* 39(2), 89-95.

Ito, H., Kojima, M., Shikata, E. and Murayama, D. (1975) Studies on legume virus diseases in Hokkaido. III. Viruses isolated from diseased azuki bean plants (in Japanese, English summary). *Memoirs of the Faculty of Agriculture Hokkaido University* (*Hokkaido Daigaku Nogaku-Bu Hobun*) 9(2), 165-175.

Ito, S. (1921) Studies on "yellow dwarf" disease of soybean (in Japanese). *Hokkaido Agricultural Experiment Station Report* 11, 47-59.

Ito, S., Kojima, M., Ohnishi, M. and Fujino, Y. (1985) Triterpenic lipids in azuki bean seeds (in Japanese, English summary). *Journal of the Agricultural Chemical Society of Japan* 59(9), 895-900.

Ito, S., Tsukidate, Y. and Fujino, Y. (1981) Separation and determination of tocopherol homologues in cereal grains and bean seeds (in Japanese, English summary). *Research Bulletin of Obihiro University, Series I* 12(2), 143-148.

Iwaki, M., Pornpod, T., Tsuchizaki, T., Honda, Y., Nonglak, S., Luck, V., Tochihara, H. and Nuanchan, D. (1980) Virus diseases of soybean in Thailand. In: *The Second Southeast Asian Symposium on Plant Disease in the Tropics.* October, 1980, Bangkok, Thailand, pp. 22-26.

Iwanami, Y. (1982) Inhibition of phototropism and gravitropism in cuttings of bean (*Phaseolus angularis*) by treatment with myrmic acid and related compounds. *Botanical Gazette* 143(4), 483-485.

Jaaska, V. and Jaaska, V. (1988) Isoenzyme variation in the general *Phaseolus* and *Vigna* (Fabaceae) in relation to their systematics: aspartate aminotransferase and superoxide dismutase. *Plant Systematics and Evolution* 159(3-4), 145-159.

Jain, H. K. and Mehra, K. L. (1980) Evolution, adaptation, relationships, and uses of the species of *Vigna* cultivated in India. In: Summerfield, R. J. and Bunting, A. H. J. (eds.), *Advances in Legume Science.* Indian Agricultural Research Institute, New Delhi, pp. 459-468.

Japan Bean Fund Association. (1977) *Legume Varieties in Hokkaido* (in Japanese). Japan Bean Fund Association, Tokyo, pp. 130-143.

Japan Bean Fund Association. (1987) *Information on Miscellaneous Beans: July 1987* (in Japanese). Japan Bean Fund Association, Tokyo.

Japan Bean Fund Association. (1988) *Report of Commercial Production Conditions of Beans Produced by Prefectures: Second Annual Report* (in Japanese). Japan Bean Fund Association, Tokyo.

Japan Bean Fund Association. (1989) *Constituents and Quality of Beans (Study on Improvement of Quality of Hokkaido Beans): February 1989* (in Japanese). Hokkaido Central Agricultural Experiment Station, 62 pp.

Japan Peas and Beans Importer Association. (1987) *Peas and Beans Statistics: June 1987* (in Japanese). Japan Peas and Beans Importer Association, Tokyo, 221 pp.

Japan Plant Protection Association. (1978) Tolerances for residues of agricultural chemicals in Japan (expressed in ppm). *Japan Pesticide Information* 35, 58-59.

Japan Plant Protection Association. (1980) Tolerances for residues of agricultural chemicals in Japan (expressed in ppm). *Japan Pesticide Information* 37, 69-70.

JETRO, Japan Trade Promotion Organization. (1991) *Trade of Agriculture, Forestry, Fisheries Products. Domestic and Foreign News of Principal 100 items.* Tokyo, pp. 186-191, 340-345.

Jiang, C. Y. (1982) Crop germplasm resources conservation system in China. *FAO/IBPGR Plant Genetic Resources Newsletter* 51(9), 2-5.

Jiang, R. and Chen, S. (1984) Study of ecological types of azuki germplasm (in Chinese). *Crop Germplasm Resources* 2, 9-12.

Jin, W. (1989a) A research on local variety resources of adzuki bean of northern China. I. Study on correlation of quantitative traits of adzuki bean in Nanjing (in Chinese, English summary). *Beijing Agricultural College* 4(3), 33-41.

Jin, W. (1989b) Behavior and potential selectivity of northern area native azuki germplasm in Nanjing (in Chinese). *Crop Germplasm Resources (Ziwupinzhongziyuan)* 3, 10-12.

Jin, W. (1990a) The study of adzuki bean variety sources of north China. III. The plurality heredity index of the local variety population of adzuki bean (in Chinese, English summary). *Beijing Agricultural College* 5(1), 47-52.

Jin, W. (1990b) The study of adzuki bean variety sources of north China. II. Application of the analysis of the principal factor to appraise source of local variety of adzuki bean (in Chinese, English summary). *Beijing Agricultural College* 5(1), 1-6.

Jin, W. and Futsuhara, Y. (1993) The induction of calli from adzuki bean and direct plant regeneration (in Chinese, English summary). *Journal of Beijing Agricultural College* 8(6), 93-99.

Jin, W. and Gu, P. (1990) Studies on the growth and development of adzuki bean. VII. The change of the leaf blade chlorophyll content of adzuki bean (in Chinese, English summary). *Beijing Agricultural College* 5(2), 17-25.

Jin, W. and Jiang, L. (1988) Azuki and cultural practices. *Jiangsu Agricultural Science* 7, 16.

Jin W. and Jiang, L. (1989a) Studies on the growth and development of adzuki bean *Vigna angularis* Ohwi and Ohashi. VI. Observation of the blooming and fruiting arrangement on an adzuki bean plant (in Chinese, English summary). *Beijing Agricultural College* 4(3), 42-48.

Jin, W. and Jiang, L. (1989b) The breeding tendency and utilization in Yangtze reaches about native azuki germplasm produced in northern areas (in Chinese). *Jiangsu Agricultural Science* 7, 12-15.

Jin, W. and Jiang, L. (1993) Research for growth and developmental rhythm of adzuki bean. IX. The distribution pattern of flower and pod of different plant. (in Chinese) *Journal of Beijing Agricultural College* 8(1), 31-36.

Jin, W., Feng, S. and Chen, X. (1993) The new adzuki bean variety of Jing Nong No. 1, 2 (in Chinese, English summary). *Journal of Beijing Agricultural College* 8(1), 74-78.

Jin W., Fong, S. and Hu, J. (1989) Studies on the growth and development of adzuki bean *Vigna angularis* Ohwi and Ohashi. V. Reaction of adzuki bean from blossoming to maturity in sowing at different times (in Chinese, English summary). *Beijing Agricultural College* 4(1), 20-26.

Johnson, D. L., Croissant, R. L., Youngman, V. E., Rubingh, J. and Shellburg, J. G. (1989) *Development of Alternative Crops in Colorado.* Colorado State University, 46 pp.

Joseph, L. S. and Bouwkamp, J. C. (1978) Karyomorphology of several species of *Phaseolus* and *Vigna. Cytologia* 43, 595-600.

Jozsef, H. and Laszlo, B. (1982) Virus susceptibility and virus resistance to bean (*Phaseolus* L.) species: new compatible and incompatible host virus relations. *Kertgazdasag* 14(2), 71-76.

Jun, T. H., Nam, M. H., Lee, S. K. and Park, W. C. (1983) Changes in peroxidase activity and its isozymes of soybean, red-bean and mungbean-bean during germination (in Korean, English summary). *Journal of the Korean Agricultural Chemical Society* 26(3), 151-156.

Jung, J. and Lee, C. Y. (1982) A flurometric assay for trypsin inhibitor soybean, red bean and mung bean, *Glycine max, Phaseolus.* (in Korean, English summary). *Journal of the Korean Agricultural Chemical Society* 25(3), 182-188.

Kaempfer, E. (1712) Amoenitatum Exoticarum Politico-Medicarum, fasciculi. V. In: *Flora Japonica.* Lemgoviae. 1983 reprint of original. Steiner, Weisbaden, pp. 837-838.

Kaga, A., Hosaka, K., Kimmura, T., Misoo, S. and Kamijima, O. (1993) Application of random amplified polymorphic DNA RAPD analysis for adzuki (in Japanese). *Science Reports of Faculty of Agriculture Kobe University* 20(2), 171-176.

Kagiwara, M. (1992) *The Situation of Beans* (in Japanese). Kokuno Marusho, Obihiro, Japan, 38 pp.

Kaiser, W. J., Wyatt, S. D. and Klein, R. E. (1991) Epidemiology and seed transmission of two tobacco streak virus pathotypes associated with seed increases of legume germ plasm in Eastern Washington USA. *Plant Disease* 75(3), 258-264.

Kakizaki, Y. (1923) Linked inheritance of certain characters in the adzuki bean. *Genetics* 8, 168-177.

Kakuta, T. and Satoh, J. (1953a) Azuki root development (in Japanese). *Tohoku Agriculture* 7(1), 10-12.

Kakuta, T. and Satoh, J. (1953b) The effect of daylength on azuki flowering (in Japanese). *Tohoku Agriculture* 7(1), 13-15.

Kamimura, T. and Maeda, T. (1991) Manufacture of low-calorie candy having sugar alcohol-treated soft center and outer shell structure. *Patent: Japan Kokai Tokkyo Koho; JP 9158750 A2; JP 0358750.* Date issued: 13 March. 5 pp.

Kamiyama, S., Tamura, N., Osada, K. and Watanuki, M. (1993) Fermentative manufacture of L-carnitine with filamentous fungi (in Japanese). *Patent: Japan Kokai Tokkyo Koho; JP 9323190 A2; JP 0523190.* Date issued: 930202. 10 pp.

Kaneta, T., Kakimoto, T. and Shibaoka, H. (1993) Actinomycin D inhibits the GA3-induced elongation of azuki bean epicotyls and the reorientation of cortical microtubules. *Plant and Cell Physiology* 34(7), 1125-1132.

Katayama, J. and Suzuki, I. (1984) Seasonal prevalence of pod borers in azuki-beans and injury caused by larval infestation (in Japanese, English summary). *Bulletin of the Kyoto Prefectural Institute of Agriculture* 12, 27-34.

Katayama, T. C. (1979) Some morphological characters of pulses collected in the Indian sub-continent. *Memoirs of the Faculty of Agriculture, Kagoshima University* 15, 29-48.

Kato, J., Hosoya, E. and Ichikawa, N. (1992) Variation of characteristics related to the seed quality of adzuki beans grown in Hokkaido and the relation between seed coat colors and ann (bean jam) colors. *Bulletin of Hokkaido Prefectural Agricultural Experiment Station* 64, 25-34.

Kato, R., Nagayama, E., Suzuki, T., Uchida, K., Shimomura, T. H. and Harada, Y. (1993) Promotion of plant cell division by an epidermal growth factor. *Plant and Cell Physiology* 34(6), 789-793.

Katsufuji, K. (1919) 'Yellow dwarf' a new nematode disease of soybean. *Annals of the Phytopathological Society of Japan* 1, 1-5.

Kaushal, R. P. (1985) *Genetics of resistance in Urdbean (Vigna mungo (L.) Hepper) to some major diseases occurring in Himachal Pradesh.* Unpublished PhD thesis, Palampur, India.

Kaushal, R. P. and Singh, B. M. (1988) Interspecific hybridization between urdbean (*Vigna mungo* (L.) Hepper) and adzuki bean (*Vigna angularis* (Willd.) Ohwi and Ohashi). *Euphytica* 39(1), 53-57.

Kawahara, E. (1962) IV. Azuki (in Japanese). In: Kawahara, E. (ed.), *Eliji Kawahara: 4th Compilation of Bean Varieties.* Yokendo, Tokyo, pp. 1-32.

Kawakami, J. (1930) Chromosome numbers in Leguminosae (in Japanese). *The Botanical Magazine* 44(522), 319-328.

Kawase, M. (1993) Wastewater treatment by anaerobic fixed-bed reactor. *Bioprocess Technology* 16(Ind. Appl. Immobilized Biocatal.), 355-376.

Kay, D. E. (1979) Azuki bean. In: Kay, D. E. (ed.), *Food Legumes.* Tropical Products Institute, London, pp. 1-10.

Kennedy, B. W. (1984a) Bacterial stem rot, a devastating disease of adzuki beans in Minnesota. *Annual Report of the Bean Improvement Cooperative* 27, 160 (Abstr.).

Kennedy, B. W. (1984b) Natural and artificial pathogenicity of *Pseudomonas adzukicola* on edible beans. In: *Proceedings of the Second Working Group on* Pseudomonas syringae *Pathovars.* 24-28 April 1984, Sounion, Greece. The Hellenic Phytopatholgical Society, Athens, pp. 65-66.

Kennedy, W. and Denny, R. (1984) Bacterial stem rot, a new disease of adzuki beans in Minnesota. *Plant Disease* 68(2), 167.

Kigasawa, K. (1978) Nematode countermeasures in azuki cultivation (in Japanese). *North Agriculture (Japan)* 518, 13-17.

Kim, B. Y., Kim, K. S. and Han, K. H. (1988) Studies on lead uptake by crops and reduction of its damage. IV. Effects of application of calcium and phosphate materials on lead uptake by upland crops (in Korean, English summary). *Journal of Korean Society of Soil Science and Fertility* 21(4), 426-433.

Kim, K. C. and Choi, H. S. (1987) Effects of temperature on the oviposition, feeding and emergence of the azuki bean weevil (*Callosobruchus chinensis* L.) in the stored beans (in Korean, English summary). *Korean Journal of Plant Protection* 26(2), 71-81.

Kim, K. J., Kim, K. H. and Kim, Y. H. (1981) Comparative studies on growth patterns of pulse crops at different growing seasons. II. Variation in distribution of flowering dates and pod setting ratio of soybean, azuki bean and mungbean (in Korean, English summary). *Korean Journal of Crop Science* 26(3), 243-250.

Kim, Y. H., Hong, E. H. and Chae, Y. A. (1993) Crossability and characteristics of F1 hybrids in interspecific hybridization among *Vigna* species (in Korean, English and Korean summaries). *Korean Journal of Breeding* 24(4), 343-350.

Kishi, K. (ed.). (1988) *Plant Diseases in Japan.* Zenkoku Noson Kyoiku Kyokai Co. Ltd. Tokyo, pp. 117-119.

Kiso, A. (1991) (B-II) Fungicides: upland crops. *Japan Pesticide Information* 58, 46.

Kitagawa, I., Wang, H. K., Saito, M. and Yoshikawa, M. (1983a) Saponin and sapogenol. XXXI. Chemical constituents of the seeds of *Vigna angularis* (Willd.) Ohwi et Ohashi. (1). Triterpenoidal sapogenols and 3-furanmethanol beta-D-glucopyranoside. *Chemical and Pharmaceutical Bulletin* (*Tokyo*) 31(2), 664-673.

Kitagawa, I., Wang, H. K., Saito, M. and Yoshikawa, M. (1983b) Saponin and sapogenol. XXXII. Chemical constituents of the seeds of *Vigna angularis* (Willd.) Ohwi et Ohashi. (2). Azukisaponins I, II, III and IV. *Chemical and Pharmaceutical Bulletin* (*Tokyo*) 31(2), 674-682.

Kitagawa, I., Wang, H. K., Saito, M. and Yoshikawa, M. (1983c) Saponin and sapogenol. XXXIII. Chemical constituents of the seeds of *Vigna angularis* (Willd.) Ohwi et Ohashi. (3). Azukisaponins V and VI. *Chemical and Pharmaceutical Bulletin* (*Tokyo*) 31(2), 683-688.

Kitazawa, K. and Yanagita, K. (1978) On the pathogenic fungus of azuki stem rot disease (in Japanese). *Annals of the Phytopathological Society of Japan* 44(1), 74-75 (Abstr.).

Kitazawa, K. and Yanagita, K. (1984) Adzuki bean wilt caused by *Fusarium oxysporum* Schl.: re-occurrence and confirmation of the causal organism (in Japanese, English summary). *Annals of the Phytopathological Society of Japan* 50(5), 643-645.

Kitazawa, K. and Yanagita, K. (1989) *Fusarium oxysporum* Schl. f. sp. *adzukicola* n. f. sp., a wilt fungus of *Phaseolus angularis* (in Japanese, English summary). *Annals of the Phytopathological Society of Japan* 55(1), 76-78.

Kitazawa, K., Suzui, T. and Yanagita, K. (1979a) The differentiation of pathogenicity of azuki phytophthora stem rot (*Phytophthora vignae*) (in Japanese). *Annals of the Phytopathological Society of Japan* 45(1), 117 (Abstr.).

Kitazawa, K., Suzui, T. and Yanagita, K. (1979b) Pathogenicity of *Phytophthora vignae* Purss. to adzuki bean and cowpea. *Annals of the Phytopathological Society of Japan* 45(3), 406-408.

Kitazawa, K., Tsuchiya, S., Kodama, F., Kamjaipai, W., Ogoshi, A. and Yanagita, K. (1978) Phytophthora stem rot of adzuki bean (*Phaseolus angularis*) caused by *Phytophthora vignae* Purss. (in Japanese, English summary). *Annals of the Phytopathological Society of Japan* 44(4), 528-531.

Kiuchi, Y., Ishikawa, H., Nitta, M. and Sasaki, T. (1987) Collection of crop genetic resources and their characteristics in Iwate Prefecture. II. Characteristics of local azuki bean varieties (*Phaseolus angularis* W. F. Wight), local kidney bean varieties (*P. vulgaris* L.), and local flower bean varieties (*P. coccineus* L.) (in Japanese). *Tohoku Agriculture Research* 40, 141-142.

Kobayashi, K. (1980) Changes of Gregatin A and Graminin A in a host plant. *Journal of the Faculty of Agriculture Hokkaido University* 59(4), 392-396.

Kobayashi, K. (1983) Soil disease injury of Hokkaido upland crops (in Japanese). In: Ui, T. (ed.), *Hokkaido Upland Crops Soil Disease Injury Publication Association.* Sapporo, Japan, pp. 203-231.

Kobayashi, K. (1984) Studies on induced mutants of *Cephalosporium gregatum*, the causal fungus of brown stem rot in adzuki beans (in English, Japanese summary). *Annals of the Phytopathological Society of Japan* 50(4), 522-527.

Kobayashi, K. and Ui, T. (1975) Isolation of phytotoxic substances by *Cephalosporium gregatum* Allington and Chamberlain. *Tetrahedron Letters* 47, 4119-4122.

Kobayashi, K. and Ui, T. (1976) The singular separation of a new wilting toxin for the azuki brown stem rot pathogenic fungus (*Cephalosporium gregatum*) (in Japanese). *Annals of the Phytopathological Society of Japan* 42(1), 94 (Abstr.).

Kobayashi, K. and Ui, T. (1977) Wilt-inducing antibiotic compounds produced by *Cephalosporium gregatum*. *Physiological Plant Pathology* 11, 55-60.

Kobayashi, K. and Ui, T. (1980a) Comparison of *Cephalosporium gregatum* isolates from adzuki bean in Japan and from soybean in U.S.A. (in English, Japanese summary). *Annals of the Phytopathological Society of Japan* 46(2), 241-246.

Kobayashi, K. and Ui, T. (1980b) Production of gregatins by an Iowa isolate of *Cephalosporium gregatum*, the cause of brown stem rot of soybeans in U.S.A. *Annals of the Phytopathological Society of Japan* 46(2), 256-257.

Kobayashi, K., Kondo, N. and Ui, T. (1981a) Appearance of a mutant from *Cephalosporium gregatum* Type A isolate in culture. *Annals of the Phytopathological Society of Japan* 47, 577-580.

Kobayashi, K., Kondo, N., Negishi, H. and Ui, T. (1984) Studies on the incidence and distribution of brown stem rot (*Cephalosporium gregatum*) of adzuki beans in Hokkaido (in Japanese, English summary). *Memoirs of the Faculty of Agriculture, Hokkaido University* (*Hokkaido Daigaku Nogaku-Bu Hobun*) 14(2), 91-97.

Kobayashi, K., Kondo, N., Tanaka, H. and Ui, T. (1980) Differences in the virulence of type A and type B isolates of *Cephalosporium gregatum*. *Annals of the Phytopathological Society of Japan* 46(2), 253-255.

Kobayashi, K., Kondo, N., Ui, T., Tachibana, H. and Aota, T. (1983) Difference in pathogenicity of *Phialophora gregata* isolates from adzuki bean in Japan and from soybean in the United States. *Plant Disease* 67(4), 387-388.

Kobayashi, K., Tanaka, F., Kondo, N. and Ui, T. (1981b) A selective medium for isolation of *Cephalosporium gregatum* from soil and populations in adzuki bean field soils estimated with the medium (in English, Japanese summary). *Annals of the Phytopathological Society of Japan* 47(1), 29-34.

Kobayashi, K., Tanaka, F., Ui, T. and Akai, J. (1979) A new strain of *Cephalosporium gregatum*, causal fungus of brown stem rot of adzuki beans. *Annals of the Phytopathological Society of Japan* 45(3), 409-411.

Kobayashi, K., Yamamoto, H., Negishi, H. and Ogoshi, A. (1991) Formae speciales differentiation of *Phialophora gregata* isolates from adzuki bean and soybean in Japan. *Annals of the Phytopathological Society of Japan* 57(2), 225-231.

Kobayashi, M. and Ohta, Y. (1983) Induction of stress metabolite formation in suspension cultures of *Vigna angularis*. *Phytochemistry* 22(5), 1257-1261.

Kobayashi, R., Michikawa, K., Tsuchibe, M. and Watanabe, T. (1992a) Comparison of processing characteristics of *Phaseolus* as materials of "An" bean jam (in Japanese, English summary). *Journal of the Japanese Society for Food Science and Technology* 39(8), 657-662.

Kobayashi, R., Michikawa, K. and Watanabe, T. (1992b) Studies on processing characteristics of *Phaseolus* as material of "An". Part III. Microscopic observation of "An" particles prepared from *Phaseolus* group and their morphological changes by treating with cellulase (in Japanese, English summary). *Journal of the Japanese Society for Food Science and Technology* 39(8), 671-677.

Kobayashi, T. (1978) Pests of grain legumes including soybeans and their control in Japan. In: Singh, S. R., van Emden, H. F. and Taylor, T. A. (eds.), *Pests of Grain Legumes: Ecology and Control*. Academic Press Inc., London, pp. 59-65.

Kobayashi, T. (1979) Control of azuki aphid and virus diseases with 'Disyston' (in Japanese). *Agricultural Chemicals Research* 89, 34-37.

Kobayashi, T., Hasegawa, T. and Kegasawa, K. (1972) Major insect pests of leguminous crops in Japan. In: *Symposium on Food Legumes: Proceedings of a Symposium on Tropical Agriculture Research*. 12-14 September 1972. Tropical Agricultural Research Series 6, pp. 109-126.

Kobayashi, T., Oku, T., Maeta, Y. and Saito, O. (1976) Studies on the soil application of insecticides. X. Control of aphid vectors and virus diseases of the azuki bean with systemic organophosphates (in Japanese, English summary). *Bulletin of the Tohoku National Agriculture Experiment Station* 53, 29-61.

Kogure, K., Naka, J. and Otoi, T. (1980) Photosynthetic and respiratory studies on the pod developing stage of azuki bean (*Vigna angularis*) (in Japanese). *Report of the Sikoku Branch, The Crop Science Society of Japan* 16, 57-61.

Kojima, M., Katagiri, Y., Furukawa, N., Mouri, H., Ohnishi, M. and Ito, S. (1991a) Sterol lipids and sphingolipids in adzuki bean *Vigna angularis* organs (in Japanese). *Research Bulletin of Obihiro University, Series I* 17(3), 235-242.

Kojima, M., Kimura, H., Ohnishi, M., Fujino, Y. and Ito, S. (1991b) Distribution and characterization of diglycosyldiacylglycerol isomers with different anomeric configuration in higher plants. *Phytochemistry* 30(4), 1165-1168.

Kojima, M., Nagasawa, T., Furukawa, N., Mouri, H., Ohnishi, M. and Ito, S. (1991c) Chemical characterization of glycerolipids sterol lipids and sphingolipids in leguminosae seeds (in Japanese). *Journal of the Japanese Society for Food Science and Technology* 38(12), 1076-1085.

Kojima, M., Ohnishi, M. and Ito, S. (1991d) Composition and molecular species of ceramide and cerebroside in scarlet runner beans (*Phaseolus coccineus* L.) and kidney beans (*Phaseolus vulgaris* L.). *Journal of Agricultural and Food Chemistry* 39(10), 1709-1714.

Kojima, M., Ohnishi, M. and Ito, S. (1991e) Fatty acid compositions in leguminosae seeds (in Japanese). *Research Bulletin of Obihiro University, Series I* 17(3), 227-233.

Kojima, M., Ohnishi, M., Ito, S. and Fujino, Y. (1989) Characterization of acylmono-, mono-, di-, tri- and tetraglycosylsterol and saponin in adzuki bean (*Vigna angularis*) seeds. *Lipids* 24(10), 849-853.

Kojima, M., Sasaki, S., Ohnishi, M., Mano, Y. and Ito, S. (1990a) Changes in molecular species of glycerolipids from developing *Vigna angularis* seeds. *Phytochemistry* 29(7), 2091-2096.

Kojima, M., Seki, K., Ohnishi, M., Ito, S. and Fujino, Y. (1990b) Structure of novel glyceroglycolipids in adzuki bean *Vigna angularis* seeds. *Biochemistry and Cell Biology* 68(1), 59-64.

Kojima, M., Seki, T., Ohnishi, M. and Ito, S. (1991f) Two digalactosyldiacylglycerols with different anomeric configuration in leguminous seeds. *Journal of the Science of Food and Agriculture* 54(1), 35-41.

Kojima, M., Yamada, S., Furukawa, N., Nakayama, O., Ohnishi, M. and Ito, S. (1992) Changes in sterollipids and sphingolipids of adzuki beans (*Vigna angularis*) during maturation (in Japanese). *Journal of the Japanese Oil and Chemistry Society* 41(3), 252-258.

Kokubo, A., Yamamoto, R. and Masuda, Y. (1993) Galactose prevents proton excretion but not IAA-induced growth in segments of azuki bean epicotyls. *Planta* 190(2), 284-287.

Kon, T. (1979) Azuki flowering, mature pod finishing numbers and rates, and features of approaching maturity during a high temperature year (in Japanese). *North Agriculture (Japan)* 46, 52-62.

Kondo, N. and Kobayashi, K. (1983a) Studies on the ecology of *Cephalosporium gregatum*, the causal fungus of brown stem rot of adzuki beans in Hokkaido. I. Seasonal variations of *Cephalosporium gregatum* populations in soil under host and non-host plant cultivation (in Japanese, English summary). *Memoirs of the Faculty of Agriculture, Hokkaido University* (*Hokkaido Daigaku Nogaku-Bu Hobun*) 14(1), 39-47.

Kondo, N. and Kobayashi, K. (1983b) Studies on the ecology of *Cephalosporium gregatum*, the causal fungus of brown stem rot of adzuki beans in Hokkaido. II. Effects of infested host residue on the number of *Cephalosporium gregatum* propagules in soil in the field (in Japanese, English summary). *Memoirs of the Faculty of Agriculture, Hokkaido University* (*Hokkaido Daigaku Nogaku-Bu Hobun*) 14(1), 48-49.

Kondo, N. and Kobayashi, K. (1983c) Studies on the ecology of *Cephalosporium gregatum*, the causal fungus of brown stem rot of adzuki beans in Hokkaido. III. The longevity and germination of *Cephalosporium gregatum* conidia in soil (in Japanese, English summary). *Memoirs of the Faculty of Agriculture, Hokkaido University* (*Hokkaido Daigaku Nogaku-Bu Hobun*) 14(1), 50-55.

Kondo, N. and Kodama, F. (1989) *Fusarium oxysporum* f. sp. *adzukicola*, causal agent of adzuki bean wilt and detection of three races of the fungus (in English, Japanese summary). *Annals of the Phytopathological Society of Japan* (*Nippon Shokubutsu Byori Gakkaiho*) 55(4), 451-457.

Kondo, N., Murata, K., Shimada, H., Shirai, S. and Kodama, F. (1990) Inheritance of resistance to three races of *Fusarium oxysporum* f. sp. *adzukicola* in adzuki bean cultivars (in English, Japanese summary). *Annals of the Phytopathological Society of Japan* 56, 677-679.

Konno, S. and Narikawa, T. (1978) Recent studies and problems on breeding and cultivation of adzuki bean in Japan. In: *Proceedings of the First International Mungbean Symposium*. Asian Vegetable Research and Development Center, Taiwan, pp. 236-239.

Koyama, Y., Nomura, N., Mori, Y. and Asahikawa, S. (1972) The new adzuki bean variety "Kotobuki-Shozu" (in Japanese, English summary). *Bulletin of the Hokkaido Prefectural Agricultural Experiment Station* 25, 81-91.

Kubota, K. (1979) Study of the rate equation for the soaking and cooking of red bean. *Journal of the Faculty of Applied Biological Science, Hiroshima University* 18(2), 161-169.

Kugimiya, M. (1987a) Solubilization of starch by treatment of ann granules with α-amylase (in Japanese, English summary). *Journal of the Japanese Society for Food Science and Technology* 34(9), 581-585.

Kugimiya, M. (1987b) A method for determination of ruptured and damaged granules in ann samples based on the solubilized starch content by α-amylase treatment (An enzymatic method for determination of ruptured and damaged granules in ann samples Part II) (in Japanese, English summary). *Journal of the Japanese Society for Food Science and Technology* 34(10), 654-658.

Kugimiya, M. (1992a) Effects of cooking conditions on rupture and damage of ann granules (denatured cotyledon cells) in cooked adzuki beans (in Japanese). *Journal of the Japanese Society for Food Science and Technology* 39(2), 167-172.

Kugimiya, M. (1992b) Effects of producing processes on rupture and damage of ann granules (denatured cotyledon cells) in Neri-Ann (bean jam made by kneading) (in Japanese, English summary). *Journal of the Japanese Society for Food Science and Technology* 39(3), 276-280.

Kugimiya, M. and Hirata, T. (1987a) Solubilization of starch by treatment of ann granules with α-amylase (an enzymatic method for determination of ruptured and damaged granules in ann samples Part I). *Nippon Shokuhin Kogyo Gakkaishi* 34(9), 581-585.

Kugimiya, M. and Hirata, T. (1987b) A method for determination of ruptured and damaged granules in ann samples based on the solubilized starch content by α-amylase treatment (an enzymatic method for determination of ruptured and damaged granules in ann samples Part II). *Nippon Shokuhin Kogyo Gakkaishi* 34(10), 654-658.

Kume, H., Tanaka, S. and Murayama, D. (1969) A strain of bean yellow mosaic virus (BYMV) isolated from mosaic-diseased red clover (in Japanese, English summary). *Memoirs of the Faculty of Agriculture, Hokkaido University* 7(1), 435-449.

Kusanagi, T. (1981) (C-II) Herbicides: upland crops. *Japan Pesticide Information* 39, 26-27.

Kusanagi, T. (1986) (C-II) Herbicides: upland crops. *Japan Pesticide Information* 49, 24-25.

Kusanagi, T. (1987) (C-II) Herbicides: upland crops. *Japan Pesticide Information* 51, 28-29.

Kusanagi, T. (1988) (C-II) Herbicides: upland crops. *Japan Pesticide Information* 53, 36-38.

Kushizaki, M., Hoshi, S. and Ito, K. (1961) Experiment on fertilizer element amounts absorbed by primary field crops in Hokkaido (in Japanese). *North Agriculture (Japan)* 28(5), 3-25.

Laurena, A. C., Garcia, V. V. and Mendoza, E. M. T. (1984) Localization and determination of condensed tannins in cowpea (*Vigna unguiculata* (L.) Walp.). *Philippine Agriculturist* 67(3), 329-336.

Lawn, R. J. (1979a) Agronomic studies on *Vigna* spp. in south-eastern Queensland. I. Phenological response of cultivars to sowing date. *Australian Journal of Agricultural Research* 30, 855-870.

Lawn, R. J. (1979b) Agronomic studies on *Vigna* spp. in south-eastern Queensland. II. Vegetative and reproductive response of cultivars to sowing date. *Australian Journal of Agricultural Research* 30, 871-882.

Lawn, R. J. (1983) Agronomic studies on *Vigna* spp. in south-eastern Queensland. III. Response to sowing arrangement. *Australian Journal of Agricultural Research* 34, 505-515.

Lawn, R. J. and Ahn, C. S. (1985) Mungbean (*Vigna radiata* (L.) Wilczek/*Vigna mungo* (L.) Hepper). In: Summerfield, R. J. and Roberts, E. H. (eds.), *Grain Legume Crops.* Collins Professional and Technical Books, London, pp. 584-623.

Lee, C. K. and Karunanithy, R. (1990) Effects of germination on the chemical composition of *Glycine* and *Phaseolus* beans. *Journal of the Science of Food and Agriculture* 51(4), 437-445.

Lee, C. Y., Shin, T. H., Kim, K. W. and Chung, C. B. (1984) A study on seed coat proteins of some legumes. I. *Journal of the National Academy of Sciences Republic of Korea Natural Sciences Series* 23, 115-123.

Lee, E. K., Lee, Y. H., Yoo, J. D. and Long, P. G. (1975) Studies on *Pythium* spp. in Korea. I. Preliminary taxonomic and physiological studies (in Korean, English summary). *Korean Journal of Mycology* 3(2), 7-12.

Lee, H. B., Lee, S. Y., Hong, C. K., Kim, D. Y., Huh, B. L. and Kim, D. R. (1992a) Studies on the intercropping system on the established sesame field (in Korean, English summary). *Research Reports of the Rural Development Administration (Suweon)* 34(1), 97-103.

Lee, I. B., Choi, K. J., Yu, K. K. and Chang, K. W. (1992b) Tocopherols and fatty acids in plant seeds from Korea (in Korean, English summary). *Journal of the Korean Agricultural Chemical Society* 35(1), 1-5.

Lee, S. C. (1989a) Survey on the population dynamics and control of *Spodoptera litura* in legumes by using sex pheromone (in Chinese, English summary). *Chinese Journal of Entomology* 9(1), 27-36.

Lee, W. H. (1989b) Adaptability trial of adzuki bean in Chia-Nan area (in Chinese, English summary). *Annual Report of Dryland Food Crops Improvement*, pp. 269-274.

Lee, W. T., Ho, C. C. and Lo, K. C. (1990) Mass production of phytoseiids. I. Evaluation of eight host plants for the mass-rearing of *Tetranychus urticae* Koch and *T. kanzawai* Kishida (Acarina: Tetranychidae). *Journal of Agricultural Research in China* 39(2), 121-132.

Lee, Y. H., Chin, M. S. and Hong, E. H. (1991) Effects of planting date on the flowering and maturing of adzuki bean (*Vigna angularis* (Willd.) Ohwi and Ohashi) and mungbean (*Vigna radiata* (L.) Wilczek) (in Korean, English summary). *The Research Reports of the Rural Development Administration* (*Suweon*) 33(3), 5-14.

Leese, B. M. Jr. (1958) Identification of Asiatic species of *Phaseolus* by seed characters. *The American Midland Naturalist* 60, 132-144.

Li, C. N. (1957) Azuki (in Chinese). In: Li, C. N. (ed.), *Selected Works of Chinese Agricultural Heritage*. Chinese Book Publishing House, Beijing, pp. 253-270.

Li, C., Sun, D. and Ma, X. (1993) Economical effects of intercropping corn with legume crops (in Chinese, English summary). In: CAAS, MAPRC and AVRDC (eds.), *The Study and Application of AVRDC Mungbean in China*. Beijing, pp. 119-125.

Li, Y. and Ding, S. (1986) Preliminary study on seed-borne mosaic virus of adzuki bean (in Chinese, English summary). *Acta Agricultural Science* (*Anhui Agricultural College*) 1, 81-83.

Liang, H. Q., An, J. and Wu, X. R. (1988) Biochemical analysis of the quality of legume seed proteins (in Chinese, English summary). *Acta Agronomica Sinica* 14(2), 117-123.

Liang, H. R. and Zhang, J. H. (1991) Mung bean and azuki (in Chinese). In: *Dry-Land Agricultural Production Technology in North China: Dry-Land Crops Cultivation*. Agricultural Publishing House, Beijing, pp. 82-88.

Liew, E. C. Y., Maclean, D. J., Manners, J. M., Dawson, D. and Irwin, J. A. G. (1991) Use of restriction fragment length polymorphisms to study genetic relationships between Australian and Japanese isolates of *Phytophthora*. *Australian Journal of Botany* 39(4), 335-346.

Lin, C. S. (1981) Resistance of beans to insects. *Chinese Journal of Entomology* 2, 1-6.

Lin, T. (1964) Laboratory test of chemicals for *Callosobruchus chinensis* control (in Chinese, English summary). *Journal of Taiwan Agricultural Research* 13(2), 55-59.

Lin, L. and Chen, Y. (1993) Nutritional characters and proper utilization of edible pulses (in Chinese, English summary). In: CAAS, MAPRC and AVRDC (eds.), *The Study and Application of AVRDC Mungbean in China*. Beijing, pp. 232-238.

Lin, W. and Dai, M. (1985) Studies on the karyotype of five species of edible legumes (in Chinese, English summary). *Journal of Sichuan Agricultural University* 3(2), 27-32.

Liu, N. (1984) Arrangement and utilization of edible germplasm resources (in Chinese). *Annual Research Report 1984: Hebei Academy of Agriculture and Forestry*, pp. 40-42.

Liu, S. D. (1991) Biological control of adzuki bean root rot disease caused by *Rhizoctonia solani*. *Plant Protection Bulletin* 33(1), 63-71.

Liu, S. E. and Wang, W. (eds.). (1959) *Botanical Key for the Northeast* (*China*) (in Chinese). Science Publishing House, Beijing, pp. 188-190.

Liu, W., Chen, G., Hsu, C. C. and Hong, A. (1981) The breeding of the new azuki cultivar Kaohsiung No. 2 (in Chinese). *Taiwan Agriculture Bimonthly* 17(6), 34-37.

Liu, X. (1987) Azuki diseases (in Chinese). In: Mao, R. (ed.), *Pests and Diseases of Major Crops in Taiwan: A Colored Pictorial Handbook.* Pesticide Commerce Association, Taipei, Taiwan, pp. B23-24.

Lu, M. S., Ge, K. L. and Yang, J. S. (1985) A preliminary report of callus induction and plantlet regeneration from cotyledon explants in *Phaseolus angularis* Wight (in Chinese, English summary). *Acta Agriculturae* (*Shanghai*) 1(4), 35-38.

Lu, T., Cao, S., Shen, Z. and Zhai, J. (1989) An identification of cucumber mosaic virus infecting adzuki bean in Hebei Province (in Chinese, English summary). *Journal of Hebei Agricultural University* 12(1), 73-79.

Lu, Y. and Ma, Z. (1985) Initial observations on the growth and development of "Tianjin" azuki (in Chinese). *Acta of Agricultural Science* (*Hebei Province*) 10(1), 76.

Lumpkin, T. A., Dean, B. B. and O'Rourke, D. A. (1993a) IMPACT and the East Asian Crops Development Program. In: Janick, J. and Simon, J. (eds.), *New Crops.* John Wiley and Sons, New York, pp. 122-126.

Lumpkin, T. A., Konovsky, J. C., Larsen, K. J. and McClary, D. C. (1993b) Azuki bean, Edamame soybean and *Astragalus*: potential new specialty crops from Asia. In: Janick, J. and Simon, J. (eds.), *New Crops.* John Wiley and Sons, New York, pp. 45-51.

MacKenzie, D. R., Ho, L., Liu, T. D., Wu, Henry B. F. and Oyer, E. B. (1975) Photoperiodism of mung bean bean and four related species. *Horticultural Science* 10(5), 486-487.

Maeda, K. (1987) *Legumes and Man: Its 10,000 Years of History.* Kokon-Shoin, Ltd., Tokyo, 379 pp.

Maeda, K. (1993) Japanese beans (in Japanese). *Genetic* 47(4) 105-108.

Maekawa, F. (1955) Topo-morphological and taxonomical studies in Phaseoleae, Leguminosae. *Japanese Journal of Botany* 15, 103-116.

Mani, A., Koshy, P. K. and Swarup, G. (1982) Nematodes associated with grain and forage legumes. In: *Technical Document No. 129, Plant Protection Committee for the South East Asia and Pacific Region.* FAO, Bangkok, Thailand, 78 pp.

Maréchal, R. (1982) Arguments for a global conception of the genus *Vigna*. *Taxon* 31(2), 280-283.

Maréchal, R., Mascherpa, J. M. and Stainer, F. (1978) Taxonomic study of one complex group of species from the *Phaseolus-Vigna* genera based on morphology and pollinology data treated to computer analysis (in French, English summary). *Boissiera* 28, 1-273.

Maréchal, R., Mascherpa, J. M. and Stainer, F. (1981) Taxonometric study of the *Phaseolus-Vigna* complex and related genera. In: Polhill, R. M. and Raven, P. H. (eds.), *Advances in Legume Systematics Part 1.* Royal Botanical Gardens, Kew (England), pp. 329-335.

Masaki, H., Murata, K., Adachi, T. and Hisanori, S. (1986) New azuki bean varieties "Beni Dainagon" and "Hatsune Shouzu" (in Japanese). *Journal of Agricultural Science* 41(3), 24-26.

Masamune, T., Anetai, M., Takasugi, M. and Katsui, N. (1982) Isolation of a natural hatching stimulus, glycinoeclepin A, for the soybean cyst nematode. *Nature* 297(June 10), 495-496.

Masuda, Y. (1985) Cell wall modifications during auxin-induced cell extension in monocotyledonous and dicotyledonous plants. *Biologia Plantarum* 27(2/3), 119-124.

Masuda, Y., Hoson, T., Yamamoto, R. and Inouhe, M. (1989) Hormone-regulated modifications to cell wall polysaccharides: relevance to cell separation. *NATO ASI (Advanced Science Institute) Series, Series H Cell Biology* 35, 139-144.

Matano, O. and Sato, R. (1963) Effect of the organophosphorus insecticide on the cholinesterase activity of adzuki been weevil (*Callosobruchus chinensis* L.) (in Japanese, English summary). *Journal of Agricultural Chemical Society of Japan* 6, 25-27.

Matlock, R. S. and Oswalt, R. M. (1963) Adsuki bean, *Phaseolus angularis* (Willd.) W. R. Wight. *Oklahoma Experiment Station Bulletin* No. B-617, 2-4.

Matsuda, K., Miyamoto, K. and Satow, T. (1992) Comparison of plant length and stem's shearing strength in crops and weeds (in Japanese, English summary). *Research Bulletin of Obihiro University, Series I* 17(4), 395-400.

Matsuda, N., Masuda, H., Komaki, M. and Matsumoto, N. (1982) Thermophilic, spore-forming, strict anaerobes isolated from spoiled canned "Shiruko" and coffee containing milk (in Japanese, English summary). *Journal of the Food Hygienic Society of Japan* 23(6), 480-486.

Matsui, M. (ed.). (1984) *Report of the Investigation and Research by a Specialist on the Promotion of Local Agricultural Enterprise* (in Japanese). Kyoto Prefecture, Agricultural/Forestry/Marketing Bureau, 52 pp.

Matsukawa, I. (1973) Differences between azuki varieties subjected to low temperature treatments. I. Effects of low temperature treatments on the flowering period (in Japanese). *North Agriculture (Japan)* 40, 1-6.

Matsumoto, H., Nishi, S. and Ishizuka, K. (1991) Action of thiocarbamate herbicide dimepiperate on plant metabolism. *Weed Research (Tokyo)* 36(3), 257-265.

Matsumoto, T. (1922) Some experiments with azuki bean mosaic. *Phytopathology* 12, 295-297.

Matsumoto, T. and Inoue, M. (1989) On a virus disease of azuki bean plants occurring on Hachirogata reclaimed land, Ohgata village (in Japanese, English summary). *Bulletin of the Akita Prefectural College of Agriculture (Akita Kenritsu Nogyo Tanki Daigaku Kenkyu Hokoku)* 15, 137-144.

Matsumoto, T. and Murayama, D. (1971) Studies on alfalfa mosaic virus. *Memoirs of the Faculty of Agriculture, Hokkaido University* 8(2), 132-174.

Matsumoto, T., Kodama, T. and Murayama, D. (1968) On the alfalfa mosaic virus strains occurring on forage crops in Hokkaido. *Journal of the Faculty of Agriculture Hokkaido University* 56(1), 1-16.

Matsuura, H. (1933) *A Bibliographical Monograph on Plant Genetics (Genetic Analysis)*, 2nd ed. Hokkaido Imperial University, Tokyo, Japan, pp. 296-99.

McCarthy, G. and Emery, F. E. (1894) Some leguminous crops and their economic value. *North Carolina Agricultural Experiment Station Bulletin* 98, 135-170.

McClary, D. C. (1990) *Azuki, Vigna angularis (Willd.) Ohwi and Ohashi: a literature review and agronomic evaluation for production in the Columbia Basin.* MSc thesis, Washington State University, 230 pp.

McClary, D. C., Hang, A. H., Gilliland, G. C., Babcock, J. M., Lumpkin, T. A., Ogg, A. G. and Tanigoshi, L. K. (1993) Herbicides for azuki production. In: Janick, J. and Simon, J. (eds.), *New Crops.* John Wiley and Sons, New York, pp. 590-594.

McClary, D. C., Raney, T. L. and Lumpkin, T. A. (1989) *Japanese Food Marketing Channels: A Case Study of Azuki Beans and Azuki Products.* Washington State University, IMPACT Center, Pullman, 41 pp.

McKern, N. M., Mink, G. I., Barnett, O. W., Mishra, A., Whittaker, L. A., Silbernagel, M. J., Ward, C. W. and Shukla, D. D. (1992a) Isolates of bean common mosaic virus comprising two distinct potyviruses. *Phytopathology* 82(9), 923-929.

McKern, N. M., Shukla, D. D., Barnett, O. W., Vetten, H. J., Dijkstra, J., Whittaker, L. W. and Ward, C. W. (1992b) Coat protein properties suggest that azuki bean mosaic virus, blackeye cowpea mosaic virus, peanut stripe virus, and three isolates from soybean are all strains of the same potyvirus. *Intervirology* 33(3), 121-134.

Me, Y., He, J., Ye, Z., Lou, N. and Feng, J. (1962) Introduction to edible legume crops (in Chinese). *Chinese Farmers' Daily* 9, 15-16.

Meiners, J. P., Gillaspie, A. G. Jr., Lawson, R. H. and Smith, F. F. (1978) Identification and partial characterization of a strain of bean common mosaic virus from *Rhynchosia minima. Phytopathology* 68, 283-287.

Melton, T. A., Edwards, D. I., Jacobsen, B. J. and Noel, G. R. (1984) The soybean cyst nematode problem. *University of Illinois Cooperative Extension Service Report on Plant Diseases* 501, 1-14.

Mengistu, A., Tachibana, H. and Grau, C. R. (1991) Selective medium for isolation and enumeration of *Phialophora gregata* from soybean straw and soil. *Plant Disease* 75(2), 196-199.

Messina, F. J. (1991) Competitive interactions between larvae from divergent strains of the cowpea weevil (Coleoptera: Bruchidae). *Environmental Entomology* 20(5), 1438-1443.

Messina, F. J. and Mitchell, R. (1989) Intraspecific variation in the egg-spacing behavior of the seed beetle *Callosobruchus maculatus. Journal of Insect Behavior* 2(6), 727-742.

Miao, S. and He, M. (1992) The experimental results of main factors in azuki yield increases. *Acta of Hua Bei Agronomy* 7(2), 18-22.

Michikawa, K., Kobayashi, R. and Watanabe, T. (1992) Comparison of physical and chemical properties of "An" (bean jam) prepared from Otebo (*Phaseolus vulgaris*) and Oshirohana (*Phaseolus coccineus*). *Nippon Shokuhin Kogyo Gakkaishi* 39(8), 663-670.

Mikami, T., Kato, S., Yamaguchi, H. and Shimamoto Y. (1991) Chloroplast DNA diversity in azuki bean and its relatives. *Japanese Journal of Breeding* 41(1), 476-477.

Mikoshiba, K. (1985) Upland crops in Japan. V. Legumes-soybeans and other legumes. *Farming Japan* 19(2), 49-57.

Milkov, E. D., Konzak, C. F. and Polle, E. A. (1992) Integrating physiological conditioning and mutagen seed treatments to reduce redrying damage in legumes. *Annual Report of the Bean Improvement Cooperative* 35, 78-79.

Mitamura, M. (1992) Stratigraphy and geologic structure of the Osaka group pliocene and pleistocene in Keihanna hills Kinki district Japan (in Japanese, English summary). *Quaternary Research (Tokyo)* 31(3), 159-177.

Mitsuhashi-Kato, M. and Shibaoka, H. (1981) Effects of actinomycin D and 2,4-dinitrophenol on the development of root primordia in azuki bean stem cuttings. *Plant and Cell Physiology* 22(8), 1431-1436.

Mitsuhashi-Kato, M., Maeda, H. and Fujii, T. (1985) Promotion of rooting in azuki cuttings by possible glycoproteins extracted from *Lentinus edodes* culture. *Plant and Cell Physiology* 26(2), 221-228.

Miyagi, M. and Yasukawa, H. (1934) Research experiments on photoperiodism in beans. I. On soybean and azuki flowering (in Japanese). *Japanese Journal of Crop Science* 6(2), 231-238.

Miyakado, M., Nakayama, I. and Yoshioka, H. (1980) Insecticidal joint action of pipercide and co-occurring compounds isolated from *Piper nigrum* L. *Agricultural and Biological Chemistry* 44(7), 1701-1703.

Miyakado, M., Nakayama, I., Yoshioka, H. and Nakatani, N. (1979) The piperaceae amides. I. Structure of pipercide, a new insecticidal amide from *Piper nigrum* L. *Agricultural and Biological Chemistry* 43(7), 1609-1611.

Mizuno, N. and Shibaoka, H. (1982) Tubulin and tubulin-associated CDR calcium-dependent regulator-like protein from azuki bean epicotyls *Vigna angularis*. In: Sakai, H., Mohri, H. and Borisy, G. G. (eds.), *Biological Functions of Microtubules and Related Structures*. Academic Press, Tokyo, pp. 1-10.

Mizuno, N. and Takao, K. (1972) Studies on the growth disorder of red bean plants grown on Usu soil. 2. Change of copper content in soils and varietal difference (in Japanese). *Japanese Journal of Science on Soil and Manure* 43(8), 302.

Mizuno, N., Suzuki, S. and Takao, K. (1971) Studies on the growth disorder of red bean plants grown on Usu soil. 1. Analysis of red bean tissue with the electron microprobe X-ray analyzer (in Japanese). *Japanese Journal of Science on Soil and Manure* 42(4), 163-166.

Montenegro, B. J. and Galindo, A. J. (1974) Bean common mosaic virus in Guanajuato state and an evaluation of varietal resistance (in Spanish, English summary). *Agrociencia* 18, 89-95.

Mori, K., Ito, T. and Tanaka, K. (1983) Synthesis and biological activity of optically active forms of (E)-3,7-dimethyl-2-octene-1,8-dioic acid (callosobruchusic acid), a component of the copulation release pheromone (erectin) of the azuki bean weevil. *Tetrahedron Letters* 39(13), 2303-2306.

Mori, T., Watanabe, K. and Fugita, I. (1966) The influences of boron fertilizers, which are favourable to sugar beet and toxic to bean crops, on the Tokachi volcanic upland soils in Hokkaido. *Research Bulletin of the Hokkaido National Agricultural Experiment Station* 90, 61-74.

Motomiya, G. and Ito, R. (1972) Domestic production, importation and utilization of food legumes and research organization in Japan. In: *Symposium on Food Legumes. Proceedings of a Symposium on Tropical Agriculture Research.* 12-14 September 1972. Tropical Agriculture Research Series No. 6, pp. 23-32.

Mou, S. and He, M. (1985) The effect of phosphorus fertilizer on azuki (in Chinese). *Soils* 17(4), 205-216.

Mukoo, H. (1975) The diseases of adzuki beans and mungbeans (in Japanese). In: Kawada, H. (ed.), *Encyclopedia of Pests and Diseases of Crops.* Yokendo, Tokyo, pp. 270-275.

Mun-Chan, B., Ho-Dzun, H. and Hammer, K. (1986) A check-list of the Korean cultivated plants. *Kulturpflanze* 34, 69-144.

Muniyappa, V., Setty, B. A. S. and Shivashankar, G. (1977) Reaction of *Phaseolus* species to horsegram yellow mosaic. *Indian Journal of Genetics and Plant Breeding* 37(3), 488-490.

Murai, M. and Fujii, K. (1970) Examination of the influences of density pressure on the pattern of adult emergence with reference to the azuki bean weevil, *Callosobruchus chinensis. Researches on Population Ecology (Kyoto)* 12(2), 219-232.

Murata, K. (1976) Explanation on new azuki bean variety, "Hayate-shozu" (in Japanese). *Journal of Agricultural Science (Nogyogijutsu)* 31(9), 410-411.

Murata, K. (1980) Explanation on new adzuki bean variety"Hokkai Shiro Shouzu" (in Japanese). *Journal of Agricultural Science (Nogyogijutsu)* 35(1), 21-22.

Murata, K. (1983a) Research related to azuki growth. VI. On leaf and pod surface temperature and frost damage (in Japanese). *Breeding and Crops Institute Hokkaido Assembly Bulletin* 23, 13 (Abstr.).

Murata, K. (1983b) Azuki. In: Nakayama, T., Osanai, S. and Nakano, H. (eds.), *Pictorial of Comprehensive Crop Management* (in Japanese). Hokkaido Prefecture Cooperative Communication Publishing House, pp. 102-115.

Murata, K. and Narugawa, T. (1978) Research related to azuki breeding. I. Differences between varieties in bean seed germination (in Japanese). *Breeding and Crops Institute Hokkaido Assembly Bulletin* 18, 38.

Murata, K., Narikawa, T., Chiba, I., Satoh, H., Adachi, T. and Matsukawa, I. (1985) New adzuki bean variety "Erimo-shozu" (in Japanese, English summary). *Breeding and Crops Institute Hokkaido Assembly Bulletin* 53, 103-113.

Murata, K., Satoh, H. and Narikawa, T. (1977) New adzuki bean "Hayate-shozu" (in Japanese, English summary). *Bulletin of Hokkaido Prefectural Agricultural Experiment Station* 38, 73-82.

Murayama, D. (1941) Research on mosaic virus seed infections in bean crops (in Japanese). *Sapporo Agriculture-Forestry Bulletin* 34, 40-70.

Murayama, D. (1966) On the witches' broom diseases of sweet potato and leguminous plants in the Ryukyu Islands (in Japanese, English summary). *Memoirs of the Faculty of Agriculture, Hokkaido University* 6(1), 81-102.

Murayama, D., Shikata, E., Shioda, H., Sekiyama, E., Sakurai, H., Waga, S., Yazu, S. and Shimamoto, Y. (1967) Studies on the purple top wilt of potatoes. I (in Japanese, English summary). *Memoirs of the Faculty of Agriculture, Hokkaido University* 6(2), 231-270.

Muszynski, S., Hiraiwa, S. and Tanaka, S. (1980) Monitoring radiation damage in the seedlings of adzuki bean (*Vigna angularis* (Willd.) Ohwi and Ohashi) after seed irradiation with gamma rays (in English, Japanese summary). *Japanese Journal of Breeding* 30(3), 225-230.

Muto, K. (1987) *Marketing Systems of Agricultural Products in Japan.* Japan FAO Association, Tokyo, p. 18.

Nagaoka, S., Hirasawa, M., Fukushima, K. and Tamura, G. (1984) Methyl viologen-linked nitrite reductase from bean roots. *Agricultural Biology and Chemistry* 48, 1179-1188.

Nagaraj, N. C., Muniyappa, V., Satyan, B. A. and Shivashankar, G. (1981) Resistance to mungbean yellow mosaic virus. *Indian Journal of Genetics* 41(3), 423-425.

Nagasawa, S. and Nakayama, I. (1968a) Joint sterilizing effect of a mixture of Hempa and Thiohema, and of Hempa and N, N, N', N'-tetramethyl-P-morpholinophosphinic diamide on the azuki bean weevil, *Callosobruchus chinensis* L. Studies on the chemosterilants of insects. XIII. (in Japanese, English summary). *Japanese Journal of Applied Entomology and Zoology* 12(4), 194-201.

Nagasawa, S. and Nakayama, I. (1968b) Studies on chemosterilants of insects. XIV. Presumption of dosage-response curve obtained by the treatment of chemosterilant for both sexes of the azuki bean weevil, *Callosobruchus chinensis* L. (in Japanese, English summary). *Scientific Pest Control* (*Botyu Kagaku*) 33, 146-152.

Nagasawa, S. and Nakayama, I. (1974) Screening of phosphorus amide and s triazine chemosterilants in house flies, fruit flies, and adzuki bean weevils (in English, Japanese summary). *Scientific Pest Control* (*Botyu Kagaku*) 39(4), 105-109.

Nagasawa, S. and Ushirokita, M. (1980) Effects of a substituted 1,3-benzodioxole on reproduction of male and female *Callosobruchus chinensis* L. (Coleoptera: Bruchidae). *Applied Entomology and Zoology* 15(4), 494-495.

Nagasawa, S. and Yamada, K. (1988) An asymptotic regression curve for describing the emergence pattern of the azuki bean weevil, *Callosobruchus chinensis* L. (Coleptera: Bruchidae). *Applied Entomology and Zoology* 23(2), 181-185.

Nagasawa, S., Kanzaki, T. and Haruyama, H. (1978) Sterilizing effects of diflubenzuron and its analogs in the azuki bean weevil, *Callosobruchus chinensis* L. (Coleoptera Bruchidae). *Bulletin of the Faculty of Agriculture, Shimane University* 12, 63-69.

Nagasawa, S., Shinohara, H. and Shiba, M. (1965) Studies on chemosterilants of insects. Sterilizing effect of Dowco-186 on the azuki bean weevil, *Callosobruchus chinensis* L., with special reference to the hatchability of the eggs deposited by treated weevils (in Japanese, English summary). *Scientific Pest Control* (*Botyu Kagaku*) 30, 91-95.

Nagasawa, S., Shinohara, H. and Shiba, M. (1966) Studies on chemosterilants of insects. IX. Differential susceptibilities in sexes of the azuki bean weevil, *Callosobruchus chinensis* L., to the sterilizing effect of HEMPA (in Japanese, English summary). *Scientific Pest Control* (*Botyu Kagaku*) 31, 108-113.

Nagayama, T., Hashimoto, S., Baba, S. and Saito, K. (1978) Experiments on red bean culture for high planting density under nursery condition (in Japanese). *Tohoku Agricultural Research* 21, 133-134.

Naidu, R. A., Sreenivasulu, P., Manohar, S. K., Ramakrishna, K. N., Reddy, D. V. R. and Nayudu, M. V. (1991) Identification of peanut green mosaic virus strains in India. *Journal of Phytopathology* 133(1), 49-56.

Naiki, T. and Ui, T. (1981) Rhizoctonia root rot of bean, soybean, and adzuki bean seedlings (in Japanese, English summary). *Memoirs of the Faculty of Agriculture, Hokkaido University* 12(4), 262-272.

Nakaseko, K. (1984a) Relationship between light energy utilization efficiency and heliotropic movement of leaves of legume, especially under canopy condition (in Japanese). *Research Foundation of Education Ministry* (*General Research B.*), *Achievement Report, Project No. 57480031*. pp. 42

Nakaseko, K. (1984b) Studies on dry matter production, plant type, and productivity in grain legumes (in Japanese, English summary). *Memoirs of the Faculty of Agriculture, Hokkaido University* 14(2), 103-158.

Nakaseko, K. and Gotoh, K. (1981) Comparative studies on dry matter production, plant type and productivity in soybean, azuki bean, and kidney bean. IV. Dry matter production of azuki bean at various population densities (in Japanese, English summary). *Japanese Journal of Crop Science* 50(3), 388-395.

Nakaseko, K. and Gotoh, K. (1982) Comparative studies on dry matter production, plant type and productivity in soybean, azuki bean, and kidney bean. VI. Differences in the early vegetative growth in relation to seed weight and character of cotyledon (in Japanese, English summary). *Japanese Journal of Crop Science* 51(1), 110-116.

Nakaseko, K. and Gotoh, K. (1983) Comparative studies on dry matter production, plant type and productivity in soybean, azuki bean, and kidney bean. VII. An analysis of the productivity among the three crops on the basis of radiation absorption and its efficiency for dry matter accumulation (in Japanese, English summary). *Japanese Journal of Crop Science* 52(1), 49-58.

Nakaseko, K. and Gotoh, K. (1983) Comparative studies on dry matter production, plant type and productivity in soybean, azuki bean and kidney bean. VII. An analysis of the productivity among the three crops on the basis of radiation absorption and its efficiency for dry matter accumulation (in Japanese, English summary). *Japanese Journal of Crop Science* 52(1), 49-58.

Nakaseko, K., Gotoh, K. and Asanuma, K. (1979a) Comparative studies on dry matter production, plant type and productivity in soybean, azuki bean, and kidney bean. I. Differences in dry matter accumulation patterns under the low population density (in Japanese, English summary). *Japanese Journal of Crop Science* 48(1), 82-91.

Nakaseko, K., Gotoh, K. and Asanuma, K. (1979b) Comparative studies on dry matter production, plant type and productivity in soybean, azuki bean, and kidney bean. II. Relationships between vertical distribution of leaf area and some morphological characteristics (in Japanese, English summary). *Japanese Journal of Crop Science* 48(1), 92-98.

Nakashima, H., Tsuda, C., Murata, K. and Narikawa, T. (1980) Histological features and inheritance of male sterile azuki bean (English summary). *Japanese Journal of Breeding* 30(3), 241-245.

Nakayama, A. and Shinya, R. (1981) A new type of flat sour spoilage of commercial canned *Shiruko*. *Journal of the Food Hygienic Society of Japan* 22(1), 37-41.

Nakayama, K. (1977) Distribution and cropping system of upland farm crops in Japan. *Japanese Agricultural Research, Quarterly* 11(3), 136-140.

Narikawa, T. (1972) Kidney bean and azuki bean in Japan with reference to breeding in Hokkaido. In: *Symposium on Food Legumes. Proceedings of Symposium on Tropical Agriculture Research.* 12-14 September 1972. Tropical Agriculture Research Series No. 6, pp. 179-188.

Narikawa, T. (1978) Miscellaneous thoughts on nomenclature of cultivated plants. IV. Adzuki bean and kidney bean (in Japanese). *Journal of Agricultural Science* 33(9), 423-427.

Narikawa, T., Miura, T. and Matsukawa, I. (1973) Studies on cool-weather tolerance in bean plants. Response to low temperature of air and soil in plant growth and seed yield (in Japanese, English summary). *Bulletin of Hokkaido Prefectural Agricultural Experiment Station* 26, 28-35.

Narita, T. (1980) *A General Survey of Hokkaido Crop Disease Damage.* Hokkaido Plant Protection Association, Sapporo, pp. 236-250.

Narita, T. and Itagaki, H. (1975) Parasitic injury by azuki circle crest fungi (in Japanese). *Northern Japan Disease and Insect Damage Research Bulletin* 26, 41.

Narita, T. and Ito, H. (1973) The fungi identified to *Ascochyta phaseolorum* Sacc. in Hokkaido. I (in Japanese, English summary). *Annual Report of the Society of Plant Protection of North Japan* 24, 6-13.

Narita, T., Akai, J. and Tsuboki, K. (1971a) Concerning azuki brown stem rot (*Cephalosporium*) (in Japanese). *Annals of the Phytopathological Society of Japanese* 37, 168-169 (Abstr.).

Narita, T., Akai, J. and Tsuboki, K. (1971b) Azuki brown stem rot and its pathogenic fungus (in Japanese). *Plant Protection* (*Shokubutsu Boeki*) 25(9), 353-358.

National Agricultural Cooperative Federation (eds). (1991) *Agricultural Cooperative Yearbook.* National Agricultural Cooperative Federation (South Korea).

National Research Council. (1989) *Alternative Agriculture.* National Academy Press, Washington D. C., 448 pp.

Nayar, E. R., Mehra, K. L., Arora, R. K. and Chandel, K. P. S. (1983) Incompatibility mechanisms in *Vigna umbellata-angularis* group. *XV International Congress of Genetics*, Part 1, 260 (Abstr.).

Negishi, H. and Kobayashi, K. (1984) Effect of soybean cyst nematode on the incidence of brown stem rot (BSR) of adzuki (in Japanese, English summary). *Annals of the Phytopathological Society of Japan* 50(4), 500-506.

Negishi, H. and Kobayashi, K. (1985) The emergence of soybean brown stem rot in field sites identified with the onset of the azuki brown stem rot disease (in Japanese). *Annals of the Phytopathological Society of Japan* 51(1), 106 (Abstr.).

Nelson, T. (1993) Adzuki beans. *Small Farm News* Jan/Feb, 3, California.

Nguyen, T. G. (1923) *Contribution a L'etude Anatomique des Tegument Seminaux des Legumineues Exotique.* Thesis, Faculty of Sciences, Marseilles.

Nishimune, A. (1984) Evaluation of the soil nitrogen supply in upland field crops in Tokachi, Hokkaido (in Japanese, English summary). *Research Bulletin of the Hokkaido National Agricultural Experiment Station* 140, 33-91.

Nishimune, A., Konno, T., Saito, G. and Fujita, I. (1983) The relationship between nitrogen fixation and grain yields of legumes on the main upland soils in Tokachi, Hokkaido (in Japanese, English summary). *Research Bulletin of the Hokkaido National Agricultural Experiment Station* 137, 81-106.

Nishimura, K. (1993) Oviposition strategy of the parasitic wasp *Dinarmus basalis* (Hymenoptera, Pteromalidae). *Evolutionary Ecology* 7(2), 199-206.

Nishitani, K. (1992) A novel method for detection of endo-xyloglucan transferase. *Plant and Cell Physiology* 33(8), 1159-1164.

Nishitani, K. and Masuda, Y. (1979) Growth and cell wall changes in azuki bean epicotyls. I. Changes in wall polysaccharides during intact growth. *Plant and Cell Physiology* 20(1), 63-74.

Nishitani, K. and Masuda, Y. (1980) Modification of cell wall polysaccharides during auxin-induced growth in azuki bean epicotyl segments. *Plant and Cell Physiology* 21(1), 169-181.

Nishitani, K. and Masuda, Y. (1981) Auxin-induced changes in the cell wall structure: changes in the sugar compositions, intrinsic viscosity and molecular weight distributions of matrix polysaccharides of the epicotyl cell wall of *Vigna angularis*. *Physiologia Plantarum* 52(4), 482-494.

Nishitani, K. and Masuda, Y. (1982a) Acid pH-induced structural changes in cell wall xyloglucans in *Vigna angularis* epicotyl segments. *Plant Science Letters* 28(1), 87-94.

Nishitani, K. and Masuda, Y. (1982b) Roles of auxin and gibberellic acid in growth and maturation of epicotyls of *Vigna angularis*: cell wall changes. *Physiologia Plantarum* 56(1), 38-45.

Nishitani, K. and Masuda, Y. (1983) Auxin-induced changes in the cell wall xyloglucans: effects of auxin on the two different subfractions of xyloglucans in the epicotyl cell wall of *Vigna angularis*. *Plant and Cell Physiology* 24(3), 345-355.

Nishitani, K., Shibaoka, H. and Masuda, Y. (1979) Growth and cell wall changes in azuki bean epicotyls. II. Changes in wall polysaccharides during auxin-induced growth of excised segments. *Plant and Cell Physiology* 20(2), 463-472.

Nishitani, K. and Tominaga, R. (1992) Endo-xyloglucan transferase, a novel class of glycosyltransferase that catalyzes transfer of a segment of xyloglucan molecule to another xyloglucan molecule. *Journal of Biological and Chemistry* 267(29), 21058-21064.

Nissin Sugar Manufacturing Company, Ltd. (1989) *Adzuki Bean Processing Plant*. Tokyo, Japan.

Nitta, T. (1991) Diversity of root fungal floras: its implications for soil-borne disease and crop growth. *Japan Agricultural Research Quarterly* 25(1), 6-11.

Nitta, T. and Matsuguchi, T. (1988) Organic amendment to upland soil as a conditioner of the rhizosphere ecosystem (Part 5): biological control of brown stem rot of adzuki beans through an organic amendment to the seedling rhizosphere. *Japanese Journal of Soil Science and Plant Nutrition* 59(2), 140-148.

Noda, A. (1952) Studies on native (primitive) varieties of crop plants: observations on a climbing variety of adsuki bean native to San'in District, Japan (in Japanese). *Japanese Journal of Crop Science* 21(2), 134-135.

Noguchi, K. (1989) (C-II) Herbicides: upland crops. *Japan Pesticide Information* 55, 27-29.

Noguchi, K. (1990) (C-II) Herbicides: upland crops. *Japan Pesticide Information* 57, 41-42.

Noguchi, K. (1991) (C-II) Herbicides: upland crops. *Japan Pesticide Information* 59, 23-25.

Nomura, N. (1962) The influence exerted by low temperature on azuki growth and development, as well as flowering-bean formation (in Japanese). *North Agriculture* (*Japan*) 31(4), 1-3.

Nomura, N. (1964) Effect of low temperature on the growth, flowering and ripening of adzuki bean (in Japanese). *North Agriculture* (*Japan*) 31(4), 1-3.

Nomura, N. (1965/66) Effect of low temperature on adzuki bean. I. The influence of low temperatures during each growth period on development and yield (in Japanese). *Japanese Journal of Crop Science* 34, 350 (Abstr.).

Nomura, N. (1968) Genotypic and phenotypic correlation of adzuki beans (*Phaseolus angularis* W. F. Wight) (in Japanese, English summary). *Bulletin of Hokkaido Prefectural Agricultural Experiment Station* 16, 114-120.

Nomura, N. (1969a) The influence of low temperature treatments on azuki finished pod rate (in Japanese). *North Agriculture* (*Japan*) 36, 41-45.

Nomura, N. (1969b) The effects of low temperatures on azuki. II. A focus on flowering and the relationship between low temperatures and average seed weight (in Japanese). *Japanese Journal of Crop Science* 38, 192 (Abstr.).

Nomura, N. (1974) Effects of low temperature on adzuki plants. II. Relations between the low temperature, treated before and after flowering, and the thickening growth of grains (in Japanese, English summary). *Bulletin of Hokkaido Prefectural Agricultural Experiment Station* 30, 9-15.

Nomura, N. and Asanuma, K. (1970) The effects of low temperature on the order and speed of flowering in the adzuki bean (*Phaseolus angularis* W. F. Wight) (in Japanese, English summary). *Bulletin of Hokkaido Prefectural Agricultural Experiment Station* 20, 73-79.

Nomura, N. and Satoh, H. (1968) Path coefficient analysis for factors influencing visual line selection of adzuki bean (in Japanese, English summary). *Bulletin of Hokkaido Prefectural Agricultural Experiment Station* 18, 24-32.

Nomura, N. and Satoh, H. (1977) Varietal differences of adzuki beans in percentage of number of pods at heights lower than 10 cm above ground surface (in Japanese, English summary). *Bulletin of Hokkaido Prefectural Agricultural Experiment Station* 36, 1-8.

Nomura, N., Asanuma, K. and Sato, H. (1965) Intervarietal difference in response of azuki, kidney bean and peas, to short day treatment (in Japanese). *North Agriculture* (*Japan*) 32(7), 7-11.

Northwest Plants Institute (eds.). (1981) *Phaseolus angularis* (Willd.) W. F. Wight (in Chinese). In: *Qinling Herbal Vol. 1: Spermatophyte Vol. 3.* Science Publishing House, Beijing, p. 112.

Noshed, S. (1991) Vertical variation of volcanic glass major element composition in a volcanic ash layer (in Japanese, English summary). *Quaternary Research (Tokyo)* 30(4), 239-250.

Nozawa, H., Yamagata, H., Aizono, Y. and Iwasaki, T. (1992) Reconstitution of the active form from the amino-and carboxyl-terminal fragments of a reactive site-modified subtilisin inhibitor of adzuki beans (*Vigna angularis*). *Journal of Biochemistry* 111(4), 456-459.

Nozawa, H., Yamagata, H., Aizono, Y. and Iwasaki, T. (1993) Temperature-dependence of the reconstitution of an adzuki bean (*Vigna angularis*) subtilisin inhibitor from its inactive two fragments. *Bioscience, Biotechnology and Biochemistry* 57(1), 154-155.

Nozawa, H., Yamagata, H., Aizono, Y., Yoshikawa, M. and Iwasaki, T. (1989) The complete amino acid sequence of a subtilisin inhibitor from adzuki beans (*Vigna angularis*). *Journal of Biochemistry* 106(6), 1003-1008.

Nuttonson, M. Y. (1951) *Ecological Crop Geography and Field Practices of Japan, Japan's Natural Vegetation, and Agro-climatic Analogues in North America.* American Institute of Crop Ecology, Washington D. C., pp. 15-190.

O'Brien, R. E. and Kohler, E. J. (1950) *Reconnaissance Soil Survey of Japan: Kyushu Area.* SCAP, Natural Resources Section Report 110-B, Washington D. C.

Obokata, J., Mikami, K., Hayashida, N., Nakamura, M. and Sugiura, M. (1993) Molecular heterogeneity of photosystem I: psaD, psaE, psaF, psaH and psaL are all present in isoforms in *Nicotiana* spp. *Plant Physiology* 102(4), 1259-1267.

Office of Rural Development, South Korea. (1981) *Annual Research Report: 1981.* Office of Rural Development, Suweon.

Ogawa, T., Abe, S. and Kugimiya, M. (1983) Gelatinization of starch in dried azuki ann granules (in Japanese, English summary). *Journal of the Japanese Society of Food Science and Technology* 30(6), 323-330.

Ohashi, H. (1979) Leguminosae. In: Hara, H. and Williams, L. H. J. (eds.), *An Enumeration of the Flowering Plants of Nepal. II.* British Museum (Natural History), London, p. 132.

Ohashi, H. (1980) Systematic position of azuki (*Vigna angularis* (Willd.) Ohwi and Ohashi) (in Japanese). In: Nippon Ikushu Gakkai (ed.), *Recent Advances in Breeding (Ikushugaku Saikin no Shimpo).* Keigaka Syuppan Co., Tokyo, pp. 73-76.

Ohashi, H. and Takahashi, H. (1981) Pollen morphology of *Vigna angularis* (Leguminosae). *Botanical Magazine Tokyo* 94(1034), 177-180.

Ohata, Y. (1988) Azuki diseases: *Rhizoctonia solani* Kühn (in Japanese). In: Kishi, K. (ed.), *Plant Diseases in Japan.* Zenkoku Noson Kyoiku Kyokai Co., Ltd., Tokyo, pp. 119-120, 123.

Ohnishi, M. and Fujino, Y. (1981) Chemical composition of ceramide and cerebroside in azuki bean seeds. *Agricultural Biology and Chemistry* 45(5), 1283-1284.

Ohwi, J. (1953) *Azukia angularis* (Willd.) Ohwi. In: *Flora of Japan.* Smithsonian Institution, Washington D. C., p. 691.

Ohwi, J. (1984) Azukia Takah. Azuki Zoku. In: *Flora of Japan.* Smithsonian Institution, Washington D. C., pp. 567-568.

Ohwi, J. and Ohashi, H. (1969) Azuki beans of Asia (in Japanese, list in English). *Journal of Japanese Botany* 44(1), 29-31.

Oikawa, K. and Ogawa, T. (1969) Influence TIBA exerts on azuki growth and yields (in Japanese). *North Agriculture (Japan)* 36, 37-41.

Oka, K. (1976) Azuki yields and plant heterogeneity (in Japanese). *Hokkaido Breeding and Crop Institute Assembly Bulletin* 16, 26.

Okiyama, A., Motoki, M. and Yamanaka, S. (1993) Bacterial cellulose. IV. Application to processed foods. *Food Hydrocolloids* 6(6), 503-511.

Oku, T., Miyahara, Y., Fujimura, T. and Toki, A. (1983) Preliminary note on *Matsumuraeses* species (Lepidoptera: Torticidae) injuring soybeans in Tohoku District (in Japanese, English summary). *Applied Entomology and Zoology* 27, 28-34.

Osaki, M., Shinano, T. and Tadano, T. (1992) Carbon nitrogen interaction in field production. *Soil Science and Plant Nutrition* 38(3), 553-564.

Oshima, K., Honda, H. and Yamamoto, I. (1973) Isolation of an oviposition marker from azuki bean weevil, *Callosobruchus chinensis* (L.). *Agricultural Biology and Chemistry* 31(11), 2679-2680.

Ozaki, K. (1985) Plant regeneration from epicotyl culture of azuki bean (*Vigna angularis* Ohwi and Ohashi) (in Japanese, English summary). *Plant Tissue Culture Letters* 2(2), 59-62.

Ozaki, K. (1986) Short communication: plantlet formation from the calli of primary leaf of azuki bean (*Vigna angularis*). *Japanese Journal of Breeding* 36, 416-419.

Ozawa, Y. (1978) On the viability of stored adzuki beans (*Phaseolus angularis*) and its hardness after cooking (in Japanese, English summary). *Report of the National Food Research Institute* 33, 37-40.

Palmer, J. (1972) Some problems with the introduction of a new bean crop to New Zealand. *Proceedings of the Agronomy Society of New Zealand* 2, 65-71.

Palmer, J. (1974) The growth habits and flowering of azuki beans in New Zealand (varieties). *New Zealand Journal of Experimental Agriculture* 2(4), 371-376.

Pandey, S. K. and Pandey, R. K. (1988) Effect of chemosterilent thiourea on the oviposition of adult azuki bean weevil, *Callosobruchus chinensis* (L) (Bruchidae: Coleoptera), by topical application. *Indian Journal of Entomology* 48(pt. 3), 264-268.

Park, K. Y. and Hong, E. H. (1970) Effect of planting date on yield of some substitutional crops for rice in paddy fields (in Korean, English summary). *Research Reports of the Office of Rural Development, South Korea, Crops* 13, 45-53.

Park, K. Y., Choi, B. H., Park, S. U., Kang, Y. K., Moon, H. G. and Lee, S. H. (1987) Effects of continuous cultivation of upland, industrial and forage crops on soil chemical composition and crop yields (in Korean, English summary). *Research Report of the Office of Rural Development, South Korea, Crops* 29(1), 232-239.

Phillips, N., Imoto, S. and Nakanishi, K. (1992) Partial purification of an endogenous elicitor from suspension-cultured cells of red bean, *Vigna angularis*. *Experientia* 48(7), 683-687.

Pierce, W. H. (1934) Viruses of the bean. *Phytopathology* 24(2), 87-115.

Piper, C. V. (1936) *Forage Plants and Their Culture*. The Macmillan Company, New York, pp. 612-613.

Piper, C. V. and Morse, W. J. (1914) Five oriental species of beans. *US Department of Agriculture Bulletin* 119, 1-32.

Planning Division, Executive Office (eds). (1990) *Statistic Yearbook of the Republic of China*. Republic of China, Taipei.

Plant Production and Protection Division. (1959) *Tabulated Information on Tropical and Subtropical Grain Legumes*. Food and Agriculture Organization of the United Nations, Rome, pp. 145-151.

Plant Protection Society. (1979) *List of Plant Diseases in Taiwan* (in Chinese). The Plant Protection Society, Taipei, Taiwan, pp. i-iv, 231-232.

Poehlman, J. M. (1987) *Breeding Field Crops*, 3rd ed. AVI Publishing Company, Inc., Westport, Connecticut.

Poehlman, J. M. (1991) *The Mungbean*. Westview Press, Boulder, Colorado, 373 pp.

Purseglove, J. M. (1974) *Tropical Crops: Dicotyledons, Vols. I and II*. John Wiley and Sons, New York, pp. 289-290.

Pushpa, G., Mahishi, D. M., Satyan, B. A. and Kulkarni, R. N. (1979) Karyomorphology of three species of *Phaseolus*. *Mysore Journal of Agricultural Science* 13, 147-151.

Qi, W. (1957) Azuki (in Chinese). *Agricultural Technology* 10, 4.

Qi, X., Chen, R., Wilson, K. A. and Tan-Wilson, A. L. (1994) Characterization of a soybean beta-conglycinin-degrading protease cleavage site. *Plant Physiology* 104(1), 127-133.

Quintero, J., Rebellon, A. and De Agudelo, F. V. (1988) Distribution and identification of host species of *Heterodera glycines* Inchinohe race 3 in Valle de Cauca, Colombia (in Spanish). *Acta Agronomica (Palmira)* 38(1), 41-52.

Ramirez, D. A. (1978) Breeding methods in legume crops. In: *Symposium on Methods of Crop Breeding: Proceedings of a Symposium on Tropical Agriculture Research*. October 1977. Tropical Agricultural Research Series No. 11, pp. 131-134.

Rands, R. D. (1925) Bean varietal tests for disease resistance. *Journal of Agricultural Research* 31(2), 101-154.

Rashid, K. A., Smartt, J. and Haq, N. (1988) Hybridization in the genus *Vigna*. In: *Proceedings of the Second International Mungbean Symposium*. Asian Vegetable Research and Development Center, Shanhua, Tainan, pp. 205-214.

Reeder, R. T., Whitmarch, J., Gray, L. E. and Peterson, R. E. (1986) Inhibition of photosynthetic electron transport by metabolites produced by *Phialophora gregata*. *Physiology and Molecular Plant Pathology* 28(3), 371-379.

Robinson, R. G. (1964) Dry field beans for Minnesota. *University of Minnesota Agricultural Extension Service Bulletin* 310, 1-12.

Robinson, R. G. (1975) Pulse or grain legume crops for Minnesota. *University of Minnesota Agricultural Experiment Station Bulletin* 513, 1-19.

Robinson, R. G. (1980) Registration of crop cultivars: registration of Minoka adzuki bean. *Crop Science* 20(4), 549.

Robinson, R. G. (1983) Yield and composition of field bean and adzuki bean in response to irrigation, compost, and nitrogen. *Agronomy Journal* 75(1), 31-35.

Rong, B. and Ma, Q. (1986) Comprehensive experimental results on azuki cultivars (in Chinese). *Achievement Report of the Shandong Agricultural Institute in 1986*, pp. 63-69.

Ruelo, J. (1980) Countering the nematode threat. *Centerpoint*, p. 4.

Sacks, F. M. (1977) A literature review of *Phaseolus angularis* - the adsuki bean. *Economic Botany* 31(1), 9-15.

Saianova, V. V. and Reva, V. A. (1977) Isolation of globulins from seeds of the adzuki bean (in Russian). *Izvestiia Akademii Nauk SSSR Seriia Biologicheskaia Khim. Nauku* 1, 11-17.

Saini, H. S. (1988) Extractability and evaluation of alpha-galactosides of sucrose in leguminous seeds. *Food Chemistry* 28(2), 149-157.

Saio, K. and Watanabe, T. (1973) Advanced food technology of soybean and other legumes in Japan. *Tropical Agricultural Research Series* 6, 209-216.

Saito, T. (1988) *Encountering Confectioneries and Fats*. Toppan Co., Osaka, Japan.

Sakai, A., Honda, H., Oshima, K. and Yamamoto, I. (1986) Oviposition marking pheromone of two bean weevils, *Callosobruchus chinensis* and *Callosobruchus maculatus* (in English, Japanese summary). *Journal of Pesticide Science (Nippon Noyaku Gakkaishi)* 11(2), 163-168.

Sakakibara, M., Aoki, T. and Noguchi, H. (1979) Isolation and characterization of 7S protein-I of *Phaseolus angularis* (adzuki bean). *Agricultural Biology and Chemistry* 43(9), 1951-1957.

Sakamoto, Y., Sasaki, H. and Kikuta, H. (1983) Ecological studies on the subfamily Apioninae injurious to leguminous crops in Japan. I. Discrimination of some Apioninae (Col. Curculionidae) species inhabiting grasses in Hokkaido. *Journal of the College of Dairying* 10, 153-160.

Sakurai, Y. (ed.). (1986) *Comprehensive Food Dictionary*, 6th ed. Dobunshoin, Tokyo, pp. 12-13, 46-47, 675-676.

Sakuratani, T. (1974) Temperature environment in the air layer over azuki bean and pasture fields (in Japanese, English summary). *Journal of Agricultural Meteorology* 30(3), 141-149.

Salmon, S. C. (1948) Crop improvement in Japan. *Journal of the American Society of Agronomy* 40(11), 1017-1035.

Sanchez, J. (1987) Oriental beans. *Rodale Organic Garden*. Emmaus, Pa. Rodale Press. 34(1), 54, 56.

Sano-Fujii, I. (1984) Effect of bean water content on the production of the active form of *Callosobruchus maculatus* (F.) (Coleoptera: Bruchidae). *Journal of Stored Products Research* 20(3), 153-161.

Sasanuma, S., Takeda, K. and Hayashi, K. (1966) Studies on anthocyanins LV. Black red pigment of "Adzuki Bean". *Botanical Magazine Tokyo* 79, 807-810.

Sathe, S. K., Deshpande, S. S. and Salunkhe, D. K. (1982) Dry beans of *Phaseolus*. A review. Part 2. Chemical composition: carbohydrates, fiber, minerals, vitamins, and lipids. *CRC Critical Reviews in Food Science and Nutrition* 21(1), 41-93.

Sato, A., Ohshima, K., Sano, T., Uyeda, I. and Shikata, E. (1987) Comparison of direct and indirect ELISA among different isolates of BYMV. *Journal of the Faculty of Agriculture, Hokkaido University* 63(3), 277-284.

Sato, T. and Sato, S. (1982) Infective ability of soybean rust to several leguminous plants. *Soybean Rust Newsletter* 5(1), 22-26.

Sato, T., Asaka, D., Harada, T. and Matsukawa, I. (1993) Plant regeneration from protoplasts of adzuki bean *Vigna-angularis* Ohwi and Ohashi. *Japanese Journal of Breeding* 43(2), 183-191.

Satoh, H. (1970) Effects of weather on azuki bean weight and pod maturity (in Japanese). *North Agriculture* (*Japan*) 37(10), 19-28.

Satoh, H. (1972) On the ripening of azuki bean (in Japanese). *Japanese Journal of Crop Science* Particular No. 2, 53-54.

Satoh, H. (1974) Responses of adzuki bean varieties to cultivating conditions (in Japanese, English summary). *Bulletin of Hokkaido Prefectural Agricultural Experiment Station* 29, 61-71.

Satoh, H. (1979) On the flowering and ripening of adzuki bean (*Phaseolus angularis* W. F. Wight): the influence of atmospheric temperature and plant density (in Japanese). *Bulletin of Hokkaido Prefectural Agricultural Experiment Station* 41, 10-20.

Satoh, H. (1982) Introduction of new varieties of summer crops (in Japanese). *Japanese Journal of Breeding* 32(1), 91-92.

Satoh, H. and Murata, K. (1978) Concerning fluctuations in principal form qualities and number of years azuki seed is in storage (in Japanese). *North Agriculture* (*Japan*) 45, 1-7.

Satoh, H., Matsukawa, I., Narikawa, T. and Ushirogi, T. (1975a) The new adzuki bean variety "Akane Dainagon" (in Japanese, English summary). *Bulletin of Hokkaido Prefectural Agricultural Experiment Station* 33, 47-57.

Satoh, H., Matsukawa, I., Narikawa, T. and Ushirogi, T. (1975b) The new adzuki bean variety "Sakae-shozu" (in Japanese, English summary). *Bulletin of Hokkaido Prefectural Agricultural Experiment Station* 33, 58-67.

Satoh, T., Takahashi, Y., Kamata, N., Sasaki, K., Furusawa, F., Kamiyama, Y., Ohno, Y., Sasaki, K., Takahashi, N. and Maita, S. (1984) A new azuki bean variety, "Beni-Nanbu" (by gamma radiation) (in English, Japanese summary). *Gamma Field Symposia, March* (*Iwate-ken Agricultural Experiment Station, Takizawa, Japan*) (21), 31-35.

Sawa, M. (1973) On the interspecific hybridization between the azuki bean, *Phaseolus angularis* (Willd) W. F. Wight and the green gram, *Phaseolus radiatus* L. I. Crossing between a cultivar of the green gram and a semi-wild relative of the azuki bean, in endemic name "Bakaso" (in Japanese, English summary). *Japanese Journal of Breeding* 23(2), 61-66.

Sawa, M. (1974) On the interspecific hybridization between the azuki bean, *Phaseolus angularis* (Willd) W. F. Wight and the green gram, *Phaseolus radiatus* L. I. On the characteristics of amphidiploid in C_1 generation from the cross green gram x rice bean *P. calcaratus* Roxb (in Japanese, English summary). *Japanese Journal of Breeding* 24(6), 282-286.

Sawada, K. (1919) Descriptive catalogue of the Formosa fungi (Part 1) (in Japanese). *Government of Formosa Agricultural Experiment Station Special Bulletin Number 19* 33, 455, 486.

Sawada, K. (1931) *List of Fungi Found in Formosa* (in Japanese). Taiwan Central Institute, Taipei, p. 77.

Sawada, K. (1958) *Government of Formosa Agricultural Experiment Station Special Bulletin Number 105.* p. 115.

Sawaguchi, M. (1986) Research on Hokkaido azuki nutritional and physiological characteristics and method of fertilizer application. *Bulletin of Hokkaido Prefecture, Tokachi Agricultural Experiment Station* 54,1-85.

Sawaguchi, M. (1987) Nutrio-physiological characteristics of adzuki-bean and methods of fertilizer application. *Japanese Agricultural Research Quarterly* 21(2), 77-83.

Sawaguchi, M. and Nomura, K. (1978) Azuki establishment techniques that increase yield. IV. The interaction of root and growth characteristics to yield as influenced by soil management practices (in Japanese). *North Agriculture (Japan)* 45, 1-12.

Sawaguchi, M. and Nomura, K. (1980) Growth and nitrogen uptake by adzuki bean varieties (in Japanese, English summary). *Bulletin of Hokkaido Prefectural Agricultural Experiment Station* 43, 1-11.

Sawaguchi, M. and Nomura, K. (1981) Changes in carbohydrates of adzuki bean in relation to its growth (in Japanese, English summary). *Bulletin of Hokkaido Prefectural Agricultural Experiment Station* 45, 59-69.

Scarbnick, C. (1979) Azuki: Minnesota's new crop. *Minnesota Science* 34(4), 6-7.

Schuff, S. (1988) Growth business. *Colorado Ranch and Farm* 42(10), 10.

Sefa-dedeh, S. and Stanley, D. W. (1979) Textural implications of the microstructure of legumes. *Food Technology* 33, 77-83.

Shaw, J. E. and Brown, R. H. (1989) Adzuki bean tolerance to herbicides. In: *Research Report Vol. 1: Expert Committee on Weeds.* Eastern Canada Section Meeting, October 31-November 2, 1989. St. John, Newfoundland, Canada, p. 345.

Shi, Y. (1988) Study of edible legume germplasm resources in Gansu province: *Phaseolus angularis* Wight (in Chinese). In: *National Cooperation Symposium of Reproduction and Selection for Edible Legume Germplasm Resources and Identification of Agronomic Characteristics.* pp. 9-11.

Shibaoka, H. (1972) Gibberellin-colchicine interactions in elongation of azuki bean epicotyl sections. *Plant and Cell Physiology* 13, 461-469.

Shibaoka, H. and Hogetsu, T. (1977) Effects of ethyl N-phenylcarbamate on wall microtubules and on gibberellin- and kinetin-controlled cell expansion (in adsuki beans, *Vigna angularis*). *Botanical Magazine Tokyo* 90(1020), 317-321.

Shibaoka, H., Hogetsu, T. and Shimokoriyama, M. (1973) Involvement of cellulose synthesis and colchicine-sensitive mechanism in gibberellin promotion and kinetin inhibition of stem elongation (azuki bean). In: *Plant Growth Substances; Proceedings of the 8th International Conference 1973* (Pub. 1974), pp. 821-827.

Shida, S. and Danjo, T. (1971) On the changes of cotyledons in the seeds of leguminous crop plants during germination. III. Plant development and the consumption of cotyledonary stored foods in azuki bean (*Phaseolus angularis* W. F. Wight) at the early stage of growth. (in Japanese, English summary). *Bulletin of the Faculty of Agriculture Miyazaki University* 18(1), 151-160.

Shih, S. (1974) *Fan Sheng Chih Shu - An agriculturalist book of China written by Fan Sheng-Chih in the first century B.C.* Science Press, Beijing, 68 pp.

Shih, S. (1979) Planting soybean and azuki (in Chinese). In: *West and East Han Dynasty Agriculture Book.* Agricultural Publishing House, Beijing, pp. 20-21.

Shikata, E. (1986) Virus diseases of rice and leguminous crops in Japan. In: *International Symposium of Virus Diseases of Rice and Leguminous Crops in the Tropics.* Proceedings of a Symposium on Tropical Agriculture Research, October 15, 1985, Yatabe, Tsukuba, Ibaraki, Japan, pp. 51-58.

Shimada, M. (1985) Niche modification and stability of competitive systems. II. Persistence of interspecific competitive systems with parasitoid wasps (in English, Japanese summary). *Research on Population Ecology* (*Kyoto*) 27(2), 203-216.

Shimizu, K. and Mitsui, Y. (1987) Varietal difference of azuki bean and kidney bean in susceptibility to the races of the soybean cyst nematode, *Heterodera glycines* Ichinohe, 1952 (in Japanese, English summary). *Annual Report of the Society of Plant Protection of North Japan* 38, 109-110.

Shimizu, S., Ikegami, M., Negishi, H., Suyama, K. and Fujii, H. (1987) Comparative studies on bean golden mosaic virus and mungbean yellow mosaic virus. I. Biological properties and serological relationships (in Japanese, English summary). *Journal of Agricultural Science, Tokyo University* (*Tokyo Nogyo Daigaku*) 31(3), 151-155.

Shinoda, K. and Yoshida, T. (1984) Relationship between adult feeding and emigration from beans of azuki bean weevil, *Callosobruchus chinensis* L. (Coleoptera: Bruchidae) (in Japanese, English summary). *Applied Entomology and Zoology* 19, 202-211.

Shinoda, K. and Yoshida, T. (1985) Field biology of the azuki bean weevil, *Callosobruchus chinensis* (L.) (Coloeptera: Bruchidae). I. Seasonal prevalence and assessment of field infestation of Aki-Azuki, an autumn variety of *Phaseolus vulgaris* (in Japanese, English summary). *Applied Entomology and Zoology* 29(1), 14-20.

Shinoda, K. and Yoshida, T. (1987) Effect of fungal feeding on longevity and fecundity of the azuki bean weevil, *Callosobruchus chinensis* (L.) (Coleoptera: Bruchidae), in the azuki bean field. *Applied Entomology and Zoology* 22(4), 465-473.

Shinoda, K. and Yoshida, T. (1991) Phytate content in fermented soybean (natto). *Tachikawa Tandai Kiyo* 24, 103-104.

Shinoda, K., Yoshida, T. and Igarashi, H. (1992) Population ecology of the azuki bean beetle *Callosobruchus chinensis* (L.) Coleoptera: Bruchidae on two wild leguminous hosts *Vigna angularis* var *nipponensis* and *Dunbaria villosa*. *Applied Entomology and Zoology* 27(3), 311-318.

Shinoda, K., Yoshida, T. and Okamoto, T. (1991) Two wild leguminous host plants of the azuki bean weevil, *Callosobruchus chinensis* (L.) (Coleoptera: Bruchidae). *Applied Entomology and Zoology* 26(1), 91-98.

Shiota, Y., Kurata, M. and Tsuchiya, F. (1983) Water absorption by adzuki beans (*Vigna angularis*). (in Japanese, English summary). *Japan Society of Home Economics* 34(12), 775-781.

Shiota, Y., Matsuura, Y. and Hatanaka, C. (1982) Studies of ann (bean jam). IX. Pectic polysaccharides in cotyledon of adzuki beans *Vigna angularis*. (in Japanese, English summary) *Journal of the Japanese Society for Food Science and Technology* 29(12), 712-719.

Shiota, Y., Matsuura, Y. and Hatanaka, C. (1991) Effect of storage temperature on the properties of pectic polysaccharides in cotyledons of adzuki beans (*Vigna angularis*) (in Japanese, English summary). *Journal of the Japanese Society for Food Science and Technology* (*Nippon Shokuhin Kogyo Gakkaishi*) 38(2), 94-101.

Shiozaki, H. and Tsuchiya, K. (1985) Differences of response to cool summer damage and wet damage among upland crops. *Japanese Agricultural Research Quarterly* 18(4), 268-274.

Shirai, S., Murata, K., Shimada, H., Adachi, T., Hara, M., Chiba, I. and Narikawa, T. (1990) A new adzuki bean variety "Kamui-dainagon" (in Japanese, English summary). *Bulletin of Hokkaido Prefectural Agricultural Experiment Station* 60, 73-78.

Shukla, D. D., McKern, N. M., Ward, C. W. and Ford, R. E. (1991) Molecular parameters suggest that cowpea aphid-borne mosaic virus is a distinct potyvirus and that bean common mosaic virus consists of at least three distinct potyviruses. *Phytopathology* 81(10), 1166.

Shukla, G. S. and Pandey, S. K. (1980) Sterilizing effect of thiourea on the oviposition of adult azuki bean weevil, *Callosobruchus chinensis* (L.) (Bruchidae: Coleoptera), by injection method. *Journal of Applied Entomology* 89(4), 419-422.

Singh, B. M., Shyam, K. R. and Sud, V. K. (1979) Web blight of *Vigna angularis* (Wight) Wilczeck caused by *Rhizoctonia solani* Kühn. *Tropical Grain Legume Bulletin* 16, 14-15.

Singh, G. and Chua, E. L. (1976) Pests in seed storage. In: Chin, H. F., Enoch, I. C. and Harun, R. M. R. (eds.), *First National Seed Symposium: Seed Technology in the Tropics*. University Pertanian Malaysia, Serdang, Malaysia, pp. 37-45.

Singh, H. P. and Singh, R. V. (1983) Methods of leaf area estimation of urdbean from linear measurement. *Legume Research* 6(1), 18-20.

Sinha, S. S. N. and Roy, H. (1979) Cytological studies in the genus *Phaseolus*. I. Mitotic analysis in fourteen species. *Cytologia* 44, 191-199.

Siriwardhane, D., Egawa, Y. and Tomooka, N. (1991) Cross-compatibility of cultivated adzuki bean (*Vigna angularis*) and rice bean (*V. umbellata*) with their wild relative. *Plant Breeding* 107, 320-325.

Smartt, J. (1971) Some thoughts on interspecific hybridization in *Phaseolus*. *Report of the Bean Improvement Cooperative* (*University of Nebraska*) 14, 109-111.

Smartt, J. (1976) Comparative evolution of pulse crops. *Euphytica* 25, 139-143.

Smartt, J. (1978) The evolution of pulse crops. *Economic Botany* 32, 185-198.

Smartt, J. (1980) Evolution and evolutionary problems in food legumes. *Economic Botany* 34(3), 219-235.

Smartt, J. (1981) Gene pool in *Phaseolus* and *Vigna* cultigens. *Euphytica* 30, 445-449.

Smartt, J. (1984) Gene pools in grain legumes. *Economic Botany* 38(1), 24-35.

Smartt, J. (1985) Evolution of grain legumes. III. Pulses in the genus *Vigna*. *Experimental Agriculture* 21, 87-100.

Smartt, J. (1990) *Grain Legumes: Evolution and Genetic Resources*. Cambridge University Press, 379 pp.

Smartt, J. and Hymowitz, T. (1985) Domestication and evolution of grain legumes. In: Summerfield, R. J. and Roberts, E. H. (eds.), *Grain Legume Crops.* Collins Professional and Technical Books, London, pp. 37-72.

Smitanond, B., Pitiyont, V., Ohsawa, K., Honda, H. and Yamamoto, I. (1991) Characterization of the ecochemical in kidney beans which inhibits the larval growth of the azuki bean weevil. *Journal of Pesticide Science* 16(2), 179-188.

Soil and Pesticide Division, Environmental Agency. (1975) Standards for withholding registration of agricultural chemicals established by the Director General of the Environmental Agency (II). *Japan Pesticide Information* 25, 34-35.

Soil and Pesticide Division, Environmental Agency. (1977) Standards for withholding registration of agricultural chemicals established by the Director General of the Environmental Agency (III). *Japan Pesticide Information* 29, 19-20.

Solomon, R. (1990) Adzukis: an alternative crop experience. *The Bean Bag* 8(4), 5, 14.

Soma, S. (1988) *Quality and Constituents of Beans.* Hokkaido Association for the Advancement of Beans, Hokkaido Agricultural Experiment Station, 69 pp.

Soma, G., Yoshimura, A., Tsukioka, D., Mizuno, D. and Oshima, H. (1992a) Lipopolysaccharides of plants and microorganisms for treatment of ulcer of digestive tract (in Japanese). *Patent: Japan Kokai Tokkyo Koho; JP 9249240 A2; JP 0449240. Date issued: 18 February.* 25 pp.

Soma, G., Yoshimura, A., Tsukioka, D., Mizuno, D. and Oshima, H. (1992b) Pharmaceuticals containing lipopolysaccharides for treatment of rheumatism. *Patent: Japan Kokai Tokkyo Koho; JP 9249241 A2; JP 0449241. Date issued: 18 February.* 26 pp.

Song, D., Zhang, X. and Lou, Y. X. (1989) Comparative experiments on azuki cultivars and breeding lines (in Chinese). *Shanxi Agricultural Sciences* 6, 15-16.

Sortland, M. E. and MacDonald, D. H. (1985) Reproduction and survival of a Minnesota population of *Heterodera glycines* Race 5 following one to three growing periods in nine crops and two weed species. *Phytopathology* 75, 1366 (Abstr.).

Sortland, M. E. and MacDonald, D. H. (1987) Effect of crop and weed species on development of a Minnesota population of *Heterodera glycines* race 5 after one to three growing periods. *Plant Disease* 71(1), 23-27.

Souma, J., Kobayashi, K. and Ogoshi, A. (1991) Studies on the decrease of the population of *Phialophora gregata,* the causal fungus of brown stem rot of adzuki beans, by corn cultivation. II. Comparisons of the rhizoplane and the rhizosphere microfloras between corn and adzuki bean (in Japanese, English summary). *Memoirs of the Faculty of Agriculture, Hokkaido University* 17(4), 479-488.

Sperbeck, J. (1981) Adzuki beans: one specialty crop with an undeveloped market. *The New Farm* 3, 48-49.

Springer, M. E. (1950) *Reconnaissance Soil Survey of Japan: Shikoku.* SCAP, Natural Resources Section Report 110-D, Washington D. C.

Srinives, P. (1990) Mungbean breeding and genetic resources in Thailand. In: Thavarasook, C. (ed.), *Proceedings of the Mungbean Meeting.* February 23-24, 1990, Chiang Mai, Thailand, pp. 31-42.

Statistical Research Department, Ministry of Agriculture and Forestry. (1951) Adzuki bean, small red bean, small bean. In: *Local Names for Field Crops.*

Agriculture and Statistics Association, Tokyo, Agriculture and Forestry Research Data Report No. 27, pp. 60-61.

Statistics and Information Department. (1992) *Statistical Yearbook of Ministry of Agriculture, Forestry and Fisheries, Japan.* Association of Agriculture and Forestry Statistics, Tokyo.

Strand, A. B. (1943) Species crosses in the genus *Phaseolus. Proceedings of the American Society for Horticultural Science* 42, 569-573.

Strang, J. (1980) Adzuki bean cv. Bloodwood. *Journal of the Australian Institute of Agricultural Science* 48, 264.

Sugano, T., Ueda, I., Senboku, T., Shikata, E., Hiwatari, M., Yashima, Y., Niizeki, M. and Kita, F. (1981) Effect of Nichino Mushicon mulch film on control of virus disease and growth of azuki bean (in Japanese, English summary). *Research Bulletin of the University Farm Faculty of Agriculture, Hokkaido University* 22, 73-79.

Summerfield, R. J. and Wien, H. C. (1980) Effects of photoperiod and air temperature on growth and yield of economic legumes. In: Summerfield, R. J. and Bunting, A. H. (eds.), *Advances in Legume Science.* Royal Botanical Gardens, Kew, Surrey (England), pp. 17-36.

Sunada, K., Nakano, H. and Narukawa, T. (1975) The course of assigned experiments and their results. V. Breeding of soybean, corn, and azuki bean in Hokkaido, Japan (in Japanese). *Journal of Agricultural Science (Nogyogijutsu)* 30(5), 227-229.

Suzuki, F. and Konno, S. (1982) Report on A. P. O. regional survey: regional report on grain legumes production in Asia. In: *Grain Legumes Production in Asia.* Asian Productivity Organization, Tokyo, Japan, pp. 15-95.

Suzuki, I. (1973) An experiment related to protection of azuki from a harmful family of bugs (in Japanese). *Bulletin of the Kyoto Prefectural Institute of Agriculture*, pp. 10-26.

Suzuki, S., Usa, M., Nagoshi, T., Kobayashi, M., Watanabe, N., Watanabe, H. and Inaba, H. (1991) Two-dimensional imaging and counting of ultraweak emission patterns from injured plant seedlings. *Journal of Photochemistry and Photobiology B: Biology* 9(2), 211-217.

Tabiki, T. and Tsuchiya, T. (1990) Field evaluation of varietal difference to Phytophthora stem rot of adzuki beans by flooding (in Japanese, English summary). *Bulletin of Hokkaido Prefectural Agricultural Experiment Station* 60, 133-142.

Tadano, T. and Sakai, H. (1991) Secretion of acid phosphatase by the roots of several crop species under phosphorus-deficient conditions. *Soil Science and Plant Nutrition* 37(1), 129-140.

Tai, F. L. (ed.). (1979) *Sylloge Fungorum Sinicorum* (in Chinese). Science Press, Academia Sinica, Beijing, pp. 320, 779.

Tai, M. (1980) Field seed selection methods for edible legumes (in Chinese). *Heilongjiang Agriculture* 7, 22.

Tajiri, T. (1979) Studies on cultivation and keeping quality of bean sprouts. II. Effect of packaging materials and storage temperatures on keeping quality of bean sprouts (in Japanese, English summary). *Journal of the Japanese Society of Food Science and Technology* 26(12), 542-546.

Tajiri, T. (1981) Studies on cultivation and keeping quality of bean sprouts. V. Effect of application of artificial light on the growth of hypocotyl and vitamin C (reduced ascorbic acid) content of bean sprouts (in Japanese, English summary). *Journal of the Japanese Society of Food Science and Technology* 28(8), 430-436.

Takabayashi, M. (1975) 1974 evaluation of candidate pesticides (C-II) herbicides: upland crops. *Japan Pesticide Information* 25, 10-12.

Takabayashi, M. (1978) (C-II) Herbicides: upland crops. *Japan Pesticide Information* 35, 47-48.

Takabayashi, M. (1980a) (B-II) Fungicides: upland crops. *Japan Pesticide Information* 37, 53-54.

Takabayashi, M. (1980b) (C-II) Herbicides: upland crops. *Japan Pesticide Information* 37, 59-60.

Takahama, U. (1993) Regulation of peroxidase-dependent oxidation of phenolics by ascorbic acid: different effects of ascorbic acid on the oxidation of coniferyl alcohol by the apoplastic soluble and cell wall-bound peroxidases from epicotyls of *Vigna angularis*. *Plant and Cell Physiology* 34(6), 809-817.

Takahashi, N. (1957) Ecological studies of fruiting behaviors of azuki beans (in Japanese, English summary). *Japanese Journal of Crop Science* 26, 44-45.

Takahashi, R. (1936) Thysanoptera of Formosa. *Philippine Journal of Science* 60, 427-459.

Takahashi, Y. (1909) Botanical studies on adzuki plant (in Japanese). *Journal of the Society of Agriculture and Forestry* 1, 140-163.

Takahashi, Y. (1917) Experimental results of azuki characterization and breeding (in Japanese). *Bulletin of Hokkaido Agricultural Experiment Station* 7, 1-181.

Takeda, K. and Shibaoka, H. (1981) Changes in microfibril arrangement on the inner surface of the epidermal cell walls in the epicotyl of azuki bean *Vigna angularis* Ohwi et Ohashi during cell growth. *Planta* 151(4), 385-392.

Takeda, K. and Shibaoka, H. (1981) Effects of gibberellin and colchicine on microfibril arrangement in epidermal cell walls of azuki bean *Vigna angularis* Ohwi et Ohashi epicotyls. *Planta* 151(4), 393-398.

Takeuchi, H., Watanabe, N. and Sonda, M. (1977) Loss in weight of adzuki-bean caused by a bruchid beetle, *Callosobruchus analis* Fabricius (in Japanese, English summary). *Plant Protection Research Experiment Report* 14, 60-63.

Takeuchi, S. (1980) (B-II) Fungicides: upland crops. *Japan Pesticide Information* 37, 53-54.

Takeuchi, S. (1984) (B-II) Fungicides: upland crops. *Japan Pesticide Information* 44, 30-32.

Takeuchi, S. (1986) (B-II) Fungicides: upland crops. *Japan Pesticide Information* 48, 38-39.

Takeuchi, S. (1988) (B-II) Fungicides: upland crops. *Japan Pesticide Information* 52, 46-47.

Talekar, N. S. and Lee, Y. H. (1988) Biology of *Ophiomyia centrosematis* (Diptera: Agromyzidae) a pest of soybean. *Annals of the Entomological Society of America* 81(6), 938-942.

Tanaka, A. (1982) Performance of crops on soils low in pH and phosphorus. *Tropical Agricultural Research Series 15*, pp. 157-166.

Tanaka, F., Kobayashi, K. and Ui, T. (1981) Azuki brown stem rot disease fungal culture medium selected and used to separate fungus from infected soil (in Japanese). *Annals of the Phytopathological Society of Japan* 47, 99 (Abstr.).

Tanaka, K. (1988) (A-II) Insecticides: vegetables, ornamental crops and industrial crops. *Japan Pesticide Information* 52, 38-39.

Tanaka, K., Ohsawa, K., Honda, H. and Yamamoto, I. (1981) Copulation release pheromone, erectin, from the azuki bean weevil, *Callosobruchus chinensis* L. *Journal of Pesticide Science* 6(1), 75-82.

Tanaka, K., Ohsawa, K., Honda, H. and Yamamoto, I. (1982) Synthesis of erectin, a copulation release pheromone of the azuki bean weevil, *Callosobruchus chinensis* L. *Journal of Pesticide Science* 7(4), 535-537.

Tanaka, N. (1968a) Studies on root system formation in leguminous crop plants. III. On classification of root systems (in Japanese, English summary). *Japanese Journal of Crop Science* 37(3), 411-415.

Tanaka, N. (1968b) Studies on root system formation in leguminous crop plants. IV. Anatomical observation of vascular system and origination of roots in hypocotyl and main root (in Japanese, English summary). *Japanese Journal of Crop Science* 37(3), 424-429.

Tanaka, N., Fujii, Y., Soejima, M. and Ueno, M. (1966) Research on leguminous crop root system formation. V. On hypocotyl root growth intercrop differences (in Japanese). *Japanese Journal of Crop Science* 35, 277 (Abstr.).

Tanaka, Y. (1991a) Individual variation of scent avoiding oviposition behavior in azuki bean weevil *Callosobruchus chinensis*. *Journal of Ethology* 9(1), 31-33.

Tanaka, Y. (1991b) Genetic variation in age-specific fecundity of the adzuki bean weevil *Callosobruchus chinensis* (Coleoptera: Bruchidae) (reared on adzuki beans (*Vigna angularis*)) *Applied Entomology and Zoology* 26(2), 263-265.

Tanaka, Y. (1993) A genetic mechanism for the evolution of senescence in *Callosobruchus chinensis* (the azuki bean weevil). *Heredity* 70(3), 318-321.

Tanaka, Y. and Konno, S. (1982) Country reports on the status of grain legumes production: Japan. In: *Grain Legumes Production in Asia*. Asian Productivity Organization, Tokyo, pp. 385-430.

Tanaka, Y. and Ohno, M. (1895) *Phaseolus radiatus* L. In: *Useful Plants of Japan, Described and Illustrated*. Agricultural Society of Japan, Akasaka, Tokyo, pp. 7-8.

Tanaka, Y., Tsunogae, Y., Hayakawa, Y., Tanaka, I., Yamane, T., Ashida, T., Ishikawa, C., Watanabe, K., Nakamura, S. and Takahashi, K. (1983) Preliminary crystallographic study of Bowman-Birk protease inhibitor (adzuki bean) and its complex with trypsin. *Journal of Biochemistry* (*Tokyo, Japan*) 94(2), 611-613.

Tanda, Y. (1958) Selecting azuki seeds (in Japanese, English summary). *Japanese Journal of Crop Science* 27(2), 319-320.

Tanigoshi, L. K. (1991) *Tetranychus urticae* Koch: its control and impact on azuki bean culture in eastern Washington. In: Dusbabek, F. and Bukva, V. (eds.), *Modern Acarology*. Academia, Prague and SPB Academic Publishing bv, The Hague, Vol. 2, pp. 737-740.

Tanigoshi, L. K. and Aguilar, H. (1993) Control of twospotted spider mite on azuki bean 1992. In: Burditt, A. K. Jr. (ed.), *Insecticide & Acaricide Tests*. Entomological Society of America, Lanham, Maryland, Vol. 18, II + 405 pp.

Tanii, A. and Baba, T. (1971) Bacterial plant diseases in Hokkaido. I. Bacterial brown spot of adzuki bean caused by *Pseudomonas syringae* van Hall (in Japanese, English summary). *Bulletin of Hokkaido Prefectural Agricultural Experiment Station* 23, 90-97.

Tanii, A. and Baba, T. (1974) Species name of the causal pathogen causing bacterial stem rot of azuki bean (in Japanese). *Annals of the Phytopathological Society of Japan* 40, 202-203 (Abstr.).

Tanii, A. and Baba, T. (1979) Bacterial stem rot of adzuki bean (*Phaseolus radiatus* var. *aurea* Prain) caused by *Pseudomonas adzukicola* A. Tanii et T. Baba nov. sp. (in English, Japanese summary). *Bulletin of the Hokkaido Prefectural Agricultural Experiment Station* 42, 29-42.

Tanii, A., Narita, T., Akai, J. and Baba, T. (1973) Bacterial stem rot of azuki bean and its causal bacterium (in Japanese). *Annals of the Phytopathological Society of Japan* 39, 151 (Abstr.).

Tanusi, S., Kasai, T. and Kawamura, S. (1972) Determination of oligosacharides in some edible legume seeds. *Journal of the Japanese Society of Food and Nutrition* 25(1), 25-27.

Tasaki, J. (1957a) The ecological classification of the varieties in azuki beans (in Japanese). *Japanese Journal of Crop Science* 25, 244 (Abstr.).

Tasaki, J. (1957b) Studies on the reproductive physiology in azuki beans (*Phaseolus radiatus* var. *aurea* Prain). II. On the germination of pollens (in Japanese, English summary). *Japanese Journal of Crop Science* 26, 275-276.

Tasaki, J. (1959) Effect of soil moisture content on growth and yield of azuki-beans, *Phaseolus radiatus* var. *aurea* Prain (in Japanese, English summary). *Japanese Journal of Crop Science* 28, 73-76.

Tasaki, J. (1960) Research on azuki response to photoperiod. II. Intervarietal differences in the influence exerted by temperature on photoperiod (in Japanese). *Japanese Journal of Crop Science* 28(3), 325 (Abstr.).

Tasaki, J. (1961) On differences between ecotype in azuki chlorophyll content and nyctinastic movement (in Japanese). *Japanese Journal of Crop Science* 29, 447-448 (Abstr.).

Tasaki, J. (1963) Genecological studies in the azuki bean (*Phaseolus radiatus* L. var. *aurea* Prain), with special reference to the plant types used for the classification of ecotypes (in English, Japanese summary). *Japanese Journal of Breeding* 13, 32-44.

Tasaki, J. (1965) Studies on varietal differences in photoperiod responses used for the classification of adzuki bean varieties (in Japanese, English summary). *Japanese Journal of Crop Science* 34, 14-19.

Tasaki, J. and Honma, H. (1965) Studies on varietal differences in relation to sensitivity to temperature of adzuki beans (*Phaseolus angularis* W. F. Wight): consideration for the method of the test of sensitivity to temperature (in Japanese, English summary). *Japanese Journal of Crop Science* 34, 20-24.

Tateishi Y. (1983) Leguminosae collected in the Arun Valley, east Nepal. In: Numata, M. (Ed.), *Structure and Dynamics of Vegetation in Eastern Nepal,* pp. 131-145.

Tateishi, Y. (1984) Contributions to the genus *Vigna* (Leguminosae) in Taiwan I. *Sci. Rep. Tohoku University, 4th series* (*Biology*) 38, 335-350.

Tateishi, Y. (1989) A revision of the Azuki bean group, the subgenus *Ceratotropis* of the genus *Vigna* (Leguminosae), Ginkgoana 7.

Tateishi, Y. and Ohashi, H. (1990) Systematics of the azuki bean group in the genus *Vigna.* In: Fujii, K., Gatehouse, A. M. R., Johnson, C. D., Mitchel, R. and Yoshida, T. (eds.), *Bruchids and Legumes: Economics, Ecology and Coevolution.* Kluwer Academic Publishers, Dordrecht, Netherlands, pp. 189-199.

Tatt, O. H. (1980) Effects of mannitol-induced water stress on the ribosomes of intact leaves of azuki bean seedlings. *Biologia Plantarum* 22(4), 249-254.

Tawaraya, K., Satoh, H. and Wagatsuma, T. (1992) Effect of the fungicide, benomyl, on vesicular-arbuscular mycorrhizal infection and phosphorus uptake of plants. *Yamagata Daigaku Kiyo, Nogaku* 11(3), 481-487.

Tazaki, K. (1979) Biosynthesis of alicyclic acids from 14C carbon isotope-labeled glucose and erythrose by isolated cells from *Vigna angularis,* adsuki bean leaves. *Plant and Cell Physiology* 20(1), 224-231.

Terai, K. and Horie, T. (1991) Comparison of yield characters among the native varieties of white adzuki bean (*Vigna angularis*) in Akita prefecture (in Japanese, English summary). *Japanese Journal of Crop Science* 60(1), 8-14.

Tevis, C. (1987) Residue-free food. *Successful Farm* 85(14), 40-41.

Thomas, T. A., Patel, D. P. and Bhagat, N. R. (1974) Adzuki bean: a new promising pulse for the hills. *Indian Farming* 23(12), 29-30.

Tian, J. (1990) Estimation of heritability and response to selection for agronomic characters of azuki (in Chinese). *Hebei Crop Research* 4, 26-28.

Tjahjadi, C. (1983) *Isolation and characterization of adzuki bean* (*Vigna angularis*) *protein and starch.* Unpublished PhD thesis, University of Minnesota, 169 pp.

Tjahjadi, C. and Breene, W. M. (1984) Isolation and characterization of adzuki bean (*Vigna angularis* cv Takara) starch. *Journal of Food Science* 49, 558-562.

Tjahjadi, C., Lin, S. and Breene, W. M. (1988) Isolation and characterization of adzuki bean (*Vigna angularis* cv Takara) proteins. *Journal of Food Science* 53(5), 1438-1443.

Tojo, M., Okamura, T. and Ishibashi, K. (1975) Freezing of bean jam "Azuki-an" fresh by cold air (in Japanese, English summary). *Research Bulletin of Obihiro University, Series I* 9(2), 357-364.

Tokitomo, Y. (1988) Odor development of azuki beans during cooking. *Bulletin of the Faculty of Education, Yamanashi University* 39, 164-167.

Tokitomo, Y. (1992) Changes in the odor constituents of "ann" (bean jam) during processing and storage (in Japanese). *Bulletin of the Faculty of Education, Yamanashi University* 43(4), 103-109.

Tokitomo, Y. and Kobayashi, A. (1988) Odor of cooked Japanese adzuki beans (in Japanese, English summary). *Journal of the Agricultural Chemical Society of Japan* 62(1), 17-22.

Tomizawa, C. (1977) Past and present status of residues of pesticides. *Japan Pesticide Information* 30, 35-42.

Tomooka, N. (1991) Taxonomic position of mungbean and its close relatives, the subgenus *Ceratotropis* (a literature review). In: *Mungbean and the Genetic Resources of the Subgenus* Ceratotropis *in Ecological Studies on Tropical Food Legumes in Relation to Adaptation to Cropping Systems in Thailand.* Tropical Agriculture Research Center, Tsukuba, Japan, pp. 25-31.

Tomooka, N., Lairungreang, C., Nakeeraks, P., Egawa, Y. and Thavarasook, C. (1991) *Vigna riukiuensis.* In: *Mungbean and the Genetic Resources of the Subgenus* Ceratotropis *in Ecological Studies on Tropical Food Legumes in Relation to Adaptation to Cropping Systems in Thailand.* Tropical Agriculture Research Center, Tsukuba, Japan, pp. 14-15.

Toms, G. C. (1981) Lectins in Leguminosae. In: Polhill, R. M. and Raven, P. H. (eds.), *Advances in Legume Systematics.* Royal Botanical Gardens, Kew, Surrey (England), pp. 561-577.

Tong, P. Y. (1981) Azuki (in Chinese). In: *Cultivation of Grain Legumes.* Agriculture Publishing House, Beijing, pp. 64-72.

Trung, B. C., Yoshida, S. and Kobayashi, Y. (1985) Influence of soil moisture stress on the nitrogen nutrition and grain productivity of mungbean (in English, Japanese summary). *Japanese Journal of Crop Science* (*Nihon Sakumotsu Gakkai*) 54(1), 72-78.

Tsai, S. L. (1982) Growth analysis of rice stubble azuki bean (*Phaseolus angularis*) (in Chinese, English summary). *National Science Council Monthly* 10(12), 967-982.

Tsai, W. F. (1988) Study on the production of adzuki, mungbean, broadbean and common beans (in Chinese). *Scientific Agriculture* 36(7-8), 173-179.

Tsuchiya, S. (1985) The ecology and emergence of the azuki phytophthora stem rot disease (*Phytophthora vignae* Purss.) (in Japanese). *Annals of the Phytopathological Society of Japan* 51(1), 106-107 (Abstr.).

Tsuchiya, S. (1988) Studies on the stem rot disease of adzuki bean caused by *Phytophthora vignae* f. sp. *adzukicola* and its control (in Japanese, English summary). *Report of Hokkaido Prefectural Agricultural Experiment Station* (*Hokkaidoritsu Nogyo Shikenjo Hokoku*) 72, 1-76.

Tsuchiya, S. and Akai, J. (1975) Infection ecology of azuki brown stem rot (in Japanese). *Annals of the Phytopathological Society of Japan* 41(3), 226 (Abstr.).

Tsuchiya, S. and Akai, J. (1977) Analysis of the origin of azuki brown stem rot occurrence (in Japanese). *North Japan Disease and Pest Research Bulletin* 28, 7.

Tsuchiya, S. and Kodama, F. (1978) Azuki phytophthora stem rot and its pathogenic fungus (in Japanese). *Plant Protection* (*Shokubutsu Boeki*) 32(9), 357-360.

Tsuchiya, S. and Kodama, F. (1979) On the pathogenicity of azuki phytophthora stem rot in azuki and cowpea (in Japanese). *Annals of the Phytopathological Society of Japan* 45(1), 116 (Abstr.).

Tsuchiya, S. and Kodama, F. (1981a) Phytophthora rot of soybean and adzuki bean (in Japanese). *Plant Protection* (*Shokubutsu Boeki*) 35(10), 439-442.

Tsuchiya, S. and Kodama, F. (1981b) On the differences in pathogenicity as it relates to azuki variety *P. vignae* (in Japanese). *Annals of the Phytopathological Society of Japan* 47(1), 100 (Abstr.).

Tsuchiya, S. and Tanaka, F. (1983) Results of foliage and seed coating chemicals for azuki stem rot control (in Japanese). *Annals of the Phytopathological Society of Japan* 49(1), 117 (Abstr.).

Tsuchiya, S. and Tanaka, F. (1984) A survey of the outbreak of *Phytophthora* stem rot of adzuki bean caused by *Phytophthora vignae* in Kamikawa District (in Japanese, English summary). *Bulletin of Hokkaido Prefectural Agricultural Experiment Station* 51, 105-112.

Tsuchiya, S., Fujio, K., Kamjaipai, W. and Akira, O. (1978) Azuki stem rot disease pathogenic fungus and circumstances of development in the Kamikawa District (in Japanese). *Annals of the Phytopathological Society of Japan* 44(1), 75 (Abstr.).

Tsuchiya, S., Yanagawa, M. and Ogoshi, A. (1986) Formae speciales differentiation of *Phytophthora vignae* isolates from cowpea and adzuki bean (in English, Japanese summary). *Annals of the Phytopathological Society of Japan* 52(4), 577-584.

Tsuchizaki, T. and Hibino, H. (1971) Seed transmission of viruses in cowpea and azuki bean plants. IV. Relations between seed transmission and virus distribution in apical meristem of flower bud (in Japanese, English summary). *Annals of the Phytopathological Society of Japan* 37, 17-21.

Tsuchizaki, T. and Omura, T. (1987) Relationships among bean common mosaic virus, blackeye cowpea mosaic virus, azuki bean mosaic virus, and soybean mosaic virus. *Annals of the Phytopathological Society of Japan* 53(4), 478-488.

Tsuchizaki, T., Yora, K. and Asuyama, H. (1970a) The viruses causing mosaic of cowpeas and azuki beans, and their transmissibility through seeds (in Japanese, English summary). *Annals of the Phytopathological Society of Japan* 36, 112-120.

Tsuchizaki, T., Yora, K. and Asuyama, H. (1970b) Seed transmission of viruses in cowpea and azuki bean plants. I. Factors influencing seed transmission (in Japanese, English summary). *Annals of the Phytopathological Society of Japan* 36, 121-126.

Tsuchizaki, T., Yora, K. and Asuyama, H. (1970c) Seed transmission of viruses in cowpea and azuki bean plants. II. Relations between seed transmission and gamete infection (in Japanese, English summary). *Annals of the Phytopathological Society of Japan* 36, 237-242.

Tsuchizaki, T., Yora, K. and Asuyama, H. (1971) Seed transmission of viruses in cowpea and azuki bean plants. III. Relationship between seed transmission and embryo infection (in Japanese, English summary). *Annals of the Phytopathological Society of Japan* 37, 11-16.

Tsunogae, Y., Tanaka, I., Yamane, T., Kikkawa, J., Ashida, T., Ishikawa, C., Watanabe, K., Nakamura, S. and Takahashi, K. (1986) Structure of the trypsin binding domain of Bowman-Birk type protease inhibitor and its interaction with trypsin. *Journal of Biochemistry (Tokyo, Japan)* 100(6), 1637-1646.

Tsutsumi, M. and Sakurai, K. (1966) Influence of root diffusates of several host and non-host plants on the hatching of the soybean cyst nematode, *Heterodera glycines* Ichinohe, 1952 (in Japanese, English summary). *Japanese Journal of Applied Entomology and Zoology* 10, 129-137.

Turkova, V. and Klozova, E. (1985) Comparison of seed proteins in some representatives of the genus *Vigna*. *Biologia Plantarum* (*Praha*) 27(1), 70-73.

Ueno, T., Kuwahara, Y., Fujii, K., Taper, M., Toquenaga, Y. and Suzuki, T. (1990) D-Catechin. An oviposition stimulant of azuki bean weevil, *Callosobruchus chinensis*, in the host azuki bean. *Journal of Pesticide Science* (*Nippon Noyaku Gakkaishi*) 15(4), 573-578.

Umega, K. and Shimizu, K. (1968) Studies on the comparative ecology of bean weevil. III. Effect of feeding on the life span and oviposition of the adult of three species of bean weevils (in Japanese, English summary). *Yokohama Plant Protection Station Report* 2(5), 34-48.

Unfried, K., Schiebel, K. and Hemleben, V. (1991) Subrepeats of rDNA intergenic spacer present as prominent independent satellite DNA in *Vigna radiata* but not in *Vigna angularis*. *Genetics* 99(1), 63-68.

Ushirogi, T., Satoh, H. and Nomura, N. (1972) The new adzuki bean variety "Akatsuki-Dainagon" (in Japanese, English summary). *Bulletin of the Hokkaido Prefectural Agricultural Experiment Station* 25, 71-90.

Ushirogi, T. (ed.), Yamakawa, T. and Hara, M. (1986) *Hokkaido Bean Production Techniques: Azuki and Kidney Bean*. Hokkaido Ebetsu, Agricultural Production Dissemination Association, 205 pp.

Utida, S. (1941a) Studies on experimental population of the azuki bean weevil, *Callosobruchus chinensis* (L.). I. The effects of population density on the progeny populations. *Memoirs of the College of Agriculture, Kyoto Imperial University* 48, 1-30.

Utida, S. (1941b) Studies on experimental population of the azuki bean weevil, *Callosobruchus chinensis* (L.). III. The effect of population density upon the mortalities of different stages of life cycle. *Memoirs of the College of Agriculture, Kyoto Imperial University* 49, 21-42.

Utida, S. (1941c) Studies on experimental population of the azuki bean weevil, *Callosobruchus chinensis* (L.). IV. Analysis of density effect with respect to fecundity and fertility of eggs. *Memoirs of the College of Agriculture, Kyoto Imperial University* 51(Entomological Series No. 8), 1-26.

Utida, S. (1941d) Studies on experimental population of the azuki bean weevil, *Callosobruchus chinensis* (L.). V. Trend of population density at the equilibrium position. *Memoirs of the College of Agriculture, Kyoto Imperial University* 51(Entomological Series No. 8), 27-34.

Utida, S. (1943) Studies on experimental population of the azuki bean weevil, *Callosobruchus chinensis* (L.). VIII. Statistical analysis of the frequency distribution of the emerging weevils on beans. *Memoirs of the College of Agriculture, Kyoto Imperial University* 54(Entomological Series No. 10), 1-22.

Uyeda, I., Kojima, M. and Murayama, D. (1975) Purification and serology of bean yellow mosaic virus. *Annals of the Phytopathological Society of Japan* 41(2), 192-203.

Uyeda, I., Sugano, T., Nakasone, T., Senboku, T. and Shikata, E. (1982) Studies on 'Top necrosis disease' of French bean cv Ofuku in Hokkaido (in English, Japanese summary). *Memoirs of the Faculty of Agriculture, Hokkaido University* 13, 69-79.

van der Maesen, L. J. G. and Somaatmadja, S. (eds.). (1989) *Plant Resources of South-East Asia No. 1: Pulses.* Pudoc, Wageningen, 117 pp.

van Oers, C.C.C.M. (1989) *Vigna angularis* (Willd.) Ohwi & Ohashi. In: Maesen, L. J. G. and Somaatmadja, S. (eds.), *Plant Resources of South-East Asia No. 1: Pulses.* Pudoc, Wageningen, pp. 67-69.

Vance, C. P., Boylan, K. L. M., Maxwell, C. A., Heichel, G. H. and Hardman, L. L. (1985) Transport and partitioning of CO_2 fixed by root nodules of ureide and amide producing legumes. *Plant Physiology* 78(4), 774-778.

Verdcourt, B. (1970) Studies in the Leguminosae-Papilionoideae for the 'Flora of Tropical East Africa'. IV. *Kew Bulletin* 24, 507-569.

Vetten, H. J., Green, S. K. and Lesemann, D. E. (1992) Characterization of peanut stripe virus isolates from soybean in Taiwan. *Journal of Phytopathology* 135(2), 107-124.

Vietmeyer, N. D. (1986) Lesser-known plants of potential use in agriculture and forestry. *Science* 232(6), 1379-1384.

Wagatsuma, T., Ikarashi, H. and Ishii, K. (1985) Lead toxicity in crops: comprehensive studies on the environmental improvement for crop productions (in Japanese, English summary). *Journal of the Yamagata Agriculture and Forestry Society* 42, 65-74.

Wallace, E. (1987) *Vegetable Growers' Handbook.* Agro-Research Enterprises Ltd., Havelock North, New Zealand, 186 pp.

Wang, C. L. (1989) Effects of endomycorrhizae on the growth and yield of adzuki bean (*Phaseolus angularis*) (in Chinese, English summary). *Journal of the Chinese Agricultural Association* (*Taiwan*) 148, 67-80.

Wang, H. (1991) An analysis of the reason for low yields and cultivation methods to increase yields (in Chinese). *Tillage and Cultivation* 2, 41-42.

Wang, J. J. (1986) The new cultivar of adzuki: "Jihong Xiaodou No. 1" (in Chinese). *Agricultural Sciences and Technology Communication* 1, 39.

Wang, J. J. (1990) An initial study on improved cultivation practices to improve yields (in Chinese). *Hebei Agricultural Sciences and Technology* 6, 7.

Wang, X. and Tian, J. (1990) Identification of saline resistance in azuki germplasm (in Chinese). *Crop Research in Hebei Province* 2, 33-36.

Wang, Z. and Zhao, Y. (1984) Seed selection for mungbean and azuki in the fields (in Chinese). *New Agriculture* 16, 23.

Watanabe, H. and Yuriko, H. (1955) Studies on "Azuki-An" (crude starch product for azuki-bean). I. The yield and the quality of the product (in Japanese). *Food Research Institute Report* 10, 89-96.

Watanabe, N. (1984) Comparative studies on ecology of the adzuki bean weevil, *Callosobruchus chinensis* (L.) and the cowpea weevil, *C. maculatus* (F.) (Coleoptera: Bruchidae). I. Effect of seed size on creeping females into the pile (in Japanese, English summary). *Japanese Journal of Applied Entomology and Zoology* 28(4), 223-228.

Watanabe, N. (1985) Comparative studies on ecology of the azuki bean weevil. *Callosobruchus chinensis* (L.) and the cowpea weevil, *Callosobruchus maculatus* (F.) (Coleoptera: Bruchidae). II. Relation between female body size and degree of creeping into the stored pile of seeds. *Japanese Journal of Applied Entomology and Zoology* 29(2), 107-112.

Watanabe, N. (1986) Comparative studies on the ecology of the azuki bean weevil *Callosobruchus chinensis* (L.) and the cowpea weevil *Callosobruchus maculatus* (F.) (Coleoptera : Bruchidae). IV. Effect of adult body size on easiness of escaping from the bottom of piled seeds. *Japanese Journal of Applied Entomology and Zoology* 30(1), 64-66.

Watanabe, S., Sato, M., Shiina, T., Usui, N. and Kato, K. (1986) Long-term storage and distribution of crop seeds for genetic resources at the Germplasm Seed Storage Center. In: *Annual Report 1985*. National Institute of Agrobiological Resources, Japan, pp. 19-20.

Watanabe, T., Kobayashi, R. and Sasaki, T. (1983) Studies of red bean ann. III. Enzyme susceptibility of starch separated from red bean ann (in Japanese, English summary). *Journal of the Japanese Society of Food Science and Technology* 30(10), 539-543.

Waterhouse, G. M. (1963) Key to the species of *Phytophthora* de Bary. *Mycology Paper 92*, Commonwealth Mycology Institute, Kew, England.

Wei, B., Yamauchi, R. and Kato, K. (1993) Chemical structure of α-glucan in the seed of adzuki bean. *Denpun Kagaku* 40(3), 299-303.

Wei, Q. (1979) Preliminary report on a distant cross between azuki and *Phaseolus vulgaris* (purple vegetable bean) (in Chinese). *Jilin City Agricultural Sciences and Technology* 1, 4-5.

Wei, S. C. and Chiu, W. F. (1964) On a mosaic disease of adzuki-bean (*Phaseolus angularis* Wight) (in Chinese, English summary). *Acta Phytopathologica Sinica* 7(1), 67-76.

Westphal, E. (1974) Pulses of Ethiopia, their taxonomy and agricultural significance. *Agricultural Research Report of Wageningen* 815, 263 pp.

White, D. (1972) Azuki bean potential being assessed. *New Zealand Journal of Agriculture* 125(3), 43-45.

Whiteside, E. P. (1950) *Reconnaissance Soil Survey of Japan: Kyoto Area*. SCAP, Natural Resources Section Report 110-C, Washington D. C.

Willdenow, K. L. (1801) *Dolichos angularis* Willd. *Species Plantarum* 3, 1051.

Wu, Y. L. and Hsu, C. C. (1986) *Growth environment and basic cultivation practices for azuki* (in Chinese). Kaohsiung Agricultural Experiment Station, Kaohsiung, Taiwan, pp. 1-16.

Wu, Y. L., Hsu, C. C. and Chen, G. (1985a) New azuki cultivar "Kaohsiung No. 3" (in Chinese). *Taiwan Agriculture Bimonthly* 21(1), 26-28.

Wu, Y. L., Hsu, C. C. and Lin, C. L. (1985b) Breeding of adzuki bean (in Chinese, English summary). *Annual Report of the Dryland Food Crops Improvement* 27, 140-143.

Wu, Y. L., Hsu, C. C. and Lin, C. L. (1986) Breeding of adzuki bean (in Chinese, English summary). *Annual Report of the Dryland Food Crops Improvement* 28, 130-134.

Xu, Z., Huang, W. and Zhang, G. (1992) Acetyl urea content in seedlings and variations of allantoic enzyme activity during azuki seed germination. *Bulletin of SubTropical Botany* 21(1), 1-5.

Xu, Z. H., Yang, L. J., Wei, W. M. and Gao, M. X. (1984) Plant regeneration in tissue culture of four leguminous species (in Chinese, English summary). *Acta Biologiae Experimentalis Sinica* 17(4), 483-486.

Yamaguchi, F. and Yoshitada, M. (1959) On the pest, 'Awano Meiga' and its control (in Japanese). *Chugoku Agricultural Research* 16, 52-54.

Yamaguchi, H. (1989) Weed adzuki bean, an overlooked representative. *Bulletin of the University of Osaka Prefecture Series B* 41, 21-27.

Yamaguchi, H. (1992) Wild and weed azuki beans in Japan. *Economic Botany* 46(4), 384-394.

Yamaguchi, H. and Kosuge, K. (1991) Evaluation of wild azuki beans (*Vigna angularis* s. lat.). 2. SDS-page profiles in wild, weed and domesticated forms of azuki bean and its related wild species. *Japanese Journal of Breeding* 41(2), 164-165.

Yamamoto, A., Yamada, K., Hayashi, C. and Tanida, N. (1991a) Partial purification and product specificity of adzuki bean (*Vigna angularis*) lipoxygenase (in Japanese, English summary). *Journal of the Japanese Society for Food Science and Technology* 38(3), 214-216.

Yamamoto, H., Kobayashi, K. and Ogoshi, A. (1990) Isozyme polymorphism in *Phialophora gregata* isolates from adzuki bean and soybean in Japan (in English, Japanese summary). *Annals of the Phytopathological Society of Japan* 56, 584-590.

Yamamoto, H., Kobayashi, K. and Ogoshi, A. (1991b) Comparison of culture characteristics of *Phialophora gregata* f. sp. *adzukicola* and *sojae* in Japan. *Transactions of the Mycological Society of Japan* 32(4), 529-533.

Yamamoto, I. and Honda, H. (1977) Population ecology of bean weevils - its chemical basis. In: *Symposium on Insect Pheromones and Their Applications.* December 8-11, 1976, Nagaoka and Tokyo, p. 163.

Yamamoto, R., Inouhe, M. and Masuda, Y. (1988) Galactose inhibition of auxin-induced growth of mono- and dicotyledonous plants. *Plant Physiology* 86(4), 1223-1227.

Yamanouchi, M. (1964a) Studies on the inorganic nutrients of the azuki bean plant. I. On the absorption of nitrogen, phosphorus, potassium and calcium (in Japanese, English summary). *Bulletin of the Hokkaido Prefectural Agricultural Experiment Station* 13, 91-98.

Yamanouchi, M. (1964b) Studies on the inorganic nutrients of the azuki bean plant. II. On the growth investigation of each leaf and the movement of nitrogen, phosphorus, and potassium in leaves at definite positions on the main stem (in Japanese, English summary). *Bulletin of Hokkaido Prefectural Agricultural Experiment Station* 14, 68-79.

Yamashita, K. (1991) Occurrence of azuki bean mosaic disease in Nanbu region of Aomori Prefecture. (*Vigna angularis*), azuki bean mosaic potyvirus (in Japanese). *Annual Report of the Society of Plant Protection of North Japan* 42, 40-41.

Yan, J. (1988) Developing new mungbean germplasm resources (in Chinese). In: *Future Research Trends in Mungbean Germplasm.* National Cooperation Symposium on

Reproduction and Selection for Edible Legume Germplasm Resources and Identification for Agronomic Characters, November, 1988, Beijing, pp. 4-6.

Yanagawa, M., Ogoshi, A. and Tsuchiya, S. (1985) Electrophoresis pattern of *Phytophthora vignae* azuki fungus and cowpea fungus (in Japanese). *Annals of the Phytopathological Society of Japan* 51(1), 107 (Abstr.).

Yang, J. and Lu, Y. (eds). (1985) Ch'i Min Yao Shu: nine kinds of lost agricultural books in Chinese history. In: *The Series of World Literature Treatises: One of Chinese Well-Known Works*. World Book Bureau Press, Taipei.

Yao, B. (1985) Cultivation methods for azuki planted in autumn (in Chinese). *Jiangsu Agricultural Sciences* 6, 7-8.

Yarnell, S. H. (1965) Cytogenetics of the vegetable crops. IV. Legumes (continued). *The Botanical Review* 31(3), 247-330.

Yasui T. (1980) Identification of a new galactosyl cyclitol from seeds of *Vigna angularis* Ohwi et Ohashi. *Agricultural and Biological Chemistry* 44(9), 2253-2255.

Yasui, T., Tateishi, Y. and Ohashi, H. (1985) Distribution of low molecular weight carbohydrates in the subgenus *Ceratotropis* of the genus *Vigna* (Leguminosae). *Botanical Magazine Tokyo* 98(1049), 75-87.

Ye, D., Chen, Z. And Pan, H. (1993) Azuki cultivation technology for high yield in Yanjiang regions in Jiangsu Province (in Chinese). *Bulletin of Agricultural Science and Technology*, 11, 9.

Yearbook Committee of China Foreign Trade (eds). (1992) *1990/1992 Yearbook of China Foreign Trade*. Yearbook Committee of China Foreign Trade, Beijing, p. 413.

Yi, G. (1961) Azuki (in Chinese). *Agricultural Technology* 3, 18.

Yokota, K., Kiyohara, T., Iwasaki, T. and Yoshikawa, M. (1983) Isolation and some properties of adzuki-bean trypsin isoinhibitors. *Science Reports of Faculty of Agriculture, Kobe University* 15, 349-356.

Yoneyama, T. and Ohtani, T. (1983) Variations of natural ^{13}C abundances in leguminous plants. *Plant and Cell Physiology* 24(6), 971-977.

Yoneyama, T., Ladha, J. K. and Watanabe, I. (1987) Nodule bacteroids and *Anabaena*: natural ^{15}N enrichment in the legume-*Rhizobium* and azolla-*Anabaena* symbiotic systems. *Journal of Plant Physiology* 127, 251-259.

Yoneyama, T., Yamada, N., Kojima, H. and Yazaki, J. (1984) Variations of natural ^{15}N abundances in leguminous plants and nodule fractions. *Plant and Cell Physiology* 25(8), 1561-1565.

Yoshida, K. (1987) Studies on seed size variation in grain legumes (in Japanese, English summary). *Memoirs of the Faculty of Agriculture, Hokkaido University* (*Hokkaido Daigaku Nogaku-Bu Hobun*) 15(4), 385-434.

Yoshida, K. and Sato, H. (1990) The extent and its source of variation for characteristics related to seed quality of adzuki beans. *Japanese Journal of Crop Science* 59(3), 450-454.

Yoshida, K. and Satoh, H. (1988) Azuki harvest curing factors on Hokkaido and differences between regions in bean size (in Japanese). *Japanese Journal of Crop Science* 57(2), 117-118 (Abstr.).

Yoshida, K., Bando, A., Yanagisawa, K. and Gotoh, K. (1986a) Effects of TIBA (2,3,5-triiodobenzoic acid) on growth and yield of azuki bean, *Vigna angularis* (in Japanese, English summary). *Japanese Journal of Crop Science* 55(3), 345-351.

Yoshida, K., Ishii, N., Ueshima, T. and Nakaseko, K. (1989a) Factors influencing azuki yields: influence of azuki branch and flower pruning (in Japanese). *Japanese Journal of Crop Science* 58(2), 37-38 (Abstr.).

Yoshida, K., Sato, H., Ueshima, H., Ishii, N. and Sato, M. (1991) The extent and source of variation of seed quality characteristics in adzuki bean. II. Variation in seed coat colour among growers batches in Hokkaido (in Japanese, English summary). *Japanese Journal of Crop Science* 60(2), 234-240.

Yoshida, K., Satoh, H., Ishii, N. and Ueshima, T. (1988) Differences in protein content rates in Hokkaido azuki (in Japanese). *Japanese Journal of Crop Science* 57(2), 119-120 (Abstr.).

Yoshida, K., Satoh, H., Ueshima, T. and Ishii, N. (1989b) Concerning Hokkaido adzuki bean coat color (in Japanese). *Japanese Journal of Crop Science* 58(1), 74-75 (Abstr.).

Yoshida, K., Satoh, H., Ueshima, T., Ishii, N. and Satoh, M. (1989c) Research related to azuki quality report #4: water imbibition characteristics of azuki (in Japanese). *Japanese Journal of Crop Science* 58(2), 277-278 (Abstr.).

Yoshida, T. and Gichuki, E. M. (1983) Use of solar heat to control stored-product insect pests: bean depth. *Scientific Reports of the Faculty of Agriculture, Okayama University* 61, 5-8.

Yoshida, T., Igarashi, H. and Shinoda, K. (1986b) Life history of *Callosobruchus chinensis* (L.) (Coleoptera, Bruchidae). In: Donahaye, E. and Navarro, S. (eds.), *Proceedings of the 4th International Working Conference On Stored-Product Protection*. Bet Dagan, Israel, pp. 471-477.

Yoshikawa, M., Yokota, K. and Hiraki, K. (1985) Purification and some properties of a subtilisin inhibitor from adzuki beans. *Agricultural and Biological Chemistry* 49(2), 367-371.

Yoshikawa, S. (1970) Azuki withering and dying occurrence-causes and countermeasures (in Japanese). *Farm and Garden* 45, 1381-1387.

Yu, T. F. (1946) A mosaic disease of cowpea (*Vigna sinensis* Endl.). *The Annals of Applied Biology* 33(4), 450-455.

Yuan, C. (1991) New cultivation technology and cropping systems of broad bean, maize, azuki and dryland rice (in Chinese). *Modern Agriculture* 1, 11-12.

Zhang, M. (1987) Production and breeding selection of edible legumes in the Baicheng region, Jilin Province (in Chinese). *Baichen Agricultural Science and Technology* 3, 19-21.

Zhang, M. and Li, H. (1988) The breeding of red beans: "White Red Number One" (in Chinese). *Bulletin of Agricultural Science and Technology* 3, 31.

Zhang, T., Yu, J., Li, A., Zhao, S. and Zhang, Y. (1981) Selective breeding of, "(Long) Dragon Azuki No. 1" (in Chinese). *Compilation of Research Achievements by the Heilongjiang Academy of Agricultural Sciences (1978-1980)*, pp. 53-56.

Zhao, G. (1990) Genesis of rust in azuki and control methods (in Chinese). *Agricultural Science and Technology Communication* 7, 28.

Zhao, G. P. and Tong, P. Y. (1993) Azuki bean. In: *Mung Bean and Azuki Bean Cultural Technology*. Jindun Publication House, Beijing, pp. 26-43.

Zheng, J. Y., Nakata, M., Irifune, K., Tanaka, R. and Morikawa, H. (1993) Fluorescent banding pattern analysis of eight taxa of *Phaseolus* and *Vigna* in relation to their phylogenetic relationships. *Theoretical and Applied Genetics* 87(1/2), 38-43.

Zheng, J. Y., Nakata, M., Uchiyama, H., Morikawa, H. and Tanaka, R. (1991) Giemsa c-banding patterns in several species of *Phaseolus* L. and *Vigna* Savi fabaceae. *Cytologia* (*Tokyo*) 56(3), 459-466.

Zheng, X. and Zhang, Y. (1983a) Research on edible legume resources in Heilongjiang Province (in Chinese). *Heilongjiang Agriculture Sciences* 2, 57-61.

Zheng, X. and Zhang, Y. (1983b) Research on ecological types of native azuki germplasm in Heilongjiang Province (in Chinese). *Crop Germplasm Resources* (*Ziwupinzhongziyuan*) 2, 25-28.

Zheng, Z. (1988) Commercial production of azuki by counties (in Chinese). In: *The Design and Construction of Commercial Production Bases for Chinese Edible Legumes*. The National Cooperation Symposium on Reproduction and Selection of Edible Legume Germplasm Resources and Identification of Agronomic Characteristics, November, 1988, Beijing, pp. 27-28.

Zhou, L. (ed.). (1982) Azuki cultivar "Dahongpao" (in Chinese). In: *Excellent Crop Cultivars in Shanghai.* Shanghai Publishing House, Shanghai, pp. 171-172.

Zhou, X. (1983) Conclusions about edible legume germplasm in Shaanxi province: final conclusions about azuki germplasm resources (in Chinese). *Journal of the Chinese Academy of Agricultural Sciences* 5, 1-21.

Zhou, H., Wang, H., Zhou, J. and Zhang, B. (1964) Initial study on native cultivars of edible legumes (mungbean, azuki and *Vigna* spp.) (in Chinese). *Journal of Anhui Agricultural Science* 3, 33-46.

Zhu, J. Q. (1991) The effect of β-Pyridin propyl alcohol on photosynthetic performance of azuki (in Chinese). *Acta of Hebei Agricultural University* 14(2), 30-34.

Zhu, Z. (1989) Breeding goals for azuki in Tianjin (in Chinese). *Tianjin Agricultural Sciences* 2, 22-24.

AUTHOR INDEX

A

B

C

D

E

F

G

H

I

J

K

L

M

P

Q

R

S

Y

Z

SUBJECT INDEX

A

C

F

G

H

I

J

K

L

M

Q

R

S

Z